JUST WAR THINKERS REVISITED

This book comprises essays that focus on a range of thinkers who challenge the boundaries of the just war tradition.

The ethics of war scholarship has become a rigid and highly disciplined activity, closely associated with a very particular canon of thinkers. This volume moves beyond this by presenting thinkers not typically regarded as part of that canon but who have interesting and potentially important things to say about the ethics of war. The book presents 20 profile essays on an eclectic cast of heretics, humanists, and radicals, from ancient Greece to the twenty-first century, who lived through and theorized about violence. The book asks how ethics of war scholars might benefit from engaging with them. Some of these thinkers engage directly with—to augment or criticize—the just war tradition, while others contribute to military thinking across the ages, pushing the boundaries of what was acceptable in war. Many proffer alternative moral frameworks regarding the legitimacy of political violence. The present volume thus invites scholars to reconsider the ethics of war in a way that challenges the standard delineation between just war theory, realism, and pacifism and to reflect on how those positions might inform our own approach to these matters.

This book will be of much interest to students of just war theory, ethics of war, war studies, and International Relations.

Daniel R. Brunstetter is Professor in Political Science at University of California, Irvine, the United States. He is the author of two books and editor of two volumes, including *Just War Thinkers* (2018).

Cian O'Driscoll is Professor of International Relations at the Australian National University (ANU), Canberra, Australia. He is the author of two books and editor of three volumes, including *Just War Thinkers* (2018).

War, Conflict and Ethics

Series Editors

Michael L. Gross
University of Haifa

and

James Pattison
University of Manchester

Ethical judgments are relevant to all phases of protracted violent conflict and inter-state war. Before, during, and after the tumult, martial forces are guided, in part, by their sense of morality for assessing whether an action is (morally) right or wrong, an event has good and/or bad consequences, and an individual (or group) is inherently virtuous or evil. This new book series focuses on the morality of decisions by military and political leaders to engage in violence and the normative underpinnings of military strategy and tactics in the prosecution of the war.

Ethics at War
How Should Military Personnel Make Ethical Decisions?
Deane-Peter Baker, Rufus Black, Roger Herbert and Iain King

Military Necessity and Just War Statecraft
The Principle of National Security Stewardship
Edited by Eric Patterson and Marc LiVecche

Warfare Ethics in Comparative Perspective
China and the West
Edited by Sumner B. Twiss, Ping-cheung Lo, and Benedict S. B. Chan

Just War Thinkers Revisited
Heretics, Humanists and Radicals
Edited by Daniel R. Brunstetter and Cian O'Driscoll

For more information about this series, please visit: www.routledge.com/War-Conflict-and-Ethics/book-series/WCE

JUST WAR THINKERS REVISITED

Heretics, Humanists and Radicals

Edited by Daniel R. Brunstetter and Cian O'Driscoll

Routledge
Taylor & Francis Group

LONDON AND NEW YORK

First published 2025
by Routledge
4 Park Square, Milton Park, Abingdon, Oxon OX14 4RN

and by Routledge
605 Third Avenue, New York, NY 10158

Routledge is an imprint of the Taylor & Francis Group, an informa business

British Library Cataloguing-in-Publication Data
A catalogue record for this book is available from the British Library

ISBN: 978-1-032-55033-6 (hbk)
ISBN: 978-1-032-55032-9 (pbk)
ISBN: 978-1-003-42868-8 (ebk)

DOI: 10.4324/9781003428688

Typeset in Sabon
by Apex CoVantage, LLC

CONTENTS

NOTES ON CONTRIBUTORS

Luke Armstrong is Lecturer at the University of Glasgow, where he also completed his Ph.D. His research looks at autonomy and free will in contemporary liberal political philosophy. He is particularly interested in how conceptions of autonomy might alter depending on the stance we take on the free will problem.

Christian Nikolaus Braun is Lecturer in the Defence Studies Department at King's College London. His research on the ethics of war and peace has been published in *Ethics & International Affairs*, *International Theory*, and the *Journal of International Political Theory*. Christian's most recent book, *Limited Force and the Fight for the Just War Tradition*, was published with Georgetown University Press (2023).

Chris Brown is Emeritus Professor of International Relations at the LSE. He is the author of numerous books, including *Practical Judgement in International Political Theory* (2010) and (with Robyn Eckersley) *The Oxford Handbook of International Political Theory* (2018). *The Politics of International Political Theory: Reflections on the Work of Chris Brown* (2018) examines his life's work.

Daniel R. Brunstetter is Professor of Political Science at the University of California, Irvine. He has written numerous articles related to the ethics of war published in *Ethics & International Affairs*, the *Journal of Military Ethics*, *Global Intellectual History*, and elsewhere. Daniel's most recent monograph, *Just and Unjust Uses of Limited Force*, was published with Oxford University Press (2021).

Yvonne Chiu is Professor of Strategy, Policy, and Warfare at the U.S. Naval War College, Newport, Rhode Island, the United States. Her book *Conspiring with the Enemy: The Ethic of Cooperation in Warfare* (Columbia University Press, 2019) won the North American Society for Social Philosophy Book Award 2020 and the ISA—International Ethics Book Award in 2021. She has been a National Fellow at the Hoover Institution and a Member of the Institute for Advanced Study.

Juan M. Floyd-Thomas is Associate Professor of African American Religious History at Vanderbilt Divinity School and Affiliated Faculty of Religious Studies. Among his publications, he is author of *The Origins of Black Humanism* (2008), *Liberating Black Church History* (2014), and *Critical Race Theology: White Supremacy, American Christianity, and the Culture Wars* (2024) and co-editor of *Religion in the Age of Obama* (2018).

Luke Glanville is Professor of International Relations at the Australian National University. His research spans past and present thinking about war, atrocities, and refugees. Recent books include *Sharing Responsibility: The History and Future of Protection from Atrocities* (2021), *Sepúlveda on the Spanish Invasion of the America* (co-edited and translated with David Lupher and Maya Feile Tomes, 2023), and *Prioritizing Global Responsibilities* (with James Pattison, 2024).

Pablo Kalmanovitz is Professor of International Studies at the Instituto Tecnológico Autónomo de México (ITAM) in Mexico City, Mexico. He specializes on the theory and practice of the laws of war and is currently co-editor of the *Yearbook of International Humanitarian Law*. His book *The Laws of War in International Thought* was published by Oxford University Press in 2020.

Rosemary Kellison is Associate Professor of Religion at Florida State University, Tallahassee, Florida, the United States, where she teaches religious ethics with a focus on ethics of war and feminist moral philosophy. She is the author of *Expanding Responsibility for the Just War: A Feminist Critique* (Cambridge University Press, 2019).

John Kelsay is the Lucius Moody Bristol Professor of Religion and Ethics at Florida State University. Author of a number of publications dealing with the ethics of war in Christian and Muslim traditions, he is currently working on a comparative study of the abolitionist John Brown and other leaders of armed resistance movements.

Anthony F. Lang, Jr. is Professor of International Political Theory in the School of International Relations at the University of St Andrews. His research sits at the intersection of politics, law, and ethics. He has written about global constitutionalism, universal values, the just war tradition, political responsibility, international law, and human rights. His publications can be found at: www.st-andrews.ac.uk/international-relations/people/al51. He tweets @ProfTonyLang.

Francisco Lobo holds an LLM in International Legal Studies from New York University, New York City, the United States, and a Master of Laws from the University of Chile, Santiago, Región Metropolitana, Chile. He recently received his Ph.D. from the Department of War Studies, King's College London. His published research explores international law and the just war tradition, including *Teoría y Práctica de la Intervención Humanitaria en la Tradición de la Guerra Justa* (2016).

David Lupher is Professor of Classics, Emeritus, University of Puget Sound, Tacoma, Washington, the United States. His research focuses on classical receptions in early modern colonial America. He is the author of *Romans in the New World: Classical Models in Sixteenth-Century Spanish America* (2003) and *Greeks, Romans, and Pilgrims: Classical Receptions in Early New England* (2017) and translator and co-editor of *Sepúlveda on the Spanish Invasion of the Americas* (2023).

Gabriel Mares is Postdoctoral Research Associate in the Department of Politics, Princeton University, New Jersey, the United States, studying the regulation of state violence and how concepts in international politics emerge "from below." His work explores how communities and states conceptualize and regulate violent threats – in particular, how colonial legacies within societies and institutions are reflected in such conceptualizations and regulations.

Andrée-Anne Mélançon is a defense and security professional. She was the top civil servant to complete the UK's Advance Command and Staff Course 26. Previously, she was Senior Lecturer in Defence and International Affairs at the Royal Military Academy Sandhurst (LOAC, field exercises, and WPS lead). She holds a Ph.D. in Politics from the University of Sheffield.

Valerie Morkevičius is Associate Professor of Political Science at Colgate University, Hamilton, New York, the United States. Her other work focuses on the intersection between power and ethics and on the applicability of traditional just war thinking to contemporary challenges, including cyber and information warfare. She is the author of *Realist Ethics: Just War Traditions as Power Politics* (2018), which explores how just war thinking is no stranger to pragmatic politics.

Cian O'Driscoll is Professor of International Relations at the Coral Bell School, Australian National University (ANU), Canberra, Australia. Cian is the co-editor of *Just War Thinkers* and the author of two monographs, the most recent of which is *Victory: The Triumph* and *Tragedy of Just War* (Oxford University Press: 2021). Cian is on the editorial team of the *Review of International Studies*.

Eric D. Patterson (Ph.D.) serves as President of the Victims of Communism Memorial Foundation, Scholar-at-Large at Regent University, and Research Fellow at Georgetown University's Berkley Center for Religion, Peace & World Affairs. He is the author of over 20 books on just war thinking, Christian Realism, and related topics including *Power Politics and Moral Order*, *Ending Wars Well*, and *Just American Wars*.

Alex Prichard is Associate Professor of International Political Theory at the University of Exeter, Exeter, the United Kingdom. He has published widely on Proudhon's international theory, the concept of anarchy in international relations, and the praxis of anarchist constitutional politics.

Gregory M. Reichberg is Research Professor at the Peace Research Institute Oslo (PRIO), where he examines the history of philosophical reflection on war and peace, just war in the Catholic tradition, contemporary military ethics, and the use of artificial intelligence in warfare. He has published *Thomas Aquinas on War and Peace* (2017) as well as numerous articles in *Ethics and International Affairs*, *Journal of Religious Ethics*, *Studies in Christian Ethics*, and *The Thomist*.

Jessica Wolfendale is Professor of Philosophy at Case Western Reserve University, Cleveland, Ohio, the United States. She is the author of two books and more than 40 articles and book chapters on topics including war crimes, torture, terrorism, military ethics, and security. Her current book project, *Torture and Terrorism in America*, examines the toleration of torture and terrorism in the U.S. criminal justice and national security contexts.

Owen Worth is Head of the Department of Politics and Public Administration at the University of Limerick, Limerick, Ireland. He writes in the areas of International Relations and Global Political Economy. Owen is the author of *Morbid Symptoms: The Global Rise of the Fall Right* (Zed Books, 2019), *Rethinking Hegemony* (Palgrave Macmillan, 2015), and *Resistance in the Age of Austerity* (Zed Books, 2013).

ACKNOWLEDGMENTS

Daniel Brunstetter would like to thank his doctoral students who listened patiently as he tried to explain the project when it was just a fledgling idea and who offered astute and constructive criticism that helped iron out many rough edges. Some of the chapters were presented at the 2023 International Studies Association conference, where the audience peppered our brave authors with great comments and questions, making my job as a discussant on the panel rather redundant. My appreciation goes out to UC Irvine's Council on Research, Computing, and Libraries, which provided funding to travel to the conference where I had many fruitful exchanges that helped to shape the book moving forward. The Center for Citizen Peace Building provided the needed funding to bring Juan Floyd-Thomas to UC Irvine to deliver the annual Hal Smith lecture, which served as the basis of his chapter on Martin Luther King Jr. I cherish the many conversations I had with Juan and the center's board members, especially in the home of the late Larry Kugelman, who was a long-time friend and dedicated peace builder. It has been fun working with each of the contributors. I've learned a great deal from their chapters and enjoyed the back and forth (maybe a little too exuberantly) as they responded to initial feedback. I appreciate their patience with my over-abundant e-mails as deadlines approached. Our editor, Andrew Humphrys, has been a joy to work with, as has the entire team at Routledge. We first chatted about the book idea at The Future of War conference held in Amsterdam. He has been ever so helpful and patient throughout all the stages it took to bring the project to fruition. Last but not least, my thanks go to Cian for agreeing to embark on this sequel to *Just War Thinkers*. Cian has been a friend through thick and thin, and the conversations we have shared over the years on war, peace, and the human condition have left an indelible mark on me as a scholar and as a person.

Cian O'Driscoll would like to thank Daniel for the pleasure of once again working together. We have traded many e-mails, Zoom calls, tracked comments, and whiskey drams over the course of compiling our two volume *Just War Thinkers* set. It has been an intellectually rewarding experience to work so closely with Dan, but, even more so, it has been a lot of fun. Thank you, Dan. Thank you, too, to our very generous and patient contributors, many of whom are dear friends. We are so grateful for the time and effort you poured into your chapters, and we hope you are as pleased with the result as we are. I also want to acknowledge the guidance of Jim Johnson. I am not sure Jim will especially like this book, but it owes its genesis to Jim's own monumental work in the field – work that is an ongoing source of inspiration for both Dan and me. Finally, echoing Dan, I would also like to express my gratitude to Andrew Humphrys and the team at Routledge, who have done so much to make the publishing process easy for us.

INTRODUCTION

Heretics, Humanists, and Radicals

Cian O'Driscoll

Introduction

The origins of this book lie in the text it follows. Published in (2018), our previous volume *Just War Thinkers* offers a one-stop shop guide to the key figures associated with the just war tradition, from Cicero to Jeff McMahan. Comprising a set of clear but detailed essays on 19 of the seminal thinkers of the tradition, it challenges the reader to think about how traditions are constituted—who is included and excluded, and how that is determined— and how they serve to enable, constrain, and indeed channel subsequent thought, debate, and exchange. So why, you may ask, just seven years later, do we need another book on just war thinkers?

The answer may not be what you are expecting. It lies both in what this book offers and in what it does not. Let's start with the latter. This book is not a straightforward sequel to the original. It does not extend the narrative traced in that collection of essays to include what we might term an outer ring of secondary, more obscure, B-list thinkers associated with the just war tradition. It is not, in other words, simply a continuation of that work or an attempt to somehow complete it. Such a book would undoubtedly be useful and, no doubt, interesting to certain scholars. But this is not such an endeavour. Rather, what this book offers is an inversion of this approach. Instead of curating a deep dive into ever more remote figures affiliated with the just war tradition, it brings together a collection of essays on political thinkers who had interesting and significant things to say about the ethics of war but who also, for one reason or another, are seldom regarded under the rubric of the just war tradition. These are heretics and outsiders whose ideas have been set aside, bracketed, overlooked, and sometimes even forgotten by

DOI: 10.4324/9781003428688-1

ethics of war scholars, all because they do not fit the mould of just war theory or align with the story we tell about it. The 'just war thinkers' featured in this volume, therefore, and perhaps ironically, are not really just war thinkers at all. Where, *Just War Thinkers* was very much an introduction to the canon of just war thought, this book is its anti-canonical counterpart. We wanted to revisit the perennial subject of ethical war by considering the range of ethical thinking about war that is not filed under just war theory, that does not speak the same moral language or use the same categories, but which tries to make sense of the war all the same. The humanists, heretics, and radicals whose positions it canvasses have something to teach us—something that one simply cannot get from mainline just war theory.

Members Only?

The just war tradition is the predominant western frame for thinking about the ethics of warfare. Often posited by its proponents as a via media between pacifism and realism, the just war idea supposes that the use of military force may, in certain circumstances, constitute a legitimate enterprise. This provokes a series of further questions that students of the just war tradition will be familiar with. For instance, what might those circumstances be, and how they should be circumscribed? Further, what if any limits should be attached to how as well as when military force is deployed? Questions such as these cut to the heart of international political theory, for they compel us to think more deeply about what values are worth fighting, killing, and dying for. They have elicited a great deal of attention in recent years, from both political and military leaders and scholars. Yet, these questions are in no way new. In fact, they, or questions very similar to them, have been a topic of debate since the sunset years of the Roman Empire. Schematised in the Middle Ages by canon lawyers, and further developed in the early modern period by jurists, these questions have latterly coalesced around three poles of analysis: the *jus ad bellum*, which considers the conditions under which the recourse to force may be justified; the *jus in bello*, which stipulates the limitations that apply to how force may be employed; and the *jus post bellum*, which stipulates the moral responsibilities of both the victors and vanquished in the wake of armed conflict. *Just War Thinkers* was designed to provide readable accounts of those who forged, enhanced, and perpetuated this tradition of thought. But we miss something if our ethical questions remain wedded to the just war canon, namely, the different questions and assumptions about war that those not trained in the canon might hold.

What is most intriguing, however, is neither the form that these questions take nor the readiness of influential actors, including presidents and prime ministers, to invoke them. Rather, it is the fact that these leaders cite the venerable history of these questions as their reason for engaging them. Leading scholars in the field (Johnson, 1975; Rengger, 2013; Bellamy, 2006; Kalmanovitz, 2020; Cox, 2023) argue that contemporary references to just war ideas draw their force from the deeper historical tradition that they evoke and from which they ostensibly derive. It is, in other words, its historical association with seminal figures in the mostly western canon—from Cicero to Hugo Grotius and beyond—that anchors just war discourse and furnishes its authority. Indeed, contemporary actors frequently invoke these figures to lend their own arguments' gravitas. If, then, we wish to make a sense of contemporary debates about the ethics of war, it is necessary to grapple with the historical canon that they invoke. This was one of the premises of *Just War Thinkers*, which we seek to revisit in this volume. Considering recent conflicts that shock the moral conscience—Russia–Ukraine, Israel–Gaza—from the perspective of just war theory, as many have done, offers some guidance, sure enough. But what if we step outside the canon? What might we learn from others who offer moral intuitions about ethical war?

The problem with the just war canon approach, of course, is that it can be construed as reproducing history as ethics. Our concern is related, albeit more oblique. Our worry is not simply that this approach reproduces history as ethics, it is that it reproduces *a very particular* history as ethics. Specifically, it casts a western and largely Eurocentric canon of political thought (which boasts of having roots in Christian political theology) as the source or point of reference from which all subsequent legitimate ethical thinking about war both proceeds and engages. With only one or two exceptions, notably Christine de Pizan, this canon is predominantly dead, white, and male. We do not raise this point to play identity politics, however. We raise it to draw attention to the worlds of thought that it occludes. Where are the figures like Frantz Fanon, Rosa Luxemburg, and Judith Butler, who all offer not only profound ethical thinking about war but also ethical thinking about war that diverges from that offered by the classical just war tradition? They have been filtered out such that they are seldom if ever cited in contemporary ethics of war scholarship. This produces what we see as some potentially very unfortunate effects. It conserves ethics of war scholarship as a members-only club for a very traditional mode of western political thought. And it does so by denying admission to and screening out ethical thinking about war that dares to be different or in any way departs from the script.

Herein lies the rub. Some of the most interesting political thought comes exactly from people who dare to be different or depart from the script. It is sometimes from outsiders and marginal figures that fresh and radical thinking often emerges. This book is designed to re-capture some of that thinking, for those interested in just war to take into consideration. The reasoning behind this is easily explained. We are keen to discover if an exposure to dissident thinking will arrest the confidence of contemporary ethics of war scholars who believe they hold all the right questions and answers, stir up novel debates, or prompt new lines of inquiry. Might, for instance, engagement with the work of Martin Luther King Jr. and John Brown lead scholars to re-evaluate the relation between the experience of institutional racism and the right to use armed force? Alternatively, could we find in the texts of Epictetus and Pierre-Joseph Proudhon the source-material for an ethical vision of war that eschews rather than assumes the authority of the state? Less provocatively perhaps, what additional insight might we glean by widening our lens to encompass the salient writings of idiosyncratic and difficult-to-pigeonhole figures like Carlos Calvo and Elizabeth Anscombe? It would seem counterintuitive to suppose that they, and other figures we discount as marginal to just war debates, have nothing of interest to say to today's ethics of war scholars.

Apart from this, there is a further ancillary benefit to reading them. By putting mainline ethics of war scholarship in dialogue with its critics and those who stand outside it, we can better probe its strengths and weaknesses. Acquiring some familiarity with rival views will enable us to get a better grasp of not only what is truly distinct and valuable about, for instance, mainstream just war theory—but also what is contingent and perhaps also idiosyncratic and even problematic about it. This will allow us to ensure that our ethical thinking about war does not succumb to complacency or a lazy conformism but instead remains fresh and vital. Here, then, is an invitation to consider an eclectic range of ethical positions on war that rarely get mentioned in ethics of war scholarship—and to reflect on how those positions might inform our own approach to these matters.

The aim, then, is to produce a text that both complements and challenges the mainstream literature on the ethics of war. Our hope is that this book will appeal to students, for whom the thinker-by-thinker format will prove a helpful resource, and advanced researchers, who will be drawn to both the selection of essays on fascinating but sometimes overlooked figures and the underlying argument regarding how we delimit the ethics of war as a field of thought.

Orphans

Who are the cast of *unusual* suspects that we will be considering? Before answering that question directly, we might say a few words about the usual suspects we will not be treating. We have decided not to pitch this book as an exercise in comparative ethics of war. That means we have not selected thinkers to represent different religious and cultural approaches to the ethics of war (e.g., Islamic, Hindu, or Buddhist approaches to the ethics of war). We see the value in this approach, but there is already an abundance of material on this topic (e.g. Reichberg, Syse, and Hartwell, 2014; Sorabji and Rodin, 2006; Morkevicius, 2018; Popovski, Reichberg, and Turner, 2009; Johnson and Kelsay, 1991). We have preferred instead to highlight some figures who occupy more liminal positions vis-à-vis ethics of war scholarship, others who are more deeply intertwined but rarely read on their own, and still others who have something in common with canonical figures but who have forged their views on different moral foundations. Our purpose, after all, is not to reify the boundaries of our ethical thinking about war but to test and expand them.

So, who exactly will be considering? One answer to this question reflects inspiration from an unlikely source. Tom Waits released a triple album in 2006 called *Orphans: Bawlers, Brawlers, and Bastards*. It comprises a collection of songs that did not fit, or make the cut for, his earlier studio albums. These are not necessarily lesser songs—some of them stand up to comparison with his very best work—they just happen not to fit any coherent album narrative; much, we wish to suggest, like many of the fascinating thinkers we feature in this volume. If we decide not to stick to the script of the just war tradition, which is in a large part historically constructed and misses many key figures who had much to say about violence but are too often eschewed (or worse, forced into just war categories), and if we instead look at thinkers who may have had different inspiration, categories, and experiences with war, where might this lead? To this end, we asked leading scholars of just war to pick a figure to write on, and it turns out that we too have our bawlers, brawlers, and bastards who compliment, challenge, and enlighten the way we—the editors—understand the heritage of just war.

In the first category are our bawlers, those thinkers who would lament or protest the wars of their time but eschew just war ethical thinking about war to make sense of them. Or indeed war itself. This includes not only Montaigne, Judith Butler, and Charles Mills but also perhaps Elizabeth Anscombe and Pope Francis. We also include Vera Cruz and Rawls in this group; they inherited just war categories and had to make them work in the context of their broader philosophical projects, though one can sense their unease with this approach. The brawlers are more forceful in their rejection of the constraints that mainstream just war theorising insists upon. We include in this bracket a diverse set of figures: Sepúlveda, Martin Luther, Reinhold Niebuhr, Pierre-Joseph Proudhon, John Brown, and Frantz Fanon. Each pushed the limits of what was acceptable in war, and challenged how we ought to view the oft-touted restraint-oriented mindset of just war thinking. We could also include those like Rosa Luxemburg and Martin Luther King Jr. who decried war in this category. Our bastards are (mischievously) so-called only because, to paraphrase the Wu Tang Clan, there appears to be no father to their style. These are one-off figures who are not easily assimilated into any single canon. We offer here Aristotle, Epictetus, Vico, Calvo, Sturzo, and Alasdair MacIntyre. This typology is, of course, intended playfully. Being more earnest, there is no one way to typologise the thinkers' profiled in this volume. Some of them reject the very idea of 'just war,' others do not; some are canonical in their own right, others less so; and some are classical, others contemporary. Some are humanists, some heretics, and others radicals. Playing off the subtitle of the book, we return in the Conclusion at the end to offer some broader reflections on what revisiting *Just War Thinkers* via these chapters can offer to those interested in the ethics of war today.

It is important to note that, in presenting essays on these figures, we do not mean to imply that they should be characterised as just war thinkers or embraced by scholars of the just war tradition as one of their own (even if they do not know it). As we have already noted, what is interesting about these figures for our purposes is precisely that most are *not* typically regarded as being part of the just war tradition and yet have staked out substantive ethical positions vis-à-vis the use of force. It may also be worth noting that, although we sequence the chapters chronologically, we are not suggesting that there is any kind of developmental narrative that connects them to one another. They do not represent a tradition, but our readers will no doubt see that their thoughts were forged in the context of many of the wars that shaped the canonical just war texts, including those of ancient Greece and Rome, the imperial conquests of the so-called New World, the religious wars in Europe, World War I, World War II, the Vietnam War, and the colonial antecedents of the War on Terror.

Structure and Design

This brings us to the structure of the book. This book has been designed in such a way as to supply the interested reader with a set of concise and accessible introductions to a range of important historical thinkers—in this case those who stake out views on the ethics of political violence, which sit outside or somehow crosscut the mainstream categories of just war theory, realism, and pacifism. Each chapter will be dedicated to a different thinker and will be written by an expert with a detailed knowledge of the figure concerned. All the authors approached to produce chapters are experts in the field; some are already established professors, and others are very much rising stars. Although the figures profiled hail from a variety of different intellectual and historical backgrounds, chapters will be presented in a manner that renders the material amenable to a contemporary international relations readership. With this in mind, each chapter will be structured in such a way as to highlight:

- The biographical, historical, and intellectual context of the thinker concerned;
- The drafting history and key points of their major texts;
- The controversies that the thinker provoked or partook in; and
- The legacy of the thinker and his/her ideas for the development of ethical thinking about war.

These focal points can be summarised in four terms: contexts, texts, controversies, and legacies. Viewed as a whole, this text will challenge to the reader to think about how traditions are constituted—who is included and excluded, and how that is determined—and how they serve to enable, constrain, and indeed channel subsequent thought, debate, and exchange about the ethics of war.

Conclusion

While there is an apparent playfulness about the composition of this book, it serves a deadly serious purpose. Since the publication of *Just War Thinkers*, Russia has invaded Ukraine to bring war to the doorsteps of Western Europe in a way reminiscent of the world wars of the twentieth centuries where ethical restraint was a footnote in military strategy, but with the adder threat of nuclear escalation. We suddenly face similar dilemmas that G.E.M. Anscombe deliberated. The long-simmering Israel–Gaza conflict has escalated, leading to untold destruction on the ground and the erection of antagonistic moral barricades across the world, with the just war stance all but untenable in face of those who would espouse the decolonial violence of Frantz Fanon. Between them, the conflicts in Ukraine and Gaza lay bare the poverty of the ethical resources at our disposal. The just war tradition, as it is classically conceived, only seems to take us so far. The question thus arises: might a wider, more eclectic take on the ethics of war be more helpful? Might, in other words, the humanists, heretics, and radicals treated in the essays that follow have something to teach us that might be useful when it comes to working through the ethical challenges of our current world? This is the wager of this book—and it is one we will return to in the Conclusion.

Works Cited

Bellamy, Alex J. 2006. *Just Wars: From Cicero to Iraq*. Cambridge: Polity.

Cox, Rory. 2023. *Origins of the Just War: Military Ethics and Culture in the Ancient Near East*. Princeton: Princeton University Press.

Johnson, James Turner. 1975. *Ideology, Reason, and Limitation of War*. Princeton: Princeton University Press.

Johnson, James Turner and John Kelsay (eds.). 1991. *Just War and Jihad: Historical and Theoretical Perspectives on War and Peace in Western and Islamic Traditions*. New York: Greenwood Press.

Kalmanovitz, Pablo. 2020. *The Laws of War in International Thought*. Oxford: Oxford University Press.

Morkevicius, Valerie. 2018. *Realist Ethics: Just War Tradition as Power Politics*. Cambridge: Cambridge University Press.

Popovski, Vesselin, Gregory M. Reichberg, and Nicholas Turner (eds.). 2009. *World Religions and Norms of War*. Tokyo: United Nations Press.

Reichberg, Gregory M., Henrik Syse, and Nicole M. Hartwell (eds.). 2014. *Religion, War, and Ethics: A Sourcebook of Contextual Traditions*. Cambridge: Cambridge University Press.

Rengger, Nicholas J. 2013. *Just War and International Order: The Uncivil Condition in World Politics*. Cambridge: Cambridge University Press.

Sorabji, Richard and David Rodin (eds.). 2006. *The Ethics of War: Shared Problems in Different Traditions*. Abingdon: Routledge.

1

ARISTOTLE (384 BC–322 BC)

Andrée-Anne Mélançon

Introduction

The legacy of Aristotle's work in Western Philosophy, and knowledge more broadly, cannot be understated. Over what would be the equivalent of 50 contemporary volumes, Aristotle, known to generations to come as simply the Philosopher, explored topics ranging from zoological studies to poetics, logic, politics, and ethics—many of which have something to do with war. Such was his influence on moral thinking that, to paraphrase the popular TV series, *The Good Place* (S1E3), it is as if Plato had "died and left Aristotle in charge of ethics." Long before Augustine—oft considered the father of the just war tradition, or indeed Cicero, his Roman antecedent who was the starting point of the first volume of *Just War Thinkers*—Aristotle contemplated just and unjust war. For war was ever-present in Aristotle's lifetime, from the stories he inherited via Homer to expeditions—undertaken by his student who came to be known as Alexander the Great—that heralded the expansion of Greek world dominance. His contemplations on war left, for better or worse, their influence for centuries to come.

Little is known of Aristotle's personality and character, but what can be confirmed is that he was driven by a deep desire for knowledge (Barnes, 2000, p. 1). This extended to knowledge of the ethics of war. In his own time, Aristotle could be read in conversation with other Greek thinkers not often associated with the just war tradition, who were concerned with the ethics of war (O'Driscoll, 2015). For later Christian generations—from the revival period with Aquinas in the thirteenth century to the discovery of the New World that implicated Vitoria, Las Casas, and Sepúlveda (see Luke Glanville's chapter in this volume) in the sixteenth century—he was both the source of

DOI: 10.4324/9781003428688-2

authority and controversy. For modern generations, his virtue ethics remains a source of inspiration, as the chapter on Alasdair MacIntyre in this volume demonstrates. Thus, although Aristotle might not be the first name often associated with the just war tradition, his body of work was fundamental in establishing the moral foundations upon which just war thinking would evolve from. If Augustine is the grandfather of just war theory, Aristotle has a claim to be the too often forgotten great-grandfather.

Contexts

Aristotle was born in 383 BC in Stagira, northern Greece. Although little is known of his early education, it can be estimated that he followed the literary and gymnastic education expected of well-connected Greek families. At 17, Aristotle moved to Athens to study in the Academy under Plato's leadership. Despite his time at the Academy, Aristotle was no Platonist. As Barnes states, "many of the doctrines central to Platonism are strongly criticized in Aristotle's treatises" (2000, p. 35). Aristotle was accused of ingratitude and nicknamed "the Foal" by Plato himself. Nevertheless, his first Athenian period lasted 20 years and ended in 347 BC, the year of Plato's death, when Aristotle left Athens hastily. It is suspected that the move was motivated by rising political issues related to Philip II of Macedonia's expansionist ambitions in the region (Haworth, 2004, p. 36).

In 345 BC, Aristotle moved to Lesbos and settled there until his return to Athens. It is during his time in Lesbos that Aristotle developed his zoological interests and wrote his two most important documents when it comes to just war thinking: the *Nicomachean Ethics* (named after his son) and *The Politics* (Haworth, 2004, p. 37). In 343 BC, Philip II of Macedonia invited Aristotle to Pella to become the tutor of his son, Alexander. Aristotle is credited with having brought Homer's *Iliad* to Alexander, a book that had a lasting impact on both young Alexander (Wu, 2022, pp. 71–72) and Aristotle himself. As Simone Weil argues,

> [T]he true hero, the true subject, the center of the *Iliad* is force. Force employed by man, force that enslaves man, force before which man's flesh shrinks away. In this work, at all times, the human spirit is shown as modified by its relations with force.
>
> *(1956, p. 3)*

Man's relationship to force is exactly what Aristotle seeks to understand—and regulate. Aristotle recognised that art, especially Homer's poetry, plays a crucial role in moral education (Cannatella, 2006, pp. 9–12). A connection that was also recognised by Rembrandt, who, when commissioned to produce a painting without a specific subject, decided to paint *Aristotle*

Contemplating a Bust of Homer in 1653. In *Poetics* (Aristotle, 1995, 1459a), where Aristotle explores the social function of art, Homer is described as having been "divinely inspired." Poems such as the *Iliad* offered a means to explore and assess various role models in action, showcasing a range of both vices and virtues, ultimately making these more recognisable or discoverable to the reader. Thus, epic poetry serves a social function; as a vehicle for moral reflection, it permits the exploration of moral dilemmas, virtue, and the tragic consequences of human action related to war. In addition to poetry, Aristotle also taught Alexander history, science, ethics, and politics – the topics that constitute what the Romans would later call *arts liberals* (Wu, 2022).

Aristotle's second Athenian period began in 335 BC when he returned and began teaching in the Lyceum until he retired to Chalcis in 322 BC. At the Lyceum, Aristotle was a public figure, lecturing both his students and the public. It is suspected that Aristotle's retirement was prompted by political turmoil. As Jonathan Barnes asserts, Aristotle allegedly stated that following Socrates' trial and execution, "he did not want the Athenians to commit a second crime against philosophy" (2000, p. 9).

Texts and Tenets

Aristotle's works are extensive. Christopher Shields (2023) classifies his surviving works into four categories: Organon, Theoretical Sciences, Practical Sciences, and Productive Sciences. This categorisation reflects Aristotle's core belief that thought is *practical* (concerned with actions and how one should act), *productive* (concerned with the production/creation of things), or *theoretical* (concerned with truth).

Aristotle sees knowledge as a science, with generalisations drawn out from perceptions as the source of knowledge (Barnes, 2000, pp. 92–94). This approach is possible thanks to Aristotle's core belief that nature does nothing in vain; hence the possibility to rely on empirical research, perceptions, and observations as sources of knowledge. Take, for example, zoology. His *History of Animals* (and *Dissections* which is now lost) catalogued creatures ranging from insects to European bison and octopuses. Alan Haworth describes Aristotle's zoological work as "so systematically constructed, and it contains so many detailed and accurate observations, that it once prompted Charles Darwin himself to remark of his own contemporaries that they were 'mere schoolboys compared to old Aristotle'" (2004, pp. 36–37). Importantly for our purposes, the legacy of this zoological study has fingerprints in Aristotle's later work on practical philosophy, where he also applies the scientific method. This leads Haworth to label Aristotle as the first *political scientist* (2004, pp. 38–39). Indeed, Aristotle builds on his classification of animals to classify people by social class in *Politics* (Aristotle, 1999, 4.1290b–4.1291a).

Social classes are based on the function someone performs. Aristotle argues that men can be divided and classified according to their birth, wealth, education, employment, etc. This implies that social class has a *function* in the same way that a body part does, which has wider implications for Aristotle's ideas on just and unjust slavery, as well as war. Broadly, this leads Aristotle to assert that man is a political animal (*Politics*, 1.1253a), that the city-state exists by nature, and that "nature, as we declare, does nothing without a purpose" (*Politics*, 1.1253a).

Aristotle's famous assertion that man is a political animal is meant literally. Man is described as the only animal to possess speech (*Politics*, 1.1253a) which grants mankind the ability to perceive and discuss right and wrong, good and bad, just and unjust. Further, men have an impulse to form partnerships, which leads to the formation of political associations. These groupings begin with the household (the family), which is responsible for fulfilling everyday needs. Households then associate together to form villages, the second form of association. The *polis* (state or political community) is the most complex form of association. It allows for the fulfilment of man's potential: "the state is one form of partnership of similar people, and its object is the best life that is possible" (*Politics*, 7.1328a). In other words, the state is an association of persons who aim for the best life possible in a situation where "the greatest good is happiness" (*Politics*, 7.1328a). Here, Aristotle uses the term "*eudaimonia,*" which is generally translated to "happiness." However, the Greek term has a more expansive meaning, which also includes flourishing (Kraut, 2022). As Aristotle explains in *Nicomachean Ethics*, happiness is more of a process: "to be happy takes a complete lifetime; for one swallow does not make spring, nor does one fine day; and similarly one day or a brief period of happiness does not make a man supremely blessed and happy" (Aristotle, 1995, 1098a.1).

Aristotle's understanding of the role of the polis as essential to the attainment of virtue, as well as his views on social class, has broad ramifications for the use of force. Rulers hold a responsibility to defend and safeguard the community from threats to the attainment of *eudaimonia*. Indeed, Aristotle argues that war is needed to establish and maintain what he understands to be the natural hierarchy of the world (*Politics*, 7.1333b). Aristotle states quite matter-of-factly that since "barbarians have no class of natural rulers" (*Politics*, 1.1252b), Greeks *should* rule over barbarians, whom he described as natural slaves. As will be explored later and in several chapters across this volume, notably those on Sepúlveda and Charles Mills, this assumption has had long-term and controversial implications for Western thought about just and unjust war.

This is not to say Aristotle was a warmonger, quite the opposite in fact. Aristotle's key contribution to just war thinking is the clear formulation of the argument that war must be fought for the just cause of creating peace:

"war must be for the sake of peace" (*Politics*, 7.1333a). Aristotle never for-mulated a doctrine on war and peace. As Martin Ostwald suggests, "this may explain why there has been no coherent and comprehensive scholarly treatment of the subject of [his] attitudes to war and peace" (1996, p. 104). Nevertheless, Aristotle's thoughts about war can be assembled from a range of statements, mainly found in *Politics*.

To begin, Aristotle sees war as somewhat inevitable since all political life is divided into war and peace: "Also life as a whole is divided into business and leisure, and war and peace, and our actions are aimed some of them at things necessary and useful, others at things noble" (*Politics*, 7.1333a). War was necessary to ensure defence but was not an end in itself:

> Experience supports the testimony of theory, that it is the duty of the lawgiver rather to study how he may frame his legislation both with regard to warfare and in other departments for the object of leisure and of peace. Most military states remain safe while at war but perish when they have won their empire; in peace-time they lose their keen temper, like iron. The lawgiver is to blame, because he did not educate them to be able to employ leisure.
>
> *(Politics, 7.1334a)*

Following this line of thought, he takes aims at warrior cultures, such as Sparta, the Scythians, and the Celts, who prepare only for war and never teach their citizens how to live in peace. He concludes that for such "warlike races," war is regarded as noble when it shouldn't: "how can that be worthy of a statesman or lawgiver which is not even lawful? and government is not lawful when it is carried on not only justly but also unjustly—and superior strength may be unjustly exercised" (*Politics*, 7.1324b). The highest pursuit of the state remains the quest for a good life. Thus, the lawgiver must dis-seminate the idea that the only noble pursuit is to live well, which he develops fully in the *Nicomachean Ethics*.

Aristotle begins *Nicomachean Ethics* by arguing that a person who lives well has the skill of acting virtuously:

> [T]he function of a good man is to perform these activities well and rightly, and if a function is well performed when it is performed in accordance with its own proper excellence—from these premises it follows that the Good of man is the active exercise of his soul's faculties in conformity with excellence or virtue, or if there be several human excellences or virtues, in conformity with the best and most perfect among them.
>
> *(1098a.1)*

Aristotle further defines virtue as a *disposition in action*, differentiating between possessing and displaying virtue (*Nicomachean Ethics*, 1094a.1).

Thus, virtue is something that is developed through practice. It is worth noting here that the Greek term "*aretē*," which is generally translated to "virtue," is broader in meaning and is closer to "skill" and "excellence" (Shields, 2023). Building on this understanding, Aristotle presents a typology of virtue and vice whereby for each sphere of action or feeling, both deficiency and excess are vices whereas the mean (mid-point) represents virtue: "There are then three dispositions—two vices, one of excess and one of defect, and one virtue which is the observance of the mean; and each of them is in a certain way opposed to both the others" (*Nicomachean Ethics*, 1108b.1). For example, the vice of rashness is an excess of confidence, the vice of cowardice is a deficiency, and the virtue/*aretē* of courage is the mean. This scale of vice–virtue–vice is sometimes referred to as Aristotle's Golden Mean.

Aristotle's virtue ethics emerges from this conception of virtue/*aretē*. He views moral behaviour as the exercise of the virtues: "it is the active exercise of our faculties in conformity with virtue that causes happiness, and the opposite activities its opposite" (*Nicomachean Ethics*, 1100b.1). Acting virtuously is not a means to something else, like courage in battle could be a means to receive glory and honour. Rather:

> [H]appiness above all else appears to be absolutely final in this sense, since we always choose it for its own sake and never as a means to something else; whereas honor, pleasure, intelligence, and excellence in its various forms, we choose indeed for their own sakes (since we should be glad to have each of them although no extraneous advantage resulted from it), but we also choose them for the sake of happiness, in the belief that they will be a means to our securing it. But no one chooses happiness for the sake of honor, pleasure, etc., nor as a means to anything whatever other than itself.
> (*Nicomachean Ethics, 1097b.1*)

This vision of happiness as being self-sufficient contrasts with the utilitarian account which sees happiness as distributive. It also raises different questions than utilitarianism. Instead of asking how to maximise good for the greatest number of people, Aristotle is more retrospective and introspective: "when you are dying, and you look back over the course your life has taken, what judgement will you form?" (Haworth, 2004, p. 51). In other words, for Aristotle, happiness is *evaluative*; it is used to evaluate the course of a life. Thus, happiness/*eudaimonia* is not a state of mind or being but rather about "the good life" or "doing well" (*Nicomachean Ethics*, 1095a.1). As Haworth (2004, p. 53) summarises, "to have lived *eudaimonically* is thus to have lived well."

To return to his example of the Spartans—or any people who are always on the war footing—waging war for the sake of war is not *living well*. It is here that Aristotle formulates his core argument that "war must be for the sake of peace"—an argument that would be picked up by Cicero and Augustine, and which remains central to the just war tradition (*Politics*, 7.1333a). This foundational argument leads him to make a distinction between just and unjust wars:

> The proper object of practising military training is not in order that men may enslave those who do not deserve slavery, but in order that first they may themselves avoid becoming enslaved to others; then so that they may seek suzerainty for the benefit of the subject people, but not for the sake of world-wide despotism; and thirdly to hold despotic power over those who deserve to be slaves.
>
> *(Politics, 7.1333b-7.1334a)*

With this, Aristotle lays out the foundations for what will eventually become the *jus ad bellum* just cause criterion. At least his version of it. His just causes could be summarised as:

i.   To defend against aggressors who unjustly try to enslave one.
ii. To establish governance for the good of the governed (and not for despotism).
iii. To establish mastery over "natural" slaves.

Aristotle also lays the groundwork for the legitimate authority criterion. In his critique of the Spartan Constitution, he argues against wars of conquest, stating that instead of being focused on dominating outsiders, legislators should establish peace and justice *within* the state (*Politics*, 7.1333b). His discussion on "barbarians" as natural slaves who do not know right from wrong, just from unjust, suggests that they could never have the authority to wage just wars (*Politics*, 1.1252b).

While often referred to as the Philosopher, Aristotle was not conjuring these ideas from thin air. Being the political scientist that he was, he supports his views, however flawed, with theory and observation:

> Experience supports the testimony of theory, that it is the duty of the law-giver rather to study how he may frame his legislation both with regard to warfare and in other departments for the object of leisure and of peace
>
> *(Politics, 7.1334a)*

Controversies

Perhaps the most significant controversy that emerges from Aristotle's views on war is the view of just war against barbarians, which emerges from his defence of natural slaves (Fernández-Santamaria, 1975; Heath, 2008). In *Politics* (1.1254b), he distinguishes between conventional slavery, like the enslavement of defeated soldiers (a common practice at the time), and the concept of natural slavery, which pertains to certain people who are by nature meant to be ruled:

> [A]ll men that differ as widely as the soul does from the body and the human being from the lower animal . . . these are by nature slaves, for whom to be governed by this kind of authority is advantageous.

Here again, Aristotle's views are greatly influenced by his (pseudo-) scientific study of the natural world and his identification of various orders of species. This is an example of how one's method and assumptions about the world can blind one to alternative ways of thinking about morality, human being, and just war. Humans, like other animals and species, are ordered in a hierarchy, and Aristotle believes that there are moral differences between natural born slaves and born-free men. These include material/physical differences, albeit physical strength and posture might have more to do with nurture than nature. Thus, Aristotle questioned the reliability of physical indicators. Instead, the soul (*psyche*) is the most important marker between free men and natural slaves. The core differentiating characteristics are the capacity for reason, the ability to govern, and self-sufficiency:

> And all possess the various parts of the soul, but possess them in different ways; for the slave has not got the deliberative part at all, and the female has it, but without full authority, while the child has it, but in an undeveloped form. Hence the ruler must possess intellectual virtue in completeness.
>
> *(Politics, 1.1260a)*

In other words, masters/rulers are capable of both active and passive reasoning whereas natural slaves cannot be virtuous on their own. Natural slaves only possess passive reason, which means that they can recognise and obey commands: "he needs only a small amount of virtue, in fact just enough to prevent him from failing in his tasks owing to intemperance and cowardice" (*Politics*, 1.1260a). Importantly, since natural slaves lack the active reasoning which allows one to understand the world and deliberate morally about actions and command (rationality), they cannot rule themselves. Considering this, Aristotle then argues that it is *advantageous* for those with only passive reason to be ruled by a master. This is the crux of his defence of natural slavery. As Gary Simpson (2011, p. 173) summarises, "having a master is for the slave's own good."

Such an argument is disturbing, and it highlights the view that Aristotle's world was one of natural inequality, which, as we saw before, Aristotle connects with just war. Aspects of this civilisational dichotomy, as the chapter on Charles Mills shows, persist today. While many just war scholars rightly point the value of Aristotle's virtue ethics (a point we return to later in the chapter), the controversial legacy of this defence of natural slavery is far-reaching. Those who would later take Aristotle as an authority could turn to his arguments and instrumentalise them for their own "moral" purposes. They would find several passages in the *Politics* where the Philosopher specifically says the civilised can wage a just war against those deemed to be "barbarians" to use as supportive evidence.

Indeed, Western European colonisers used a variation of this argument on multiple occasions. As Simpson (2011, p. 174) highlights,

> [A]lready in the fifteenth century the moral rational for slavery began to move dramatically from Soul to body in the form of racism based on skin color and other bodily markers, which were thought to indicate more reliably the quality of reason.

In the sixteenth century, Juan Gínes de Sepúlveda (see Luke Glanville's chapter in this volume) uses natural slavery to justify the wars of subjugation of indigenous populations in the Americas at the hands of the Spanish. Moreover, Aristotle's argument for natural slavery also inspired the justification of racial slavery (Campbell, 1974; Dobbs, 1994; Monoson, 2011). He thus found many proponents in the Antebellum South, with authors such as George Fitzhugh and Thomas R. Dew being inspired by Aristotle to support racial slavery (Wish, 1949; Haynie, 2022). In contemporary times, scholars have drawn parallels with the Bush administration's justifications for the U.S.-led War on Terror in the immediate post-9/11 era (Brunstetter and Zartner, 2011).

As other chapters discuss this in details, we will leave it at this: Aristotle's inegalitarianism and lack of conception of human dignity remain one of the most controversial aspects of his work to this day, and one is forced to ask: are such views morally reconcilable with his virtue ethics? More metatheoretically, which is a question to consider when reading any authoritative moral figure: how do we reconcile their moral flaws with their virtues?

Legacy

It is difficult to overstate the importance of Aristotle's legacy. Aristotle is one of the founders of Western Philosophy. His central position, walking next to Plato and carrying a copy of his *Nicomachean Ethics* in Raphael's *The School of Athens,* is just one illustration of his foundational importance. He was the first political *scientist*. He believed that the scientific methods used to study natural phenomena could also be applied to the study of social phenomena and politics, a belief still found in contemporary positivism. Moreover, his views on ethics have had a long and illustrious shelf-life.

A key legacy of Aristotle's virtue ethics can be found in the way it shaped Thomas Aquinas' thinking in the thirteenth century, leading to the development of Scholasticism (see Reichberg, 2018 for more on Aquinas and the just war tradition). Aquinas' approach sought to reconcile classical philosophy, especially Aristotle's work, with Christian theology leading to a movement that emphasises dialectical reasoning to resolve questions and contradictions. Building on Aristotle's virtue ethics, Aquinas also supported that virtues were the path to moral excellence and that the key to understanding human action lies in understanding virtue. Renowned Aquinas scholar Leo Elders summarised Aristotle's legacy as follows:

> Thomas established the general conformity of the principles of Aristotle's thought with tenets of the Christian faith, except for the creation of the world at the beginning of time and divine providence. . . . [W]e must realize that it was Thomas's intention to defend Aristotle and to show the conformity of his philosophy with Christian doctrine.
>
> *(2013, p. 745)*

This was not a passing fad. The legacy of Aristotle's work can also be seen in contemporary debates about moral philosophy. The field of Neo-Aristotelian virtue ethics continues to explore the concept of *eudaimonia* and human nature (Hirji, 2019). Notable authors include Philippa Foot (1978), John McDowell (1979), Rosalind Hursthouse (1999), David McPherson (2020), and Alasdair MacIntyre (studied in this volume).

Aristotle's legacy as a moral philosopher impacted the field of just war thinking. His core contribution is the articulation that war is not an end in itself, but a means to an end, namely that war ought to be waged for the sake of peace. This rejection of the idea that war has any inherent value—as other Greek city-states such as Sparta believed—lays the foundation for the development of Western just war thinking. Importantly, Aristotle's early formulations of what will develop to become the just cause criterion laid the fundamental groundwork of recognising that war can *sometimes* be morally just and, consequently, affirmed that there could be unjust wars as well. These are important observations for any civilisation whose existence is structured around the arts of the military. Perhaps more importantly, he put virtue ethics at the core of just war.

More recently, Aristotle's influence is seen as being overt in Grady Scott Davis' *Warcraft and the Fragility of Virtue: An Essay in Aristotelian Ethics* (2011). Here, Davis criticises the over-reliance of the traditional just war thinking of Michael Walzer and James Childress on rules, which simultaneously neglects the foundation of said rules. Davis is particularly critical of the supreme emergency exception and rejects the language of necessity. Instead of rules, Davis' approach, which he calls "orthodox Aristotelianism," argues that war ought to be assessed in terms of virtues:

> In undertaking any activity, the Aristotelian thinks of herself as a crafts-man and acts as shaped by the virtues. Warcraft, on this account, pre-supposes the cardinal virtues and requires me to reflect on what further particulars of skill and knowledge I need to acquire, what conditions need to be satisfied in order to create a worthy product.
>
> *(Davis, 1987, p. 487)*

David Fisher (2011), who served in senior positions in the UK Ministry of Defence, continued this trend of integrating virtue ethics into debates about the ethics of war by blending Aristotelian virtue ethics with consequential-ism. Fisher's *Morality and War: Can War be Just in the Twenty-First Century?* turns to Aristotle to answer in the affirmative. Written in the throes of the just war controversies of the post-9/11 era, but casting its gaze on the wars of the twentieth century too, he argues that just war ethics needs to be refashioned to better provide practical guidance to politicians, generals, and ordinary service people. His virtuous consequentialism combines Aristotelian virtue ethics with consequentialist arguments, paying a special attention to the cor-rupting effects of the use of force, especially acts such as torture. In doing so, Fisher judges actions by their impact on human flourishing, or Aristotelian *eudemonia*. Fisher further argues that professional military education ought to be grounded in both just war theory and virtuous consequentialism in order to ensure that armed forces are prepared not only to face modern war and its challenges but also to lessen its horrors. Aristotelian virtue represents a source, *the source*, with which to teach morality to the military as well as to statesmen, so that ethical conduct are habits of both thought and action.

The idea that the military need to be schooled in virtue is essential to armies that seek to hold the high moral ground. A final legacy is the importance of cultivating Aristotelian virtue, found in the various moral codes armed forces adopt to enshrine in the service members a set of core values (Olsthoorn, 2014).[1] Teaching military ethics is, at its core, a way of ensuring the excel-lence of (moral) character through Aristotelian habit formation, which is not only essential to act justly in the face of adversity, such as in war, but also essential to cultivate in times of peace. Commenting on the fractured U.S. military ethos after the withdrawal from Afghanistan in 2022, Colonel Thomas J. Gordon makes the point explicitly:

> It was not Athens' defeat in the Peloponnesian Wars or Rome's failed excursions in Gaul that led to the collapse of their Republics, but the moral vacuum that followed. Sophists, the pundits of the day, filled this

void with political speech—unanchored in truth—to persuade people to act against their best interest. It was in this milieu that Aristotle's moral philosophy was born. Aristotle's moral theories are ethics for the real world and remain just as valid today as at their inception.

(2022)

For Gordon, the need to instil military virtue is essential to the function of democracy. In a world in which behaviour is moulded by social media, including in the military, he double downs on teaching ethics:

> Developing principled leaders—men and women of virtue and character— to bridge this political divide is a national security imperative. Military professionals need to turn away from social media feeds and turn off cable news' incessant "infotainment," and rise above the fray.

They should be reading Aristotle instead.

In closing, it is clear that Aristotle is a just war thinker. Though some aspects of his legacy remain controversial today, such as his defence of natural slavery and his limited view of human dignity, he remains a greatly influential thinker, especially in the field of moral philosophy. His belief that there are moral objective principles, which could be empirically observed, laid the groundwork for the development of just war theory's criteria or principle-based framing. In addition, his departure from the celebrated militarism of ancient Greece in favour of a more nuanced position that "war must be for the sake of peace" (*Politics* 7.1333a) set an important precedent. Without this initial articulation of the idea that war must have a just cause, perhaps just war thinking would not have developed in the way it did. For this statement alone, Aristotle should be seen as one of the forbearers of just war thinking.

Note

1 For example, the British Army uses the acronym CDRILS to instil its values in all soldiers: courage, discipline, respect of others, integrity, loyalty, and selfless commitment.

Works Cited

Aristotle. 1995. *Poetics*. Edited and Translated by Stephen Halliwell and W. H. Fyfe, with Donald Russell and Doreen C. Innes. Cambridge, MA: Harvard UP.

Aristotle. 1999. *Politics*. Translated by Benjamin Jowett. Kitchener: Batoche Books.

Aristotle. 2004. *The Nicomachean Ethics*. Translated by J. A. K. Thomson. New York: Penguin Classics.

Barnes, Jonathan. 2000. *Aristotle: A Very Short Introduction*. Oxford: Oxford University Press.

Brunstetter, Daniel R. and Dana Zartner. 2011. Just War against Barbarians: Revisiting the Valladolid Debates between Sepúlveda and Las Casas. *Political Studies* 59(3), pp. 733–752.

Campbell, Mavis. 1974. Aristotle and Black Slavery: A Study in Race Prejudice. *Race & Class* 15(3), pp. 283–301.

Cannatella, Howard James. 2006. Plato and Aristotle's Educational Lessons from the Iliad. *Paideusis* 15(2), pp. 5–13.

Davis, Grady Scott. 1987. Warcraft and the Fragility of Virtue. *Soundings: An Interdisciplinary Journal* 70(3/4), pp. 475–494.

Davis, Grady Scott. 2011. *Warcraft and the Fragility of Virtue: An Essay in Aristotelian Ethics*. Eugene: Wipf & Stock Publishers.

Dobbs, Darrell. 1994. Natural Right and the Problem of Aristotle's Defense of Slavery. *The Journal of Politics* 56(1), pp. 69–94.

Elders, Leo J. 2013. St. Thomas Aquinas's Commentary on Aristotle's 'Physics'. *The Review of Metaphysics* 66(4), pp. 713–748.

Fernández-Santamaria, José A. 1975. Juan Ginés de Sepúlveda on the Nature of the American Indians. *The Americas* 31(4), pp. 434–451.

Fisher, David. 2011. *Morality and War: Can War be Just in the Twenty-First Century?* Oxford: Oxford University Press.

Foot, Philippa. 1978. *Virtues and Vices*. Oxford: Blackwell.

Gordon, Colonel Thomas J. 2022. Washington and Aristotle Can Restore the Military's Professional Ethos. *Proceedings* 148. [Accessed 14 May 2024]. Available at: www.usni.org/magazines/proceedings/2022/february/washington-and-aristotle-can-restore-militarys-professional.

Haworth, Alan. 2004. *Understanding the Political Philosophers: From Ancient to Modern Times*. New York: Routledge.

Haynie, Olivia. 2022. Aristotle and the Argument for American Slavery. *Discentes*. Available at: https://web.sas.upenn.edu/discentes/2022/11/27/aristotle-and-the-argument-for-american-slavery/.

Heath, Malcolm. 2008. Aristotle on Natural Slavery. *Phronesis: A Journal for Ancient Philosophy* 53(3), pp. 243–270.

Hirji, Sukaina. 2019. What's Aristotelian About neo-Aristotelian Virtue Ethics? *Philosophy and Phenomenological Research* 98(3), pp. 671–696.

Hursthouse, Rosalind. 1999. *On Virtue Ethics*. Oxford: Oxford University Press.

Kraut, Richard. 2022. Aristotle's Ethics. In N. Zalta and Uri Nodelman. eds. *The Stanford Encyclopedia of Philosophy Edward*. Available at: https://plato.stanford.edu/archives/fall2022/entries/aristotle-ethics/.

McDowell, John. 1979. Virtue and Reason. *The Monist* 62, pp. 331–350.

McPherson, David. 2020. *Virtue and Meaning: A Neo-Aristotelian Perspective*. Cambridge: Cambridge University Press.

Monoson, S. Sara. 2011. Recollecting Aristotle: Pro-Slavery Thought in Antebellum America and the Argument of Politics Book I. In Richard Alston, Edith Hall, and Justine McConnell. eds. *Ancient Slavery and Abolition*. Oxford: Oxford University Press, pp. 247–278.

O'Driscoll, Cian. 2015. Rewriting the Just War Tradition: Just War in Classical Greek Political Thought and Practice. *International Studies Quarterly* 59(1), pp. 1–10.

Olsthoorn, Peter. 2014. Virtue Ethics in the Military. In Stan van Hooft. ed. *The Handbook of Virtue Ethics*. Abingdon: Routledge, pp. 365–374.

Ostwald, Martin. 1996. Peace and War in Plato and Aristotle. *Scripta Classica Israelica* 15, pp. 112–118.

Reichberg, Gregory M. 2018. Thomas Aquinas. In Daniel R. Brunstetter and Cian O'Driscoll. eds. *Just War Thinkers: From Cicero to the 21st Century*. Abingdon: Routledge, pp. 50–63.

Shields, Christopher. 2023. Aristotle. In Edward N. Zalta and Uri Nodelman. eds. *The Stanford Encyclopedia of Philosophy*. Available at: https://plato.stanford.edu/archives/win2023/entries/aristotle/.

Simpson, Gary. 2011. For Their Own Good: Moral Slavery 101—The Aristotelian Cantus Firmus. *Word & World* 31(2), pp. 166–174.

Weil, Simone. 1956. *The Iliad or the Poem of Force*. Wallingford PA: Pendle Hill Pamphlet no 91.

Wish, Harvey. 1949. Aristotle, Plato, and the Mason-Dixon Line. *Journal of the History of Ideas* 10(2), pp. 254–266.

Wu, Yuchen. 2022. The Relationship Between Aristotle and Alexander the Great. *Advances in Social Science, Education and Humanities Research* 638, pp. 71–75.

2

EPICTETUS (C. 50–135 AD)

Luke Armstrong

Introduction

When faced with the question of how to conduct oneself in cases of violent conflict, Epictetus asks us to consider only what we are able to do with complete certainty. On finding that there is not much we can do while being sure of the result, he argues we must learn to contain our ambitions. However, while we must avoid grandiose ambitions, we still hold duties in the world; we must act as good citizens and resist tyrants. There may be little we can do to resolve conflicts, but Epictetus does not lead us towards unadulterated realism; if there is something we can do within our power to resist the unjust use of force, we have a duty to do it, even if our own death is the result. In Epictetus, we find a parallel to revisionist just war theory: both the revisionists and Epictetus are concerned with the duties individuals hold in conflicts. However, whereas the revisionists debate exactly how to define these duties, there is a simplicity to Epictetus' approach which leads to quite straightforward conclusions. Rather than question the duties of individuals from an external vantage point, Epictetus asks us to look inwards: whatever is within our power to resist unjust aggression is what we must do.

Epictetus saw self-understanding as our foremost task. To understand ourselves, we must perfect the will and realise human reason's fullest potential. For Epictetus, it is when we misunderstand ourselves and what is within the scope of our own power that we come into conflict with others. To avoid or attempt to resolve conflicts, whether peacefully or through force, we must understand ourselves. According to Epictetus, an introspective account of the just use of force can be found, which offers us something otherwise missing from the just war tradition. To assess whether the use of force is warranted,

DOI: 10.4324/9781003428688-3

we must look at what is within our own power. We find very little; the world outside of us contains much that is unexpected which could thwart the aims of even the most powerful person. In what follows, I interpret Epictetus as a realist but one who retains a hope for peace. To this end, however, it is left for individuals to determine what is within the scope of their rather limited power. Epictetus, then, speaks to the revisionist concern with the individual within the just war tradition, but he asks the individual to look inwards for solutions to the revisionists' problems.

Contexts

Though little is known of Epictetus' life, he is understood to have been born into slavery, sometime between AD 50 and 60, in the Graeco-Roman city of Hierapolis (Long, 2002, pp. 10–11). Epaphroditus, the secretary of Emperor Nero and Domitian, acquired Epictetus as a slave, allowing him to attend the lectures of the Stoic Musonius Rufus and eventually allowing him his freedom (Matheson, 1916, p. 13). A.A. Long thinks it is likely that Epictetus began his teaching under Musonius (Long, 2002, p. 10). When Domitian expelled the philosophers from Rome in AD 89, Epictetus went to live and teach in Nico-polis in Epirus. It was here that Epictetus taught Arrian, due to whom Epic-tetus' philosophy survived (Matheson, 1916, p. 14). Gaining a reputation for his eloquence, Epictetus' public lectures attracted students from afar, and many of them were high-ranking officials (Brunt, 1977, p. 21). P.E. Mathe-son tells us that Epictetus is thought to have known Hadrian, whom Arrian later served. Epictetus is thought to have lived until Marcus Aurelius' reign, though Matheson thinks this unlikely. As he was already teaching by AD 89, Matheson claims it is unlikely that he lived beyond the reign of Pius (Mathe-son, 1916, pp. 14–15). In later life, Epictetus is thought to have been physi-cally disabled in some way (Long, 2002, p. 10). He is referred to as a "lame old man" in the *Discourses* (Epictetus, 1916a, pp. 93–94). This is about the extent of what is known about the details of Epictetus' life.

We know more about the intellectual environment in which he was raised. Taught by Musonius, Epictetus was educated in the Stoic tradition. Though the Stoics thought of Zeno as the founder of their school, it is Chrysippus who Epictetus cites most frequently (Long, 2002, pp. 57–58). From Chry-sippus, Epictetus takes the importance of improvement and moral progress (Hershbell, 1993). However, Epictetus does not find Chrysippus beyond criticism, who he argues wrote obscurely. According to Long, Chrysippus' Stoicism was criticised for its "rigidity, paradoxicality, esoteric terminol-ogy," and the distinction drawn between virtuous human beings with perfect rationality and those who are irrational fools (Long, 2002, p. 32). Epictetus breaks with several tenets of this form of Stoicism. However, while Epictetus may not draw such a sharp distinction between the virtuous and those who are fools, he retains rationality as an ideal. It is Socrates who Epictetus and the Stoics find as their intellectual ideal, in particular his strength of mind during his trial and death (Long, 2002, p. 24); but even Socrates, the Stoics admit, may have fallen short of their ideal of perfect rationality. According to Susanne Bobzien, Epictetus also rejects aspects of Chrysippus' understanding of metaphysics (Bobzien, 1998, p. 332). Thus, despite Chrysippus' influence on Epictetus, Epictetus finds his philosophical ideals in the lives of other philosophers. Long lists Zeno, Chrysippus' teacher Cleanthes, Socrates, and the Cynic Diogenes as those for whom he has the most praise (Long, 2002, pp. 57–58).

The Stoics looked back to Zeno as the founder of their school (Long, 2002, pp. 57–58). However, Zeno saw his own philosophy as inspired primarily by Socrates and Diogenes. Socrates' life represents the ideal for the Stoics due to his strength of mind (Long, 2002, p. 68). Diogenes' role is that of promoting "kingship and castigation," as Malcolm Schofield puts it (Schofield, 1997). In this role, the philosopher is divinely appointed, living and teaching by example rather than through speech. The philosopher must head off into the world to understand its workings and his self within this world, before returning to assure others that the world is safe (Schofield, 1997, pp. 78–79). He must then teach others by his example, rebuking others where necessary. Philosophy is a way of life, not just a way of thinking. It is this lesson which Epictetus and the Stoics take from Diogenes and Cynicism. Zeno's main contribution to Stoicism was his theory of how we come to understand the world through our mental impressions (Long, 2006, p. 223). We then base our judgements on these impressions, though through our impressions of the world, it seems to us that our judgements cannot be wrong. Our impressions can, however, be mistaken. From this, Epictetus derives the importance of revising our impressions to form better judgements. After Zeno, Stoicism was further refined by Cleanthes and Chrysippus, but as much of the work of these early Stoics was lost, it is Epictetus who is most well known today (Long, 2002, pp. 7–8).

Epictetus' Stoicism has a complex relationship with other schools of philosophy. Montaigne (see Chapter 6 of this volume) held Epictetus in high regard; the influence of Epictetus is witnessed in Montaigne's choice of ideals. For Montaigne, the Greek general and statesman Epaminondas was the greatest of human beings for his ability to combine the extremes of gentleness and courage (Montaigne, 1958, pp. 572–574), fulfilling the Stoic ideal of becoming the perfect citizen through self-reflection and realising reason's potential. While the influence of Epictetus on Montaigne is clear, Pascal contrasted Montaigne's scepticism with Epictetus' faith in reason (Pascal, 1657; Force, 2005, pp. 29–31). For Montaigne, human reason had its limits, a view Pascal saw as incompatible with Epictetus' view that reason can transcend all external obstacles. Despite his influence on Montaigne, Epictetus stands opposed to Montaigne's form of scepticism. There are two further schools of philosophy which Epictetus opposed. Against the Aristotelians (see Chapter 1 of this volume), Epictetus holds that the satisfaction of the will is not dependent on the community to which the individual belongs; our will is dependent on nothing external to us (Jensen, 2017). It was the Epicureans who were Epictetus' main philosophical opponents in his own time. Whereas the Epicureans prized the flesh, Epictetus saw the flesh as inconsequential; it was reason and the will that mattered for Epictetus. It is through understanding how Epictetus thinks about the will that we can see what he offers the just war tradition, a sense of where the individual should lie within this tradition.

Texts

The first book of Epictetus' *Discourses* begins with a chapter setting out Epictetus' view on what is within the ambit of human agency. This is perhaps the central question within his philosophy, and the answer to this question is key to determining where Epictetus lies within the just war tradition. While the Stoics tended towards belief in universal determinism, Long suggests that Epictetus' argument for the freedom of the human mind shows that he does not remain entirely committed to Stoic metaphysics (Long, 2002, p. 162). Nevertheless, Epictetus sees much in the world which is outside of human control. It is only the mind, through its faculty for reasoning, that allows for human freedom through offering us control over our judgements (Epictetus, 1916a, pp. 43–47). Many events in our lives may be beyond our control; we could be imprisoned, tortured, or executed. Yet, whatever is done to us, our reason remains free:

> I must die. But must I die groaning? I must be imprisoned. But must I whine as well? I must suffer exile. Can any one then hinder me from going with a smile, and a good courage, and at peace?
>
> *(Epictetus, 1916a, p. 46)*

As Stoics believed in universal determinism, they considered it irrational to wish our lives had panned out differently (Long, 2002, p. 22). If we are dying, imprisoned, or exiled, these are necessary events for which there was no escaping. It would be useless to resent our current condition or wish it could be otherwise. In themselves, these events are neutral; it is only our minds that judge such events to be good or bad (Long, 2002, p. 28). As the mind remains free, it is for us to learn how to make good judgements whatever circumstances we face. We may not choose the circumstance, but we can always choose how to respond to the circumstance.

For Epictetus, as for Stoics more generally, the circumstances which we face are the consequence of divine power. While Long argues that Epictetus speaks of the divine power with the same reverence as St Paul or Augustine, he also notes that Epictetus does not view divinity as transcendent as is the Christian God (Long, 2002, pp. 144–146). Instead, Epictetus speaks of "the God within." Perfect rationality is identical with this God. As our minds are free, we must train our rationality in accordance with "the God within," or we will remain emotionally disturbed.

This is the metaphysical and theological ground on which Epictetus bases his moral outlook. We are rational creatures and can only tolerate that which is in accord with reason (Epictetus, 1916a, pp. 47–51). Once we understand that an act is in accord with reason, we understand it is this act which we must perform, even if this act is suicide. It is in this way that we learn to tolerate what may appear intolerable. As Epictetus recognises, however, this does not answer how we determine what is rational. To learn what is rational requires education, but it also means acquiring an understanding of ourselves. We can only know what is rational in relation to our own character. Understanding our character, we come to know what is preferable for us. Epictetus illustrates this with the example of the athlete threatened with death if he did not sacrifice his virility; the athlete chose death over shame. What is against our character is not within reason and must be rejected. However, if we must understand our character to understand rationality, we must know how we come to understand our character. Epictetus views this task as being intuitive. On seeing a lion approach, the bull who stands to protect his herd understands the power he possesses (Epictetus, 1916a, p. 51). If we possess a power, we know we possess this power. It is this knowledge on which we base an understanding of our characters.

Our rationality, as understood in relation to our character, allows for moral progress. In Chapter IV of the first book of the *Discourses*, Epictetus tells us that moral progress leads us to peace of mind (Epictetus, 1916a, pp. 53–57). Moral progress cannot be passive. We cannot learn peace of mind from reading the works of others; instead, we must actively train our will if we are to attain a tranquil mind. We must bring our will into "perfect harmony with nature—lofty, free, unhindered, untrammelled, trustworthy, self-respecting" (Epictetus, 1916a, pp. 55–56). If we will beyond what is within our ability to control, we cannot be trustworthy or free; we are living irrationally and not in accord with nature. Learning to live according to these principles—keeping our will in accord with nature—we will be on the "path of progress" and will not have "travelled to no purpose."

Living so that our will is in accord with nature and the divine order of the universe, we come to be, like Socrates, "citizens of the universe" (Epictetus, 1916a, p. 70). As our reason is part of this divine nature, we are connected with the government of the universe through reason, being governed by God. This supersedes any territorial loyalties we may have. Epictetus acknowledges this may bring us into tension with earthly authorities.

However, in our relations with others, living in accord with reason, we must learn what is within the power of our will. That which is within this power should concern us. What falls outside of this power should be a matter of indifference (Epictetus, 1916a, pp. 140–141). The judgements other people make belong to the latter category. We can conquer the lives of others and come to control their bodies, but we cannot conquer their will nor the judgements they make (Epictetus 1916a, p. 132). Our minds remain free, whatever the circumstances in which we exist; there is no way to enslave the mind of another. Thus, as the minds of others exist outside of the power of our will, they should be a matter of indifference to us. If we find the judgements other people make go against our own, we should not be angry with them; they are merely mistaken in their impressions of the world, basing their judgements on these mistaken impressions (Epictetus, 1916a, pp. 126–128). We cannot alter their impressions, and so we should not feel anger when others make poor judgements in response to these impressions.

Instead, we must focus on what is within the power of our will. Once we learn to concern ourselves only with what is within our will, we will live more content lives, but to do this, we need philosophy to tell us how to make judgements. Epictetus sees us as coming into the world already equipped with a degree of moral knowledge (Epictetus, 1916a, pp. 175–178). Even without education, people have a sense of duty, fortune, and fairness. However, we also recognise that we disagree with others over the constitution of these values. Philosophy's first task is to establish a standard according to which we can judge in order to resolve such disagreements. Our task in life is to learn our own character, our capacity for reason, and understand how our reason coheres with "the God within." It is in this task that we find the standards necessary for making good judgements.

Though we come into the world with a degree of moral knowledge, we must continue to learn throughout our lives if we are to overcome our mistaken impressions of the world around us. We must learn to correct our impressions through practice. Thus, if we find we struggle with anger, we must focus our energies on not being angry. Through training, we must learn to resist our inclinations (Epictetus, 1916b, pp. 38–40). Whatever we find most difficult to resist is avoided by training our will in the opposite direction. Training our will in this way, we come to perfect the will.

In this sense, Epictetus' stoicism may appear a philosophy of the self—a doctrine according to which we master ourselves, while considering all that is external to us a matter of indifference. This philosophy of the self would have little to offer to the just war tradition. However, Epictetus views us as social beings. While the will is the highest faculty—with it being "the God within"—we have a duty to cultivate our other faculties, developing our faculty of speech in order to live successfully with others (Epictetus, 1916a, pp. 232–238). Though we may be "citizens of the universe," Epictetus still finds us to have duties to the places to which we belong. Wandering freely from place to place would lead us to forget our purpose. We must return where we belong and fulfil our duties as citizens: marry, have children, and hold office when required (Epictetus, 1916a, p. 237).

In our place of belonging, others might assess us harshly, and we may fall into conflict. Yet, the judgements others make of us should not concern us, as these judgements fall outside of our will (Epictetus, 1916b, pp. 51–52). Though we have duties in society, we should only concern ourselves with what is within our will in our relations with others. The poor judgements of others are not our concern. How, then, should we resolve our conflicts with those with whom we share a society? Epictetus tells us that we cannot capture their power of judgement, nor can the task of philosophy be to secure for us what is outside of ourselves (Epictetus, 1916a, pp. 90–91). Conflict resolution takes time. We must foster our relationships over time, learning to deal only with what is within our will; good relations with others will only come once we all learn this lesson. When others harm us, we must not seek revenge, which only harms us further by bringing evil into our will (Epictetus, 1916a, p. 174). They harm us and act against us because they believe in their mistaken impressions (Epictetus, 1916a, pp. 127–128). If we correct our impressions and live true to our will, we will not live in conflict. People trust those who live true to their own will and whom they can build a relationship with based on honesty (Epictetus, 1916a, p. 230). While Epictetus may not flesh out a comprehensive position within the just war tradition, there is nevertheless an idea of how to determine the root of conflict; it consists in a misunderstanding of the self and what lies within the power of the self. Correct this misunderstanding, and we cut out this root.

In sum, Epictetus tells us to live in accord with "the God within." Beginning by learning what falls within the power of our will, we learn to master our will, living in harmony with the divinity of nature. It is only that which is within our will that should concern us; all else should be considered indifferently. Through learning this lesson, we can learn to live well whatever our circumstances; we might be enslaved by someone stronger than us, but nevertheless learn to live well. This person is stronger than us, and all our attempts to resist fail. If there is nothing that can be done, resenting our situation is futile. As Epictetus argues, it is only the person living against the will who is imprisoned (Epictetus, 1916a, p. 85). Rather than desire the impossible, we must learn to master what is within the power of our will. We do this through learning and habituation throughout our lives, learning what is essential to our character, and understanding how to best develop our character within the circumstances we face. It is through doing so that we become exemplars of "the God within."

Controversies

The Epicurean school was Epictetus' main target of criticism. There are two central reasons why Epictetus rejected Epicureanism. First, Epicurus' advice to us is to withdraw from social and political relations (Epictetus, 1916a, pp. 111–112). For Epictetus, living in accord with our nature means acknowledging our social nature along with the duties that derive from our nature. Epicurus told us that wise people would avoid entering into politics, as they understand what is expected of the politician. We should also avoid raising children, according to Epicurus. For Epictetus, it is in realising our social nature that we bring our will into accord with nature; this means we have a duty to hold office when required and to marry and raise children. Second, Epicureanism, grounded as it is on the pleasure we can find in the flesh, does not provide us with stable values on which we can base good judgements. For Epictetus, it is the will that governs us, connects us with "the God within," and is thus our highest faculty, whereas for Epicurus, the highest good consists in our flesh (Epictetus, 1916a, p. 107). Holding that the good of the human consists in the flesh is akin to saying the good of the snail consists in its shell, according to Epictetus. Living according to Epicureanism would lead us into nihilism, as we recognised no Gods or social virtues (Epictetus, 1916a, pp. 219–220).

In his own time, it was the Epicurean school which Epictetus contended with, primarily concerning the tension between a doctrine emphasising pleasure and another prioritising the will. In the years since Epictetus' time, tensions have been found in his thought. One tension is found in how to interpret Epictetus' conception of freedom. As a fundamental component of Epictetus' philosophy rests on the independence of the will from the rest of the universe, it could appear to modern readers that Epictetus believed in the libertarian formulation of free will. That is, our will must be independent of external causes if it is to be free; we must be able to decide for ourselves. Susanne Bobzien argues that this is to misunderstand Epictetus and the Stoics. Freedom, for Stoics, is not an unhindered ability to do whatever we want but a virtuous state of mind (Bobzien, 1998, p. 342). This is part of the reason why Epictetus rejected Epicureanism. If freedom consists only in the ability to satisfy bodily desires, we will be slaves to our passions while thinking ourselves free (Bobzien, 1998, p. 343). Instead, what matters is that our minds remain free; an integral part of the mind's freedom is its ability to determine what is dependent on the mind. Bobzien notes that this marks an important distinction between Epictetus and Chrysippus. Whereas Chrysippus held that any event caused by us was dependent on us, Epictetus found a much smaller domain of things dependent on us (Bobzien, 1998, p. 332). Any event outside of us could be stopped by something over which we have diminished control. If we decide to go for a walk, there are many factors outside of our control which could prevent us from taking this walk. Therefore, our ability to go for a walk is not within the domain of things dependent on us. It is only that which is dependent on us with absolute certainty that we consider belonging to this domain. Essentially, it is only our will and our ability to make judgements that we can consider as firmly belonging to this domain.

Both Long and Myrto Dragona-Monachou agree with Bobzien's interpretation of Epictetus' account of freedom (Long, 2002, p. 230; Dragona-Monachou, 2007, p. 114). However, despite his focus on what is within the power of our will, Epictetus still views us as having duties within society: we ought to marry, have children, hold office when necessary. Against the objection that concerning ourselves only with our own will would lead to anarchism, as people obeying their own will would have no reason to obey the law, Epictetus claimed that his doctrine would lead to a firmer conviction in the law (Epictetus, 1916b, pp. 171–172). Concerning ourselves only with what is our own, we will not trespass against others. People acting in this way will uphold rather than undermine the law. Epictetus cares deeply about our social roles and the importance of the law. There are two possible contradictions here. First, our ability to be good citizens depends on us acting within the world; we cannot sit passively, letting things happen to us without acting. This appears to sit in tension with Epictetus' position that we should only care about that which is dependent on us. Acting as a citizen will necessarily involve stepping beyond what is within our will. There are many things outside of our control that could prevent us from acting as a good citizen. Second, if we should only care about what is within the power of our will, where can we find the motivation to be a good citizen? Surely whether we are a good citizen is dependent on the judgements others make of us? Moreover, if other citizens are committed to Stoicism, should they not be content as their city burns to the ground due to our inaction as citizens?

As John M. Cooper notes, Epictetus is not just advising us what we should think; he is advising us how to apply this way of thinking in our lives (Cooper, 2007, p. 15). Epictetus tells us how to train our will and overcome our negative inclinations. Nancy Sherman also appears to agree with this reading; she argues that Epictetus encourages us to master ourselves, and not our will alone (Sherman, 2007, pp. 3–4). This suggests that Epictetus is concerned with our actions in the world; we cannot just learn how to think, we must act, too. Perhaps the domain of that which is dependent on us is larger than Bobzien posits. If Epictetus is concerned with how we conduct ourselves in the world, this brings us back to the free will problem. Epictetus could not hold that all events external to us are outside of our control while simultaneously claiming we have a duty to act in this world. Julie Annas points to the importance of our social roles to Epictetus (Annas, 2007). We cannot live morally while ignoring the roles we have within our families and communities. Thus, it seems Epictetus does not expect us only to consider our judgements in relation to the external world but to act in this world, too.

What this points to is an unresolved question within Epictetus' thought: how wide is the scope of that which depends on us? On the narrow interpretation, we sink back into our own lives, caring only about the judgements we make when witnessing the events around us. There is little room for thinking about duty within this interpretation; there is nothing we can do within the world, as the events within it are beyond our control. On the wide interpretation, we care not only about our judgements but also about our acts within the world. Through philosophy, we must train our will, learning how to make good judgements, and apply these lessons to our acts in life. It seems the latter interpretation is the most appropriate, given the importance Epictetus places on applying the lessons of philosophy to our lives and the necessity of our social roles.

Legacy

With his faith in human reason allowing the individual to triumph against adversity, Epictetus' influence can be witnessed in philosophers from Montaigne, Spinoza, and Kant through to Deleuze. However, Epictetus has been a neglected thinker within the just war tradition. Nancy Sherman draws heavily on Epictetus in her work; yet, her focus is on the individual experiences of soldiers (Sherman, 2007); she does not attempt to use Epictetus' thought to determine a stance on the just use of force. Nor have other scholars made this attempt. Mark N. Jensen argues that, while individuals in war may learn from Epictetus, if we are to understand military virtues, we are better off looking to Aristotle (Jensen, 2017). In the previous volume of this work, there is no mention of Epictetus. Perhaps this is due to Epictetus' focus on the self. Though Epictetus sees us as socially rooted beings, he is preoccupied with what we can do for ourselves. Greater political ambitions are regarded as folly; within any such ambition, there are many circumstances beyond our control that could thwart our hopes, whether personal or social. These sorts of ambitions take us away from a focus on what is within our will. This preoccupation with the self may lead people to think that Epictetus has nothing to say about conflict resolution or how to solve problems beyond the self. However, it is precisely this preoccupation from which I think we can learn. Part of the interest lies in the central tension within Epictetus' philosophy: how do we determine what is within our will?

If we make the scope of what is within our will small, as Bobzien suggests, we retreat into ourselves. In the international realm, this might lead us to think of Epictetus as a realist inclined to pacifism; the problems of others are their concern, and there is little we can or should do to help. Political leaders should focus on their own state and what is within their control. However, this sort of reading of Epictetus is at odds with another aspect of his doctrine: his insistence that we are "citizens of the universe" and his disdain for tyrants. This side of Epictetus' thought makes him appear more sympathetic to the liberal tradition. How, then, should we interpret Epictetus within the just war tradition?

A sensible place to begin is in determining who we are. Political leaders with a good understanding of themselves will not step beyond what is within their control, nor attempt to be what they cannot. Likewise, states that understand themselves will not reach beyond themselves. On this view, the cause of conflict lies in misunderstanding of the self. A state must understand its own interests and what is within its control. It is states overreaching themselves that cause problems. However, this does not tell us what states are to do when other states do overreach themselves. In most cases, the behaviour of other states will be beyond the ability of most states to control. This suggests that Epictetus would encourage political leaders towards pacifism and resignation. What other states do is not within their control; they should therefore let happen what happens, dealing only with the consequences for their own state, insofar as these consequences are within their control. At least at the level of the state, Epictetus appears to be a realist, though with pacific tendencies.

This perspective shifts at the level of the individual. Along with their duties as citizens, people have duties as "citizens of the universe." As Annas notes, the notion that we are citizens of the universe might be thought to conflict with the idea that we are socially rooted beings with duties towards our families and communities (Annas, 2007). If a person is a citizen of the universe, why should specific relations with others matter? Surely these people should hold no more moral significance than any others? However, as Annas posits, we cannot separate our duties within our own social worlds from those as citizens of the universe. We would be failing in our latter duties if we did not act as good family members and citizens of our communities. Perhaps this could be summed up in the expression popular with environmentalists, "think global, act local." Our duties to the rest of the world are fulfilled by how we act within our own worlds.

As citizens of the universe, there is a duty on us to do what is within our power to promote peace. If we fail in this duty, either by pursuing aggression or by supporting the cruelty of others, we are not exemplifying Stoic virtues. This begins first within our communities, within which we have a duty to resist aggressive tactics where it is in our power to do so. It is not immediately clear how far we ought to go in our resistance, whether withdrawing our support is enough or if we are justified in using force against aggressors. An answer to this, I think, can be found by assessing our own power. If we understand ourselves and know that it is well within our power to stop an aggressor, then we have a duty to act against this aggressor. Like the bull defending the herd from the lion, if we understand ourselves, we will know how capable we are of resisting an opponent. However, if we know that we are powerless against the aggressor, we must still withdraw our support.

This focus on the individual's role is perhaps why Sherman draws on Epictetus in her analyses of the military. Sherman recounts the story of James B. Stockdale, an American Navy pilot shot down over Vietnam in 1965. Knowing that he was about to be captured by the enemy, Stockdale thought to himself that he was now "entering the world of Epictetus" (Sherman, 2007, pp. 1–5). Through his understanding of Epictetus, Stockdale sought to preserve his own agency while captured. Against Bobzien's narrow account of what is within our will, Sherman draws agential empowerment from Epictetus. Though we must primarily master our will, we must also master ourselves, pushing our sense of agency to its limits.

Perhaps this understanding of agency allows us respond to some of the dilemmas in contemporary just war theory concerning individuals. David Rodin argues against the position that soldiers fighting an unjust war hold the same privileges and prohibitions as do those fighting just wars (Rodin, 2008). In doing so, Rodin finds himself confronted with several dilemmas: are soldiers fighting unjust wars deserving of punishment? Can soldiers fighting just wars go further in their use of force than the current norms of *jus in bello* would allow, perhaps including the targeting of non-combatants? If we follow Epictetus, we are offered a simpler response to these dilemmas. It is always in the power of a soldier to resist the orders of a superior. If the war is unjust, and even if the consequence is death, this is what the solider must do. When faced with soldiers who failed this task and fought unjust wars, it remains within our power to refrain from retributive justice; seeking revenge only allows evil into our own will. Keeping our acts within the scope of what is within our will, we would go no further than necessary in ensuring our own defence.

While this partly clears up the duties of individuals, an issue is left unresolved at state level. For most leaders, the affairs of other states exist outside of their state's control. However, this is not true of all leaders. It would be perfectly within the power of the United States to prevent many states from acting as they wish. Assessing the power of the state, the President of the United States could conclude that it is well within their power to act. However, it seems to me that Epictetus would warn against such actions. As Bobzien argues, there are many obstacles which could prevent us from doing even the most mundane of everyday tasks. Within the interactions of states, there are countless obstacles and unknown possibilities that could thwart even the most powerful leader's intentions. Though I think Sherman's interpretation of a wider sense of agency in Epictetus is more apposite than Bobzien's narrow account, we must still be careful when we step out into the world; even on the wide account, there remains much that is beyond our control.

Thus, Epictetus is best understood as a realist when thinking about states— but a realist whose pacific inclinations mean he finds a role for the individual within the just war tradition. States will often commit atrocities in the pursuit of their own interests. When states do this, they are reaching beyond themselves, failing to act in accordance with Stoic virtues. We can find a standard in Epictetus against which we can assess conflicts as unjust. A standard for identifying the just use of force is more difficult to find. In the complexity of any conflict, there are so many factors beyond a state's control that it seems almost any use of force would be considered unjust. It appears resistance is, then, best left to individuals. From an understanding of themselves and their own capacities, it is for individuals to determine what can be done to resist unjust aggression.

Works Cited

Annas, Julie. 2007. Epictetus on Moral Perspectives. In T. Scaltas and A. S. Mason. eds. *The Philosophy of Epictetus*. Oxford: Oxford University Press.

Bobzien, Susanne. 1998. *Determinism and Freedom in Stoic Philosophy*. Oxford: Oxford University Press.

Brunt, Peter Astbury. 1977. From Epictetus to Arrian. *Athenaeum 55*, pp. 19–48.

Cooper, John M. 2007. The Relevance of Moral Theory to Moral Improvement in Epictetus. In T. Scaltas and A. S. Mason. eds. *The Philosophy of Epictetus*. Oxford: Oxford University Press.

Dragona-Monachou, Myrto. 2007. Epictetus on Freedom: Parallels between Epictetus and Wittgenstein. In T. Scaltas and A. S. Mason. eds. *The Philosophy of Epictetus*. Oxford: Oxford University Press.

Epictetus. 1916a. *The Discourses and Manual, Vol. 1*. Translated by P. E. Matheson. Oxford: Oxford University Press.

Epictetus. 1916b. *The Discourses and Manual, Vol. 2*. Translated by P. E. Matheson. Oxford: Oxford University Press.

Force, Pierre. 2005. Innovation as Spiritual Exercise: Montaigne and Pascal. *Journal of the History of Ideas* 66(1), pp. 17–35.

Hershbell, Jackson P. 1993. Epictetus and Chrysippus. *Illinois Classical Studies* 18, pp. 139–146.

Jensen, Mark N. 2017. Epictetus vs. Aristotle: What is the Best Way to Frame the Military Virtues? *Naval War College Review* 70(3), pp. 101–120.

Long, Anthony Arthur. 2002. *Epictetus: A Stoic and Socratic Guide to Life*. Oxford: Oxford University Press.

Long, Anthony Arthur. 2006. *From Epicurus to Epictetus: Studies in Hellenistic and Roman Philosophy*. Oxford: Oxford University Press.

Matheson, Percy Ewing. 1916. Introduction. In Epictetus. ed. *The Discourses and Manual, Vol. 1*. Translated by P. E. Matheson. Oxford: Oxford University Press.

Montaigne, Michel de. 1958. Of the Most Outstanding Men. In Michel de Montaigne. ed. *The Complete Essays of Montaigne*. Translated by Donald M. Frame. Stanford: Stanford University Press.

Pascal, Blaise. 1657. Entretien avec M. de Sacy sur Epictète et Montaigne. In Léon Brunschvicg (ed.). *Les Provinciales, Pensées et Opuscules*. Paris: Hachette.

Rodin, David. 2008. The Moral Inequality of Soldiers: Why Jus In Bello Asymmetry is Half Right. In D. Rodin and H. Shue. eds. *Just and Unjust Warriors*. Oxford: Oxford University Press.

Schofield, Malcolm. 1997. Epictetus on Cynicism. In T. Scaltas and A. S. Mason. eds. *The Philosophy of Epictetus*. Oxford: Oxford University Press.

Sherman, Nancy. 2007. *Stoic Warriors: The Ancient Philosophy Behind the Military Mind*. Oxford: Oxford University Press.

3

JUAN GINÉS DE SEPÚLVEDA (1490–1573)

Luke Glanville and David Lupher

Introduction

Juan Ginés de Sepúlveda is best known as the antagonist of the Dominican friar Bartolomé de las Casas. The two of them famously engaged in a bitter debate about the justice of Spain's activities in the Americas at Valladolid in 1550–1551. Some popular renderings of this debate present it as a dispute about the humanity of Indigenous peoples in the Americas or the right of Spaniards to enslave these Amerindians or the permissibility of forcibly converting Amerindians to the Christian faith. While the debate did touch on these issues, it was centrally concerned with the just reasons for war and specifically whether Spain had just reasons for war against the Amerindians.[1]

Sepúlveda was a just war theorist. He had written several treatises on war, including *Democrates Part Two, On the Just Reasons for the War against the Indians*, composed around 1544. His debate with Las Casas at Valladolid is perhaps most neatly understood as a dispute about the validity of the four just reasons for Spain's wars that Sepúlveda advanced in that treatise.[2]

In treatments of the Valladolid debate, Sepúlveda is sometimes reduced to an absurd figure whose arguments for Amerindian subjugation can be easily dismissed as focus is given instead to the passionate and seemingly modern defenses of Amerindian rights advanced by Las Casas. But Sepúlveda was a learned and agile thinker. His just war arguments are nuanced and carefully made. And they are all the more disturbing for this fact. He developed a sophisticated account of just war to justify the brutal subjugation of Indigenous peoples, and it is an account that is especially disquieting since we can still feel its reverberations in some of the ways that we theorize just war today.

DOI: 10.4324/9781003428688-4

Contexts

Sepúlveda was born in Pozoblanco, near Córdoba, around 1490.[3] He studied philosophy at the University of Alcalá de Henares and theology at the University of Sigüenza before leaving Spain in 1515 to pursue further study at the Colegio de San Clemente, the Spanish college in Bologna. He displayed his theological training proudly thereafter, deploying the title "doctor of arts and theology" in his publications, penning a 1526 treatise against Luther, *On Fate and Free Will (De fato et libero arbitrio)*, joining his mentor Alberto Pio, prince of Carpi, in a theological dispute with Erasmus in the 1520s and 1530s and demonstrating his mastery of the church fathers, medieval theology, and canon law in his defenses of Spain's activities in the Americas in the 1540s and 1550s.

While at Bologna, Sepúlveda had become the pupil and protégé of the great Italian Aristotelian Pietro Pomponazzi. Pomponazzi's enthusiasm for Aristotle inspired Sepúlveda to produce a series of translations of Aristotle into Latin, with commentaries, from 1522 to 1548. Aristotle, as the first chapter in this volume shows, had a vast impact on just war inheritance. Sepúlveda would make use of Aristotle's doctrine of "natural slavery" when justifying Spain's American wars, though other canonical just war thinkers tended to overlook that aspect of the Philosopher's thought. More broadly, Sepúlveda drew from Pomponazzi a humanist determination to understand Greek philosophy on its own terms, rather than filtering it through medieval scholastic theology (as Thomas Aquinas and others were wont to do) and a conviction that natural reason did not depend on divine revelation. This too would feature in Sepúlveda's arguments about the Americas.

Sepúlveda had also met Giulio de' Medici, the future Pope Clement VII, while at Bologna. In 1526, he joined Clement VII's papal court in Rome, where he based himself for much of the next decade. In 1529, he was sent by Clement VII to greet Charles V upon his landing at Genoa and to accompany Charles to Bologna, where the pope crowned him emperor. In 1535, Charles V offered Sepúlveda the post of official chronicler, with an additional honorary title of "chaplain." Sepúlveda gladly accepted. Thus, in 1536, having spent two decades immersed in Italian intellectual culture, Sepúlveda returned to Spain. He was by now a remarkably protean intellectual with credible claims to expertise in Christian theology, Aristotelian philosophy, and Renaissance humanism. He would weave these together deftly in defense of Spain's American "just" wars.

Texts and Tenets

Sepúlveda had already written several treatises on war while in Italy. While in Pio's court in the early 1520s, he composed a riposte to the pacifist arguments of Erasmus and Juan Luis Vives in the form of a dialogue, *Gonzalo—A Dialogue in Defense of the Pursuit of Glory* (*De appetenda gloria dialogus, qui inscribitur Gonsalus*), printed in Rome in 1523. To mark the occasion of the pope's crowning of Charles V, he wrote an *Exhortation to Charles V to Make War upon the Turks* (*Cohortatio ad Carolum V ut bellum suscipiat in Turcas*), printed in Bologna in 1529. A few years later, he composed another dialogue, *A Dialogue entitled* Democrates, *on the Compatibility of Military Training with Christian Religion* (*De convenientia militaris disciplinae cum christiana religione dialogus, qui inscribitur Democrates*), known today as *Democrates Part One (Democrates primus)*. He indicated in the prologue that he had found that many young Spanish noblemen serving as knights in the train of Charles V were troubled by their awareness of "certain views according to which the profession of a noble warrior . . . did not fit well with the mandates of Christian philosophy" (Castilla Urbano, 2013, pp. 95–96). Sepúlveda was alarmed by such claims, particularly given Spain's ongoing wars with the Turks, and so he wrote the dialogue to explain that Christians were not prohibited from engaging in wars waged for just reasons. He published the dialogue in Rome in 1535, returning to Spain the following year.

Back in Spain, he became embroiled in the controversy of the Americas. As Sepúlveda (2023b, §2) himself tells it, the issuance of the contentious "New Laws" by Charles V in 1542 generated a renewed controversy within the royal court about the justice of Spain's invasion of the Americas and its subjugation of Amerindians. Francisco García de Loaysa, the archbishop of Seville and president of the Council of the Indies, heard that Sepúlveda was going around telling people that he could prove that Spain's wars in the Americas were just. Loaysa asked him to write something about the matter. And so, sometime around 1544, Sepúlveda composed another treatise, his most controversial work, titled *Democrates Part Two, on the Just Reasons for the War against the Indians* (*Democrates secundus, sive de iustis belli causis apud Indos*).[4]

Democrates secundus comprised two books. In the first, Sepúlveda (2023a, I.1–4) recapitulated his argument about why war was not prohibited to Christians and his explication of three just reasons for war, which he had advanced in *Democrates primus*. These three reasons—to avenge wrongs, to punish injuries, and to recover things wrongly seized—were essentially the same as those that Augustine had offered in book VI of his *Questions on the Heptateuch*, that Isidore of Seville had brought together in his *Etymologies*, and that Sepúlveda had read in Gratian's *Decretum*.[5] Sepúlveda then proceeded to offer four additional reasons for a just war—reasons that he said were less widely acknowledged and that arose less often but which applied to Spain's invasion of the Americas (Sepúlveda, 2023a, I.4). Sepúlveda devoted most of the first book of *Democrates secundus* to expounding these four just reasons and their application to the Americas (detailed later in the chapter). In the much shorter second book, he addressed the question of how subjugated Amerindians ought to be ruled. His answer featured a lengthy dialogue about whether individual Amerindians should be punished for participating in an unjust war—a dialogue that previewed key elements of the debate between traditionalist and revisionist just war theorists that would emerge more than 400 years later (Sepúlveda, 2023a, II.4–7).

In *Democrates secundus*, Sepúlveda put on full display his talent for bringing together the doctrines and methods of Aristotelianism, humanism, and theology. But far from settling the controversy within Spain about its activities in the Americas, his treatise amplified it. Spanish theologians schemed to ensure that he was denied a license to publish it. He therefore wrote a defense, this time not in the form of a dialogue but in the scholastic manner preferred by the theologians. Here, he made a flamboyant show of his mastery of the church fathers, medieval theology, and canon law. He published this *Defense (Apologia)* in Rome in 1550. But this did little to ease tensions or settle consciences.

Charles V called a halt to further wars by Spaniards in the Americas pending the outcome of a *junta*, or meeting, of leading theologians and jurists to be held in Valladolid. Here, in 1550–51, Sepúlveda and Las Casas debated, albeit it seems without at any stage being in the same room as each other. The junta took place around the same time as Alonso de la Vera Cruz—the first chair of philosophy at the recently founded University of Mexico and the subject of the next chapter in this volume—grappled with applying Vitoria's just war principles in New Spain. In 1552, Las Casas printed without license a summary of the opening sessions of the Valladolid deliberations written by one of the presiding judges, Domingo de Soto, accompanied by Sepúlveda and Las Casas' respective responses. Sepúlveda retaliated by penning a furious response, *Outrageous, Scandalous, Heretical Notions. . .*, sometime around 1553. He accused Las Casas of an array of heresies, sought to have Las Casas' text publicly banned by the Inquisition, and continued to appeal for permission to publish his book. But *Democrates secundus* would remain unpublished until 1892. Complete English-language editions of Sepúlveda's writings on the Spanish invasion of the Americas—his *Democrates secundus*, his *Apologia*, the record of the Valladolid debate compiled by Las Casas, and Sepúlveda's enraged riposte—were not published until 2023.

Controversies

Each of the four "just reasons" for Spain's wars in the Americas that Sepúlveda developed in *Democrates secundus* proved controversial, and each would be subjected to sustained debate between Sepúlveda and Las Casas during the Valladolid junta.[6]

Natural Slavery

The first justification for war rested on Aristotle's claim, found in the first book of his *Politics*, that some people are "slaves by nature."[7] Sepúlveda's application of this doctrine to justify Spain's American wars has become the primary source of his notoriety. The uncivilized nature and customs of the Amerindians, coupled with their idolatry and rite of human sacrifice, constituted more than sufficient evidence that they were slaves by nature, he claimed. They are surpassed by the Spaniards in every respect, "as children are by adults, as women are by men, as savage and fierce people are by the most gentle people," and, as he infamously put it in an early version of the text, "I might also say: as apes are by human beings" (Sepúlveda 2023a, I.9.1). It is thus, he claims, not only just but also beneficial for such uncivilized people to be subjected to the rule of civilized Spaniards.

Sepúlveda claimed that their condition of natural slavery *itself* constituted a just reason to subjugate them through war, should they refuse to submit themselves to the rule of natural masters. He composed *Democrates secundus* at the same time as he was at work translating Aristotle's *Politics* into Latin (printed in Paris in 1548), and he had no trouble finding a passage that made his case: "for the art of hunting is a part of warfare which it is fitting to use sometimes against beasts, and sometimes against those humans who, though they have been born to obey, reject rule" (Aristotle, *Politics*, 1.3.8, quoted in Sepúlveda, 2023a, I.11.1, I.19.3).

This was an odious and inflammatory justification for war. It became a feature of the "Black Legend" regarding the evils of the Spanish imperialism, and it is an argument for which Sepúlveda has been repeatedly condemned through the ages. However, it is worth noting that Sepúlveda drew back from the argument soon after making it. He came to realize that a key reason why Spanish theologians at the Universities of Alcalá and Salamanca, when asked by the Council of Castile, recommended against licensing *Democrates secundus* was their opposition to his use of Aristotle's controversial doctrine. Sepúlveda noted that, when he had invoked the doctrine in a meeting at Salamanca, one theologian, Melchor Cano, who would go on to serve as a judge at Valladolid, had responded that Aristotle had offered the doctrine to gratify Alexander the Great, who was waging war against barbarians. And when Sepúlveda had attempted to defend the authority of Aristotle, Cano had replied, "Oh no, he was wicked and flawed" (Sepúlveda, 1997a, §24, our translation).

And so, when Sepúlveda came to write his *Apologia*, he changed tack. He offered only a brief defense of this first justification, and rather than deploying Aristotle's language of "natural slavery," he claimed merely that the Amerindians were barbarian in their customs and nature. While he continued to insist that such people should be made subject to the rule of superior people, he emphasized that this was the position of theologians like Augustine and Aquinas, as much as Aristotle (Sepúlveda, 2023c, III.1). When Las Casas argued at Valladolid that the Amerindians were not barbarians of the sort that Aristotle described as "slaves by nature," Sepúlveda responded only briefly, and he appealed not to Aristotle but to Aquinas in insisting that it was right that the Amerindians be forced to submit to the more rational and civilized Spaniards (Las Casas, 2023, Sepúlveda's Eighth Objection, §2).[8]

The centrality of Aristotle's doctrine for Sepúlveda's defense of Spain's wars in the Americas has been overstated in much Anglophone scholarship. This unfortunate tendency can be traced to Lewis Hanke's (1959) in many ways excellent study, *Aristotle and the American Indians*. In reality, after composing *Democrates secundus*, Sepúlveda backed away from relying on the Philosopher to defend his first justification for war. Moreover, when advancing his three other justifications, even in *Democrates secundus*, he leant much more heavily on the authority of Scripture and theologians such as Augustine.

Punishing Violations of Natural Law

The second justification for war was that it was a just punishment for the Amerindians' violations of natural law, specifically those laws prohibiting idol worship and human sacrifices. The Amerindians' impious and wicked crimes, Sepúlveda claimed, violate not only divine law but also natural law, which applies to all peoples. It was for such crimes that God meted out punishment against the pagan peoples inhabiting the Promised Land, at the hands of the Israelites, before the coming of Christ. Christ, in turn, commanded his Apostles not only to preach the Christian faith to all peoples but also to teach them to observe the laws of nature as contained in the Decalogue and the love of one's neighbour. And since all people are subject to these laws, it is just that violators be punished on the authority of the church (Sepúlveda, 2023a, I.11–12).

Vitoria had denied that the church had either spiritual or temporal juris-diction over the infidel Amerindians, invoking Paul's words in 1 Corinthi-ans 5:12: "What business is it of mine to judge those outside the church?" (Vitoria, 1991, pp. 258–264, 272–275). But Sepúlveda interpreted Paul's verse differently. Paul meant merely that the church should not waste time passing judgement *futilely* on the unbelief of pagans, since people cannot be forced to believe against their will, Sepúveda claimed. And yet the church is authorized, and indeed obliged, to do all it can to bring about the conver-sion of unbelievers, including, where possible, by punishing and correcting those who violate natural law. As Augustine had declared in a letter to the Donatist bishop, Vincentius, the church "corrects those whom it can and tolerates those whom it hasn't the power to correct" (Letter 93.9.34, quoted in Sepúlveda, 2023a, I.12.2).

Upon reading the manuscript of *Democrates secundus*, Cano (1997, §19) criticized Sepúlveda for advancing arguments for war that were contrary to those presented by such a revered figure as Vitoria.[9] It is perhaps surprising, then, that when defending the punishment of sins as a just reason for war in his *Apologia*, Sepúlveda (2023c, III.2.xii) added to the scriptural passages and theological authorities that he had previously invoked the opinions of medieval canonists Pope Innocent IV and Hostiensis, whose arguments Vitoria had explicitly rejected. These canonists had argued that the pope was authorized to punish any people who violate natural law. Vitoria had complained that this was "tantamount to saying that the barbarians may be conquered because of their unbelief, since they are all idolaters" (1991, p. 274). But Sepúlveda appealed to the canonists' authority anyway.[10]

In his remarks and written replies at Valladolid, Las Casas rejected Sepúlveda's argument for the same reason that Vitoria had offered: regardless of what certain canonists had claimed, neither the church nor Christian princes had jurisdiction over the pagan Amerindians. As for Sepúlveda's rendering of 1 Corinthians 5:12, Las Casas replied, "just as the doctor readily offers that interpretation off the top of his head, so too can it be just as readily dismissed" (Las Casas, 2023, Las Casas' Fifth Reply, §1). Sepúlveda's argument for a right to punish violations of natural law never stood much chance of winning over the influential Dominican theologians who presided at Valladolid. He later claimed, though, that it did help convince several of the jurists who presided alongside them of the merits of his defence of Spain's wars (Sepúlveda, 1997b, §8, 2023b, §5).

Protecting Innocents

Sepúlveda supplemented his argument for punitive war with a third justification for war: to ward off injuries to innocent people and specifically to protect innocent Amerindians from the rite of human sacrifice. In both *Democrates secundus* and the *Apologia*, he defended this right of war only briefly, appealing to well-known scriptural and theological proof-texts for support (Sepúlveda, 2023a, I.15.8–9). It is surprising that he did not make more of this justification for war since, of the four that he advanced, this was the one that had been most consistently endorsed by Spanish theologians and jurists (Schwartz, 2007). Vitoria had drawn a clear distinction between wars of punishment and wars of protection, insisting that, while the church lacked the power to authorize the punishment of the Amerindians for violations of natural law, including human sacrifice, Christian princes could justly wage war to protect the victims of this nefarious rite (Vitoria, 1991, pp. 287–288). In contrast, Sepúlveda presented the duty to protect innocents as little more than an addendum to his defense of punitive wars. It reads almost as an afterthought. And, yet, Las Casas would need to labour, and even flirt with heresy, to rebut it at Valladolid.

Las Casas accepted that the church was entrusted by divine law with the protection of all innocents. He argued, however, that if the duty of protection can only be accomplished by means of war, it is better to refrain from discharging it. This was for several reasons. First, it is right to choose the lesser of two evils. Allowing a few innocents to be killed for the purpose of sacrifice, as heinous as that is, is preferable to resorting to war, which not only kills a larger number of innocents but also leads those who survive to hate the Christian faith. Second, the negative commandment to not kill the innocent is more binding than the positive duty to protect the innocent, and so one should not act to fulfil the latter if it cannot be done without violating the former. Las Casas' third argument was more controversial: the Amerindians act with invincible ignorance and indeed with remarkable piety. Not only are they excused by the fact that they are following the teachings of their priests and rulers and carrying out a practice that was widespread in Greco-Roman antiquity, but in performing their depraved rite they are also piously sacrificing the best thing that people have to offer—human life—to that which they understand to be God. For these reasons, while they are ultimately accountable to God, they cannot be held to account by men (Las Casas, 2023, Soto's Summary, §13).

Sepúlveda perceived an opportunity to go on the attack. In justifying human sacrifice, he declared, Las Casas had offered an argument "far removed from Christianity." The notion that the ignorance of the Amerindians can excuse such a monstrous violation of natural law is ludicrous, for theologians and canonists agree that "ignorance of natural law does not serve to exonerate anyone" (Las Casas, 2023, Sepúlveda's Eleventh Objection, §§3–5). Las Casas doubled down on his position, insisting that, while the Amerindians may be in error, they were justified in taking up arms to defend their religion against the Spaniards. Indeed, they would violate natural law if they did not defend their rites, since all people are obliged to love God, or that which they consider God, even to the point of death (Las Casas, 2023, Las Casas' Eleventh Reply, §§4–9). Sepúlveda amplified his condemnation. In his *Outrageous, Scandalous, Heretical Notions*, Sepúlveda argued that all theologians are in agreement that the precepts of the Decalogue are natural laws, and human sacrifice violates the precepts prohibiting both idolatry and murder. Las Casas' suggestion that the Amerindians did not breach natural law in sacrificing innocents to false gods, he declared, was impious and heretical (Sepúlveda, 2023b, §16).

Spreading the Christian Faith

The fourth justification for war, Sepúlveda claimed, was also the most just: war is rightly waged to guide the Amerindians by the shortest and most direct path to the truth of the gospel. While pagans cannot be compelled to believe, he argued, Christians are obliged at least to pull them back from the precipice and lead them towards the true path. And this can be done most safely and most effectively in the Americas by first bringing the Amerindians under Spanish rule (Sepúlveda, 2023a, I.15–16.)

To support his argument, he turned again to the weighty authority of Augustine and specifically Augustine's letters endorsing the use of compulsion to lead heretical Donatists back to the true faith. Among numerous passages that proved of use was an interpretation of Jesus' parable of the banquet (Luke 14:15–24) that Augustine developed to explain why the church in his day was just in using compulsion to lead people towards the faith even though the first apostles did not. In the parable, the host had his slave first invite and then compel people to come to his banquet. Augustine suggested this was meant to symbolize that, whereas the early church, lacking the power to do more, merely invited outsiders into the faith, the church's power was now reinforced by the power of Christian rulers and so it was now right to "compel them to come in." The same remained true in his time, Sepúlveda averred. Pagans are rightly subjugated to the rule of Christians so that they can be led by Christian teaching to worship God (Sepúlveda, 2023a, I.17.1–3, citing Augustine, Letter 173.10, and Luke 14:15–24).

Sepúlveda then drew on another of Augustine's letters to push the argument further. When Christian teaching is accompanied by "useful terror," Augustine had claimed, not only are listeners persuaded by divine testimonies and rational arguments but also "the power of fear would break the bonds of evil custom," leading to their acceptance of the gospel. Sepúlveda likewise argued that the Amerindians should be reduced to Spanish rule not only so that they could be forced to listen to preachers but also so that they might be made to fear Christians more than their own priests and rulers—a fear that would encourage them to embrace the true faith. That this was the most expeditious means of converting people to the faith had been confirmed both by the experiences of Augustine with the Donatists and of Spaniards with the Amerindians, he declared (Sepúlveda, 2023a, I.18.3–4, quoting Augustine, Letter, 93.5.16, 93.1.3).

Sensing, perhaps, that he had a particularly strong argument in hand, Sepúlveda devoted more space in his *Apologia* to expounding this fourth justification for war than he did for the other three combined. He claimed that the experience of Spaniards had been that, once Amerindians are reduced to Christian rule, they respond to Christian preaching so readily that more are converted in the space of a few days than would be converted by mere preaching in 300 years. This point alone should settle the entire controversy, he declared (Sepúlveda, 2023b, III.4).

Las Casas firmly rejected this argument at Valladolid. In addition to disputing Sepúlveda's interpretations of Augustine and other theological sources, Las Casas firmly rejected his opponent's claim that the subjugation to Christian rule and the instilling of fear made Amerindians more easily instructed in the faith and more inclined to receive it. Always eager to remind his audience that he had spent many years in the Americas whereas Sepúlveda had not, Las Casas insisted that his experiences demonstrated that the model of spreading the gospel that Sepúlveda recommended actually led the Amerindians to either reject the faith, or accept it merely out of fear, and so in a way that was ultimately hollow and insincere (Las Casas, 2023, Soto's Summary, §10).

The Outcome

Our knowledge of the judges' opinions of Sepúlveda and Las Casas' arguments at Valladolid relies to a large extent on the disputants' own accounts. Sepúlveda claimed that, while the theologians opposed him, the jurists, while disputing some of his arguments, agreed with him that the medieval canonists were right to justify war against infidels on account of their idolatry and other violations of natural law, and they accepted that Spain's wars in the Americas were just for that reason (Sepúlveda, 1997b, §8, 2023b, §5). Las Casas offered a rather different account. He asserted that he had proved conclusions before the junta "that no person before me had dared to touch or write about," including that it is not against natural law to sacrifice humans to a god that one believes to be true. All the theologians and jurists were "quite satisfied," he declared, some were even "struck with admiration," and the junta pronounced Spain's conquests unjust (Gutiérrez, 1993, p. 183; Las Casas, 1992, p. 9).

Ultimately, the junta did not issue a collective judgement. The Council of the Indies seems to have decided that the weight of opinion lay with Las Casas. Writing to Charles V in 1554, it asserted that the junta had considered Spain's wars "dangerous to Your Majesty's conscience for many reasons . . . and mainly because of the difficulty of excusing the damages and grave sins that are committed in such conquests" (quoted in Castilla Urbano, 2013, p. 245). Nevertheless, Charles soon cast aside his moratorium on conquests, authorizing new wars in Peru in 1555 (Adorno, 2007, p. 83).

Legacy

Despite many efforts, Sepúlveda never succeeded in being granted a license to print the *Democrates secundus*. It remained unpublished until the late nineteenth century. Meanwhile, his arguments became known via the record of the Valladolid debate that Las Casas compiled, which was translated in whole or in part into several European languages and published in numerous editions in the sixteenth and seventeenth centuries. Sepúlveda became a central figure in the "Black Legend" about the cruelties of Spanish imperialism and earned a reputation as one of the most brazen defenders of Spain's wars. As philosophers, jurists, and theologians writing in support of other European imperial projects sought to mark their distance from the Spaniards and to develop distinct justifications for wars and conquests, it seems likely that they would have taken some care to avoid certain claims and arguments for which Sepúlveda was known. Nevertheless, there is little indication that they engaged directly with his writings. He goes unmentioned in the just war writings of Alberico Gentili and Hugo Grotius, for example, as well as Samuel Pufendorf and Emer de Vattel.[11]

We would be mistaken, though, if we were to suppose that Sepúlveda's arguments about the just reasons for war are merely disturbing but irrelevant artefacts of the early years of European colonialism. As alluded to in the final chapter in this volume on Charles Mills, troubling echoes of the justifications that he offered in defense of Spain's imperial wars can be heard in justifications that continue to be offered for certain wars today—particularly those wars fought by Western powers against weaker others. The four just reasons for war that he advanced have had long afterlives, surviving both the secularization of the law of nations in the early modern period and even the codification of positive international law since the eighteenth century.

A notion of punishment continues to be insinuated in justifications for the resort to force, especially when great powers take up arms against "rogue" and "outlaw" regimes, whose behaviours are said to violate universal notions of morality. As seen in cases like the U.S.-led invasion of Iraq in 2003, such punishment is framed as justified for the purposes of disciplining wrongdoing and motivating the embrace of the true and the good. Wars of protection are fought in the name of the Responsibility to Protect (R2P). Whatever the differences between R2P as constructed by states in recent decades and the duty to protect as presented by Sepúlveda half a millennium earlier—and these differences are significant—the danger of doing more harm than good when resorting to force to protect innocents endures, as the horrors that have followed NATO's 2011 intervention in Libya clearly show. Wars are also fought to spread beliefs, especially belief in liberal democracy. In cases such as the 20-year war waged by Western powers in Afghanistan, regimes and institutions are forcibly torn down, populations are quelled and made subject to new rulers, outside experts are deployed, and people are taught to embrace the true doctrines of good government. And distinctions between the civilized and the barbarian continue to be invoked in support of all such wars, whether to help justify the punishment of the barbaric, the protection of their innocent victims, or the propagation of civilized virtues and institutions.

The identification of Sepúlvedan antecedents of certain justifications for war may not in itself warrant giving up on all such justifications today. At the very least, though, they point to a need to reckon carefully with his arguments for the resort to force—arguments that in Sepúlveda's unabashed voice are plainly dehumanizing and marked by obvious hubris and myopia, but whose fundamental claims and assumptions may not be so different from those that we continue to use to reason about the justice of war today.

Notes

1 This chapter's arguments build on a lengthier treatment in Glanville, Lupher, and Feile Tomes (2023).
2 On Las Casas contribution to just war thinking, see Brunstetter (2018).
3 For an English-language overview of Sepúlveda's life and career, see Bell (1925). See also Losada (1949/1973); Castilla Urbano (2013).
4 One early manuscript gives the title of the book as *Democrates alter*. The four other known manuscripts are entitled *Democrates secundus* (Coroleu Lletget, 1997, p. xxxi).
5 On Augustine and Gratian's contributions to just war thinking, see Johnson (2018) and Cox (2018).
6 For further discussion, see Glanville, Lupher, and Feile Tomes (2023); Hanke (1959); Lantigua (2020); Lupher (2003); and Pagden (1986).
7 On Aristotle's contribution to just war thinking, see Andie Melançon's chapter in the present volume.
8 On Aquinas's contribution to just war thinking, see Reichberg (2018).
9 On Vitoria's contribution to just war thinking, see Bellamy (2018).
10 Vitoria's relection on the Indies was in circulation in the 1540s. Sepúlveda presumably would have had access to it if he wished. But, while he at some point added an ambiguous marginal note to a manuscript of *Democrates secundus* that mentioned the relection, he gave little indication that he had read it closely.
11 On the echoes of Sepúlveda's arguments in the writings of these just war theorists, though, see Glanville, Lupher, and Feile Tomes (2023, pp. 56–62) and Lantigua (2020, pp. 189–250).

Works Cited

Adorno, Rolena. 2007. *The Polemics of Possession in Spanish American Narrative.* New Haven: Yale University Press.
Bell, Aubrey F. G. 1925. *Juan Ginés de Sepúlveda.* Oxford: Oxford University Press.
Bellamy, Alex J. 2018. Francisco de Vitoria (1492–1546). In Daniel R. Brunstetter and Cian O'Driscoll. eds. *Just War Thinkers: From Cicero to the 21st Century.* New York: Routledge, pp. 77–91.
Brunstetter, Daniel R. 2018. Bartolomé de las Casas (1484–1566). In Daniel R. Brunstetter and Cian O'Driscoll. eds. *Just War Thinkers: From Cicero to the 21st Century.* New York: Routledge, pp. 92–104.
Cano, Melchor. 1997. Letter 81, from Melchor Cano to Sepúlveda, undated. In Ignacio J. García Pinilla and Julián Solana Pujalte. eds. *Obras Completas, vol. 9.1: Epistolario.* Pozoblanco: Ayuntamiento de Pozoblanco, pp. 214–225.
Castilla Urbano, Francisco. 2013. *El pensamiento de Juan Ginés de Sepúlveda: Vida activa, humanismo y guerra en el Renacimiento.* Madrid: Centro de Estudios Políticos y Constitucionales.
Coroleu Lletget, Alejandro. 1997. Introducción filológica. In Juan Ginés de Sepúlveda. ed. *Obras completas, vol. 3: Demócrates Segundo.* Edited by Alejandro Coroleu Lletget. Pozoblanco: Ayuntamiento de Pozoblanco, pp. xxxi–xxxvii.
Cox, Rory. 2018. Gratian (Circa 12th Century). In Daniel R. Brunstetter and Cian O'Driscoll. eds. *Just War Thinkers: From Cicero to the 21st Century.* New York: Routledge, pp. 34–49.
Glanville, Luke, David Lupher, and Maya Feile Tomes. 2023. Introduction. In Luke Glanville, David Lupher, and Maya Feile Tomes. eds. and trans. *Sepúlveda on the Spanish Invasion of the Americas: Defending Empire, Debating Las Casas.* Oxford: Oxford University Press, pp. 1–62.

Gutiérrez, Gustavo. 1993. *Las Casas: In Search of the Poor of Jesus Christ.* Maryknoll, NY: Orbis.

Hanke, Lewis. 1959. *Aristotle and the American Indians: A Study in Race Prejudice in the Modern World.* Chicago: Henry Regnery.

Johnson, James Turner. 2018. St. Augustine (354–430 CE). In Daniel R. Brunstetter and Cian O'Driscoll. eds. *Just War Thinkers: From Cicero to the 21st Century.* New York: Routledge, pp. 21–33.

Lantigua, David M. 2020. *Infidels and Empires in a New World Order: Early Modern Spanish Contributions to International Legal Thought.* Cambridge: Cambridge University Press.

Las Casas, Bartolomé de. 1992. *In Defense of the Indians.* Edited and Translated by Stafford Poole. DeKalb, Ill.: Northern Illinois University Press.

Las Casas, Bartolomé de. 2023. Contained Herein Is a Debate or Disputation. In Luke Glanville, David Lupher, and Maya Feile Tomes. eds. and trans. *Sepúlveda on the Spanish Invasion of the Americas: Defending Empire, Debating Las Casas.* Oxford: Oxford University Press, pp. 236–346.

Losada, Ángel. 1973 [1949]. *Juan Ginés de Sepúlveda: a través de su 'Epistolario' y nuevos documentos.* Madrid: Consejo Superior de Investigaciones Científicas.

Lupher, David A. 2003. *Romans in a New World: Classical Models in Sixteenth-Century Spanish America.* Ann Arbor: University of Michigan Press.

Pagden, Anthony. 1986. *The Fall of Natural Man: The American Indian and the Origins of Comparative Ethnology.* Cambridge: Cambridge University Press.

Reichberg, Gregory M. 2018. Thomas Aquinas (1224/5–1274). In Daniel R. Brunstetter and Cian O'Driscoll. eds. *Just War Thinkers: From Cicero to the 21st Century.* New York: Routledge, pp. 50–63.

Schwartz, Daniel. 2007. The Principle of the Defence of the Innocent and the Conquest of America: 'Save Those Dragged Towards Death'. *Journal of the History of International Law* 9(2), pp. 263–291.

Sepúlveda, Juan Ginés de. 1997a. Letter 82, from Sepúlveda to Melchor Cano, July 15, 1549. In J. García Pinilla and Julián Solana Pujalte. eds. *Obras Completas, vol. 9.1: Epistolario.* Pozoblanco: Ayuntamiento de Pozoblanco, pp. 226–247.

Sepúlveda, Juan Ginés de. 1997b. Letter 95, from Sepúlveda to Martín Oliván, October 1, 1551. In Ignacio J. García Pinilla and Julián Solana Pujalte. eds. *Obras Completas, vol. 9.1: Epistolario.* Pozoblanco: Ayuntamiento de Pozoblanco, pp. 267–271.

Sepúlveda, Juan Ginés de. 2023a. Democrates Part Two, on the Just Reasons for the War against the Indians. In Luke Glanville, David Lupher, and Maya Feile Tomes. eds. and trans. *Sepúlveda on the Spanish Invasion of the Americas: Defending Empire, Debating Las Casas.* Oxford: Oxford University Press, pp. 86–179.

Sepúlveda, Juan Ginés de. 2023b. Outrageous, Scandalous, Heretical Notions. In Luke Glanville, David Lupher, and Maya Feile Tomes. eds. and trans. *Sepúlveda on the Spanish Invasion of the Americas: Defending Empire, Debating Las Casas.* Oxford: Oxford University Press, pp. 366–384.

Sepúlveda, Juan Ginés de. 2023c. The Defence of the Book, On the Just Reasons for War. In Luke Glanville, David Lupher, and Maya Feile Tomes. eds. and trans. *Sepúlveda on the Spanish Invasion of the Americas: Defending Empire, Debating Las Casas.* Oxford: Oxford University Press, pp. 198–224.

Vitoria, Francisco de. 1991. On the American Indians. In Anthony Pagden and Jeremy Lawrance. eds. and trans. *Political Writings.* Cambridge: Cambridge University Press, pp. 231–292.

4

ALONSO DE LA VERA CRUZ (1507–1584)

Francisco Lobo

Introduction

Alonso de la Vera Cruz epitomizes the essence of the Spanish spirit of the six-teenth century. In this rather obscure figure coalesced many of the traits we usually associate with this historic period marked by European territorial and ideological expansion, as encapsulated by the motto Charles V adopted for his growing empire: *Plus ultra* ("Further beyond"). It was an age of explora-tion and probing, a time for testing limits and pushing boundaries—not only geographical but also political and ethical. Alas, it was also an age of war—for no boundary can be pushed without encountering friction, especially in the case of imperial expansion, which more often than not involves imposing one's worldview to replace that of others. Yet, it was also a time when discus-sions about war, its rightfulness and necessity, were as important as its execu-tion, not unlike so many other historical periods covered by this volume.

In order to navigate the tumultuous waters of this new age, our adven-turers were aided by the full panoply of tools, knowledge, and capital the Renaissance had to offer. As part of such an essential wherewithal we find not only scientific advancement but also cultural and philosophical conceptions that accompanied explorers wherever they went, even "to the uttermost parts of the earth" (Koskenniemi, 2021).

One such traveler was Alonso de la Vera Cruz, a true "missionary of knowl-edge" (Lazcano, 2007) driven by a unique desire to spread his doctrine to benefit the whole of humanity, standing up for justice and right wherever—and just as importantly, against whomever—he saw fit. And to assist him in this ethical quest of global proportions, he wielded a powerful tool: the teachings of a towering intellect, his master Francisco de Vitoria. Drawing on

DOI: 10.4324/9781003428688-5

the lessons received at Salamanca, Vera Cruz harnessed this precious knowledge as the mariner uses a compass to help them navigate through agitated waters carrying them into the unknown.

But the compass can only take the mariner so far before they realize these navigation devices sometimes prove ineffective when faced with the challenges found in uncharted waters. Having been *plus ultra* himself, Vera Cruz had a firsthand knowledge of what happens when old beliefs are confronted with new realities. As a result, his lifework and legacy were also colored by some controversy and problematic claims that introduce nuance to this all but forgotten historical figure.

Contexts

1507 was the year of two important baptisms: it was the moment when the New World was reportedly called "America" for the first time in Martin Waldseemüller's world map. It was also the year when Vera Cruz was born[1] and christened Alonso Gutiérrez in Caspueñas, a small town located in Castille, Spain. Vera Cruz was thus born into a society that was undergoing unprecedented change as a burgeoning world power engaged in the continuous conquest of vast territorial possessions that would come to dwarf his native Iberian Peninsula.

He received a full education in the humanities (including rhetoric and Latin) at the University of Alcalá de Henares, after which he pursued further studies in philosophy ("arts") and theology at the prestigious University of Salamanca between 1528 and 1532. There, he would become a student to one of the most celebrated scholars of the time, Francisco de Vitoria (Gómez Robledo, 1984, pp. 29–30).

Upon completion of his superior studies, he tutored the children of a Spanish Duque for a few years and worked as a reader of philosophy at his *Alma Mater* until 1535 (Aspe, 2021, p. 295; Heredia, 2004a, p. 25). As he approached the then advanced age of 30, probably eager to see the world and apply in practice the invaluable theoretical knowledge received at Salamanca, Alonso was inspired by a cleric who had recently returned to Spain from America to set sail on a new adventure. He signed up for the next available ship departing from Seville, carrying a crew of Augustine missionaries. They finally landed in the port of Villa Rica de la Veracruz (located in present-day Mexico) in the summer of 1536. Inspired by this new setting, Alonso decided to join the Augustine religious order, adopting thenceforth the name Alonso de la Vera Cruz[2] (Gómez Robledo, 1984, p. 30; Heredia, 2004a, p. 9)—literally "Alonso of the True Cross".

By the time of his arrival, Mexico had been under Spanish control for 15 years, which means that the Spanish *Conquista*, a feat that would span across three centuries, was still in its infancy and would continue to unfold for many generations. Before the invasion, Central and North America were characterized by a vast network of trade among tribes of differing sizes and power, as well as periodic armed conflict in order to secure the tributes demanded by the regional hegemon, Tenochtitlán. Thus, pre-Hispanic Mexico was a place plagued by war and inequality (Ávila Sandoval, 2003, p. 331), where force in numbers rather than military technology facilitated victory (Carballo, 2022, p. 4). Upon their arrival, the conquistadores imposed Catholicism as the official religion of New Spain, reinforced in 1530 by the infamous Spanish Inquisition. In their unsatiable appetite to gain more land and indigenous servants through the model of indentured servitude known as *encomienda*, these *encomenderos* did not hesitate to weaponize the Inquisition by accusing each other of heresy (Lima, 2015, p. 148). At the same time, they were tasked with the Christian instruction and pastoral care of their indigenous servants, a task for which they were assisted by the local clergymen, as the *encomenderos* were mainly in it for the profits not for the salvation of souls.

Against this backdrop, the remaining years of this born-again theologian would bear fruit to an impressive record of applying just war theory in the New World. Suffice it to say for now that, until his death in 1584, our adventurer-philosopher endeavored to bring knowledge and the promise of salvation where he thought it was most needed, speaking truth to power and standing up for justice, such that, to paraphrase Kipling's poem *If*, Vera Cruz truly managed to "walk with kings and not lose the common touch".

Texts and Tenets

The lifework of an archetypal Renaissance man such as Vera Cruz can be measured, first, by his many contributions to the advancement of knowledge in the New World, including setting up the continent's first library, holding the first chair of philosophy at the recently founded University of Mexico (1551), and founding several local monasteries and schools spread across New Spain (Gómez Robledo, 1984, pp. 33–38; Burrus, 1963, pp. 226–227; Heredia, 2004a, pp. 9–10, 25–26). True to his humanist character and erudite spirit, he also endeavored to learn the language of some of the local tribes in Mexico.

As an accomplished scholar trained at one of the best universities of his time, Vera Cruz was knowledgeable in a wide range of topics, including philosophy, theology, logic, law and ethics and even physics, astronomy, natural science, and agriculture (Gómez Robledo, 1984, p. 38; Heredia, 2004b). Alongside his impressive record as a missionary, the intellectual output of this veritable "Mexican Socrates" (Cerezo de Diego, 2015, p. 51) comprises commentaries on the Aristotelian canon as an integral part of his teaching, including such titles as *Recognitio summularum* (1554), *Dialectica resolutio*

(1554), and *Phisica speculatio* (1557). It also includes reflections on more practical issues such as the legal implications of marriage in his work *Speculum coniugiorum* (1556) and the fairness of religious taxation in *De decimis* (1555–1557) (Heredia, 2004a, pp. 31–38; Cerezo de Diego, 2015, p. 49).

But without a doubt, Vera Cruz's most renowned contribution—and the one that is most relevant for this volume on the ethics of war—is a *Relectio* or lecture he gave in 1554–1555 at the University of Mexico, following in the footsteps of his master Vitoria, titled *De dominio infidelium et iusto bello*, or "On the rule over unbelievers (or Indians) and just war". It is worth noting that this *Relectio* was delivered immediately after the famous Valladolid debates on the conquest of the New World that took place in Spain between Sepúlveda and Las Casas in 1550–1551, which have a storied place in just war tradition (Brunstetter and Zartner, 2011). As no winner was officially declared in the dispute, the rightfulness of the Spanish conquest remained an open question, one to which Vera Cruz sought to provide some answers of his own. The manuscript containing the original transcript of the lecture was lost to history until the twentieth century when it was unearthed by an American Jesuit scholar, Ernest J. Burrus. The manuscript was kept in a private library until 1958, when it was entrusted to Burrus for him to release only an English translation of its contents (Cerezo de Diego, 2015, p. 86; Velasco, 2019, p. 1036; Godinas, 2007, p. 267). The task would be eventually completed in 1968, when Burrus finally published *The Writings of Alonso de la Vera Cruz* in five volumes, including *De dominio* (Burrus, 1968). In 2004, as the original lecture turned 450 years old, the Universidad Nacional Autónoma de México (UNAM) issued a special Spanish edition of this precious piece of Hispanic–American scholarship (Heredia, 2004a).

De dominio was taught as part of the *Prima* chair in theology at the School of Philosophy and Literature of the new Royal University of Mexico (Burrus, p. 7). As expected in such a fledgling academic institution, Vera Cruz's pupils included young clerics, as well as laymen of all origins (Spaniards, *criollos*, and natives) (Heredia, 2004a, p. 9). In this sense, his teachings had a direct practical impact on people connected to the day-to-day reality of the New World in a unique way that was absent in the hallowed halls of Salamanca where his master taught.

True to its scholastic form (Aspe, 2021, p. 310), *De dominio* begins with an inquiry into the many implications of a single passage from the Bible[3]: in this iteration, Vera Cruz chose the famously pragmatic formula: "Render to Caesar the things that are Caesar's; and to God the things that are God's" (Mt 22:21) (Burrus, 1963, p. 230; Aspe, 2021, p. 310). Thereupon follow 11 *dubia* or *questiones* ("doubts" or "questions") bearing on the meaning of the passage.

In order to unfold his reasoning on each doubt, Vera Cruz presents the question, followed by an appropriate line of reasoning, culminating in an alternation of conclusions and corollaries. This scholastic method, honed over the centuries by such masters as Aquinas and Vitoria (Aspe, 2021, p. 313), offers a dialectic style that is apposite for the exploration of such thorny moral issues as the conquest of new populations and the waging of war against them (Velasco, 2019, pp. 1031, 1036; Braun, 2023).

The practical importance of this kind of exploration at the time is two-fold: On the one hand, the *Conquista* was ongoing; therefore, asking about its legitimacy continued to be as timely as it was pertinent (Carillo Cázares, 2000). On the other hand, the Valladolid debate that preceded this lecture did not end with a clear winner, thereby additional analytical effort and soul-searching exercise by the conquerors, on either side of the Atlantic, were warranted. The fact that Vera Cruz took care to learn the *Tarasca* language of his native pupils (Heredia 2004a, p. 10) guaranteed that his contribution to the debate would be received in a much more open way by those directly affected, arguably, than if it came all the way from Europe formulated in obscure Latin aphorisms and inaccessible scripture.

De dominio considers 11 doubts or questions concerning the validity of the Spanish *Conquista* (Burrus, 1963, pp. 231–252; Heredia, 2004a, p. 43).[4] Vera Cruz largely followed in the footsteps of his master Vitoria—although he inexplicably did not cite him directly in *De dominio*, referencing him elsewhere (Heredia, 2004b; Lupher, 2009, p. 160)—to examine the just and unjust titles for the European conquest of the Indies. For the purposes of this chapter, we will focus on only the last two doubts, both bearing on *jus ad bellum*; unlike Vitoria (Bellamy, 2018, p. 84), Vera Cruz did not venture into *jus in bello* matters. That said, it is worth keeping in mind that Vera Cruz did not focus only on the causes and regulation of war. He also endeavored to look at the larger picture of statecraft as a complex phenomenon comprising taxation, *dominium* (i.e., ownership and jurisdiction), as well as the administration of public force as a tool for governance, paving the way for more contemporary authors to raise the same points centuries later (Scott, 2020; Koskenniemi, 2021). Likewise, as we shall see in the last section, the essential driver behind the entirety of Vera Cruz's thinking is a rudimentary conception of self-determination.

Regarding Question 10 ("*Whether the emperor or the King of Castile could have declared just war against the natives of the New World*"), Vera Cruz concludes following Vitoria, that no power, whether it be temporal or spiritual, can declare war on the unbelievers to dispossess them of their property on the sole ground of their different religions (Scott, 2020; Koskenniemi, 2021, p. 314).

Furthermore, relying on the authority of Aquinas, Vera Cruz asserts that *dominium*—that is, ownership and/or political power or jurisdiction (Koskenniemi, 2021)—stems from natural reason, such that faith, deriving itself from divine law, cannot override the natural law. On this point, thus, natural law trumps divine law, and therefore a different religion is not enough justifica-tion for taking property away from people (Heredia, 2004a, p. 315). A key corollary to this is the bold claim that the war waged against the natives as authorized by the Catholic Kings, the emperor or the pope, was an unjust one. Consequently, the emperor and every single Spaniard who wronged the natives must compensate them (Heredia, 2004a, pp. 316–317). In this sense, Vera Cruz becomes a precursor of *just post bellum* as he was confronted with the *fait accompli* of the first wave of the Spanish conquest in Mexico but sought ways to right injustices done to certain natives. However, his work understandably focuses mostly on *jus ad bellum* rather than *jus post bellum*, as the violence of the *Conquista* was still unfolding before his eyes during his time in the New World. This position stands in stark contrast with that of Sepúlveda, whose contribution is also studied in the previous chapter. For the imperialist champion of the Valladolid debates, who all but ignored the authority of Vitoria, the natural serfdom of the Indians was a matter of fact. Vera Cruz's lecture would have offered a different authoritative stance that heightened the sense of native rights and equality in a time when Sepúlveda's paradigm might still have won the day.

Continuing with just cause, Vera Cruz believed that if the infidels attack the Christians first and cause them *iniuria*,[5] then it is lawful to punish them with war and deprive them of their *dominium* (Heredia, 2004a, p. 319). Likewise, if unbelievers do not welcome missionaries and attack them without giving them a chance to preach the Gospel, then war against them would have been justified under the authority of the pope; yet, since the first European envoys were in fact armed soldiers and not peaceful preachers, then waging war against the unbelievers is unjust, for "no one can be forced to believe" (Here-dia, 2004a, pp. 320–323). An additional conclusion that is indicative of Vera Cruz's "anti-imperial humanism" (Velasco, 2019, p. 1025) is his defense of the natives' rational agency. He believes that, however childish and underde-veloped they may seem to Europeans, they are indeed endowed with the same natural reason as every other human being, having their own customs and governance institutions and being capable of great accomplishments (Here-dia, 2004a, pp. 327–336).

Moving on to Question 11 ("*Whether there is any motive to justify war against the inhabitants of this New World*"), Vera Cruz continues to probe the potential just causes for Spaniards to wage war against the natives, focus-ing this time on more plausible rationales. This admittedly moves his theory closer to Sepúlveda's position in the Valladolid debate and away from Vito-ria's, especially as the former saw just war as a means to pave the way for spreading Christianity, something a practicing missionary like Vera Cruz val-ued in particular. He also shares with Sepúlveda a proclivity to draw directly on Aristotle for building his arguments instead of merely relying on the scho-lastic distillation of the Philosopher's ideas (Gómez Robledo, 1984, p. 40), although Vera Cruz did not go as far as Sepúlveda as to use Aristotle's postulate of natural slavery to justify the *Conquista*. In this sense, he remained a true *Vitoriano*.

After engaging dialectically with a full battery of arguments and counter-arguments, Vera Cruz reaches a number of conclusions, with evidence to support each. Among the foremost conclusions is the admissibility of using force after proper preaching has been ineffective, which is admittedly a departure from his master Vitoria (conclusions 1 and 2). More in line with Vitoria, overthrowing tyranny and the stopping of cannibalism are just causes for war (conclusions 4 and 5), as is providing military assistance to allies (conclusion 6) and using force against the natives if they prevent the Spanish from peacefully settling, negotiating, mining, or traveling (conclusions 12–15). While these just causes might be interpreted as justifications for European domination that solidified a hierarchy in international relations, which continues to legitimize Western just wars (as explored in the chapter on Charles Mills), it is also important to note that the Spanish had many native allies, including the Tlaxcaltecas, who were also *conquistadores* shaping local international relations and pushing the limits of just war norms (Matthew and Oudijk, 2007; Carmagnani, 2011, p. 275; Brunstetter and Lobo, 2024).

Controversies

Despite being a man of the cloth, Vera Cruz did not always see eye to eye with the hierarchy of the Catholic Church. In fact, his life's work is evidence that he preferred to remain a man of the people, a grassroots philosopher who would always take the side of the oppressed and the downtrodden over the powers that be. Accordingly, and even though he proved his managerial aptitude when he was left in charge of the diocese of Michoacán as a deputy while the bishop traveled to Europe to attend the Council of Trento, he turned down three times the promotion to the official position of bishop for the dioceses of León de Nicaragua, Puebla/Tlaxcala, and Michoacán. The first of these was offered to him by emperor Charles V himself, to which Vera Cruz's first reaction reportedly was to say "save me, oh Lord, from the lion's mouth"[6] (Gómez Robledo, 1984, pp. 31, 34; Burrus, 1963, p. 227).

But his complicated history with Catholic authorities does not arise solely from declining a few job offers. His political preference for grassroots communities and local governance also led Vera Cruz to actively lobby for the exemption from the Trento Decree for the religious orders he worked closely with in Mexico. The Decree aimed at tightening control that central episcopal authority held over local religious privileges. No bishop or missionary from the New World was in attendance at this momentous congressional event. With his hands-on experience from the Americas, Vera Cruz was convinced that this was a mistake that would harm native converts who maintained some ties to traditional ways. He traveled back to Spain in 1562, where he argued his case before King Philip II, who held the cleric in high intellectual esteem (Heredia, 2004a, p. 27). As a testament to his political acuteness, Vera Cruz finally managed to get the Spanish King to intercede with the pope. As a result, the overseas dominions of the Spanish Empire were exempted from the Tridentine legislation, and local religious orders could preserve their autonomy and privileges (Burrus, 1963, p. 229; Gómez Robledo, 1984, pp. 36–37).

Beyond his clerical activism, Vera Cruz also clashed with religious authorities due to the contents of his teachings. As mentioned earlier, he delivered his *Relectio De dominio* within the context of the greater Valladolid debate that took place right before his inaugural lecture at the University of Mexico. Insofar as his efforts to check attempts by *Encomenderos* to conquer more lands caused a stir, he was, like Las Casas, not always looked upon favorably. Moreover, his penchant to speak truth to power, what one scholar calls such a "paradigm of academic freedom" (Heredia, 2004b), earned him the animosity of Catholic authorities in the New World. Although admittedly more the result of his professional differences with the archbishop of Mexico over the best way to evangelize the natives than a consequence of its actual contents, the fact is that Vera Cruz's *De dominio* was prevented from being published by order of the latter (Heredia 2004a, p. 41; Burrus, 1963, p. 252; Godinas, 2007, p. 267), which explains why it was lost to history until the twentieth century. Admittedly, Sepúlveda's work was also banned at some point, which speaks as much to the even-handedness of the censorship applied by the Spanish crown as to the complexity of theorizing amidst controversy no matter the position defended.

Perhaps the greatest irony is that the actual contents of *De dominio* are not all that subversive or threatening to the powers that be as the censorship it was subjected to might suggest. On the contrary, it included some claims that could have very well been used to bolster the case for Spanish expansion.

The most puzzling of such claims is Vera Cruz's view on forced conversions, a matter in which he definitively departed from the teachings of his master Vitoria, who was against forcing people to embrace the Christian faith if they did not truly believe (Aspe, 2021, p. 316). Even Vera Cruz himself seems to subscribe to this opinion when he claims that "no one can be forced to believe" (Heredia, 2004a, pp. 320–323). Yet, Vera Cruz refers in that passage to the specific practice of sending in armed soldiers first instead of preachers. On the other hand, as stated in the first conclusion to Question 11 of *De dominio*, the theologian of the True Cross is persuaded that if the Christian faith has been sufficiently and peacefully presented to the natives and they still refuse to embrace it, then the emperor would be justified to wage war against them to make them believe: "that they may wholeheartedly want what they did not accept before" since they would no longer be excused by their prior invincible ignorance of the true faith (Heredia, 2004a, pp. 340–341). Reminiscent of Rousseau's perplexing "forced to be free" formula, Vera Cruz calls this kind of coercion "indirect", namely the type of coercion that the emperor can also apply against unbelievers within its European domains, including Jewish and Muslim populations (Heredia, 2004a, p. 341). Vera Cruz is not specific as to whether this use of coercion within the confines of imperial jurisdiction qualifies as just war or merely as governance by a Christian sovereign.

Another rather controversial claim found within *De dominio* relates to the ideas of popular sovereignty and self-determination that permeate Vera Cruz's thinking. Undoubtedly a precedent for the rules-based international order, as we shall see in the next section, Vera Cruz's doctrine of self-determination was not necessarily synonymous with values held dear to the international community of today, such as democracy and respect for human dignity. His concept of self-determination could just as well accommodate some rather illiberal contents so long as the will of the people mediated in their implementation. Thus, he concludes about the rule of Moctezuma:

> Perhaps that which seems tyrannical to a nation may be convenient and appropriate from the standpoint of this barbarous people, as if it was better for them to be ruled by their own lords with fear and dominance and not with love.
>
> *(Heredia, 2004a, pp. 357–358)*

This malleability in Vera Cruz's understanding of self-determination confirms that this principle is usually affected by an "in-built ambivalence" turning it into a battleground for different ideologies to compete (Mckenna, 2023, pp. 11–13).

Finally, an additional potential point of contention to be found in *De dominio* is its treatment of *post bellum* issues. Pragmatic as ever, Vera Cruz's starting point is the Spanish conquest as a *fait accompli*. Consequently, he does not argue for the complete expulsion of Europeans from the New World and restoration of native sovereignty, as Las Casas purportedly did at the end of his life (Cerezo de Diego, 2015, p. 587, Brunstetter 2018, p. 241). He differentiates between questions of justice at the beginning of the war and those bearing on the justice of preserving a realm after the war is won. Thus, even if the war suffers from a poor *ad bellum* justification, or a complete lack thereof, there might be still justice in retaining what has been gained (Heredia, 2004a, p. 340). At the same time, he believes that it is only fair for the emperor and the Spanish soldiers who have committed injustice to compensate the victims "as it is the case with robbery" (Heredia, 2004a, p. 317).

Compensations notwithstanding, Vera Cruz also reaches a final conclusion building on the *ius communicationis* stemming from natural law and articulated as international law (Burrus, 1963, p. 252), in the sense that Europeans are allowed to peacefully travel, settle, negotiate, and exploit the natural resources of the land, and if they are forcefully prevented from doing so by the natives, they are allowed to reply in kind and take their property as due compensation (Heredia, 2004a, p. 386). However, when referring to the perennial issue (especially in the Americas) of land ownership and cultivation, Vera Cruz takes the stance that the natives are true owners of their land, even if they do not always farm it, and therefore the Spanish are not allowed to take it away from them but only exact tribute, a view he managed to reconcile with the subsistence of the scheme of exploitation known as the *encomienda* (Heredia, 2004a, pp. 49–50, 151–153). Some might say he thus laid the foundation for what has been called a true "economy of solidarity" whereby the sovereign rights of the natives are recognized and respected as the basis for a fair distribution of wealth and social peace (Pereña, 1998, p. 55). The fact that the Tlaxcala nobles did reportedly preserve some form of nominal autonomy after the fashion of European medieval custom may have helped in making such exploitation more palatable for local elites (Navarrete, 2019). But not so much so for native exploited on the *encomiendas* and in the mines.

This nuanced and seemingly sensible approach to *jus post bellum* in the context of the Spanish conquest of the Americas has even been characterized as a "colonial ethic" that is both practical and realistic, thus enabling two different communities to coexist even after armed confrontation (Heredia, 2004a, p. 66). At the same time, such an attempt at sanitizing European imperial expansion can be construed as self-serving and even disingenuous or delusional. It is arguably in the transcendence of Vera Cruz's work that we may find a yardstick to ascertain the ultimate measure of his legacy.

Legacy

The shores of Veracruz in Mexico did not witness the rebirth of Alonso Gutiérrez as Alonso de la Vera Cruz in 1536 alone. Almost two decades prior, near that same harbor, Hernán Cortés decided to sink his own fleet in order to signal to his troops his uttermost resolve and that henceforth there would be no turning back (to Cuba, or Spain) until they have achieved their goal of conquering the inland that lay *plus ultra*. Cortés thus burned his bridges with Europe and its political establishment to pursue his ambitious path of personal glory unhindered.

Alonso de la Vera Cruz, on the other hand, did not so drastically cut all ties with the Old World. On the contrary, our mariner did not bury his compass in the sands of Veracruz but kept it in order to harness all the neo-scholastic knowledge enclosed within, a wealth he was eager to share with the inhabitants of this *Terra Nova*.

Further, Vera Cruz is one of a few Salamanca scholars who "talks about America from America" (Cerezo de Diego, 2015, p. 580), and as a result no one can accuse him of being one of the "armchair theologians" weighing in on the Affair of the Indies without firsthand knowledge of the situation on the ground (Lantigua, 2011; Aspe, 2021, p. 322), or of exaggerating in his account of the atrocities, like Las Casas admittedly did. He was undoubtedly an exceptional witness (Pereña, 1998, p. 55). In Vera Cruz's own words: "I speak from experience" (Heredia, 2004a, p. 52). This, naturally, does not confer watertight validity to every single argument he made and does not shield his work from justified criticism. But it is an important consideration when assessing the impact of his legacy.

This legacy is a mixed one. On the one hand, he has been praised for being an advocate of the rights of indigenous populations, much like Las Casas was at the time. In the words of his most famous translator: "The lifework of Veracruz may be summed up in one phrase: the defence of the natives" (Burrus, 1963, p. 229). This fabled "anti-imperial humanism" (Velasco, 2019, p. 1025) was attributable to not only Vera Cruz but also to an entire "Iberian School of Peace" comprising the Salamanca scholars (Vitoria, Suárez, and Vera Cruz among them), and also thinkers from the rest of Spain and Portugal whose arguments helped lay the groundwork for modern theories and institutions promoting human rights and the rule of law, as pointed out by late Judge Antônio Cançado Trindade (Trindade, 2014, pp. 43, 59, 89).

Furthermore, like Las Casas, Vera Cruz believed in a restricted version of just war that must be held in reserve only for the most extreme of circumstances or the most recalcitrant of enemies, such as the Turks (Brunstetter, 2018, pp. 231–233; Burrus, 1963, p. 245). Also like Las Casas, Vera Cruz spoke from experience as a witness of the facts on the ground. Yet, an important lesson to be drawn when comparing both thinkers is that even accounts from the ground can differ as everyone interprets reality in different ways, and that includes Las Casas and Vera Cruz. The fact that both of them were "in country" and witnessed the facts they were reflecting on does not render either of their works infallible but at best more or less plausible.

Vera Cruz's contributions to the development of modern international law build on the Vitorian doctrine that, since there is no single ruler of the world, the principles governing international relations must rest on the "authority of the entire globe" (*"totius orbis auctoritate"*) (Vitoria, 2010, pp. 40, 252–258; Heredia, 2004a, pp. 219–250; Gómez Robledo, 1989, pp. 32–33). This horizontal scheme presupposes the agency of the peoples comprising such a multipolar community, with self-determination and popular sovereignty always being at the core of Vera Cruz's doctrine (Heredia, 2004a, p. 45). Growing strong roots in the Western Hemisphere particularly during the age of national wars of liberation (Mckenna, 2023, p. 44; Velasco, 2019, pp. 1042–1043), the principle of self-determination is today a well-established rule of conventional and customary international law (Sands, 2022)—one that has even been elevated to the rank of a peremptory norm or *jus cogens*. As a result of this interplay between Western and non-Western sovereigns, a veritable "*mestizo* international law" was enabled to eventually develop, one that finds parallels in the work of Carlos Calvo explored later in this volume (Becker Lorca, 2016, p. 11).

Vera Cruz's acceptance of military assistance to allies as a plausible just cause for war, also predicated on the premise of equal agency and self-determination, is another important precedent for the development of the doctrine of collective self-defense at both the regional and the UN levels (Lobo and Silvestre, 2020, pp. 199–233).

On the other hand, if we take his lifework to amount to no more than an attempt at whitewashing or sanitizing an otherwise nasty endeavor of colonial expansion and oppression, Vera Cruz could be accused of being a shameless apologist of empire and of the violence it inevitable engenders—one of the "barbarous philosophers" Rousseau famously called out (Coker, 2010, p. 3) or the "miserable comforters" Kant likewise despised (Koskenniemi, 2009). Accordingly, it could be said that his pragmatism and firsthand knowledge of life in the New World led him to justify not only the continuous presence of Europeans on the continent but also the perpetuation of the exploitative *encomienda* system and even a highly debatable defense of forced conversions. Similar to modern-day imperial "nation-building" (Ignatieff, 2003), there seems to be an unresolved tension between the recognition of the agency and rights of a group and the promotion of its subordination to a larger power and so-called universal values.

Was Vera Cruz being disingenuous then? Conceivably so, but then no more than Vitoria and other thinkers trying to rationalize the march of history unfolding before their eyes. Even if not, and if he was assisted by the purest of intentions, was not Vera Cruz's fight against the inexorable advance of the Spanish Empire (and the West) a merely Quixotic gesture whereby, as he put it himself, he might have just been "taking swings in the dark" (Heredia, 2004a, p. 171)? Could he not also be accused of suffering from a messianic disposition by giving one of the earliest performances on record of the infamous "white savior" role?

These are all questions that cannot, and should not, be answered in this short account, for more research and debate are needed around this obscure but highly influential actor in the Spanish conquest of the Americas. In such an endeavor, it is always salutary to venture beyond the traditional just war canon to look at those who tried to apply its principles, and this chapter can hopefully become a first step in that direction.

Notes

1 There are some disputes around this date, with some arguing Vera Cruz was actually born in 1504 (Heredia, 2004a, p. 7).
2 Or "Veracruz" as it is commonly found in the literature written in Spanish.

3 Vitoria's verse of choice in *De Indis* was: "Go ye therefore, and teach all nations, baptizing them in the name of the Father, and the Son, and the Holy Spirit" (Mt 28:19) (Vitoria, 2010, p. 233).

4 The 11 doubts/questions are:

1) "Whether those who have taken over natives without any title to them may justly exact tribute from them or whether they must make restitution of the tribute received and free the natives".
2) "Whether owners with a just title to the Indians are bound to attend to their instruction".
3) "Whether the encomendero who had just dominion over his charges by virtue of a royal grant, might arbitrarily occupy such lands of the natives as were untilled but destined for cattle-grazing or planting".
4) "He who is the real owner or lord (verus dominus) may levy and exact such tribute. But the Spaniard who has these natives entrusted to his care is the real owner or lord. Therefore, he may levy and exact such tribute".
5) "Whether the natives who ruled in the New World prior to the arrival of the Spaniards were true lords; and if they were, whether they could be justly deprived of their dominion and whether they are now actually so deprived".
6) "Whether the Spaniards who purchase land from the Indians can do so conscientiously regardless of the price they pay for it".
7) "Whether the emperor is the lord of the world".
8) "Whether the emperor is the owner of all things".
9) "Whether the pope has supreme power as lord of the world".
10) "Whether the emperor or the King of Castile could have declared just war against the natives of the New World".
11) "Whether there is any motive to justify war against the inhabitants of this New World".

5 Since the writing of Aquinas, the just war canon has considered *iniuria* to be the epitome of the just cause for war. *Iniuria* has been translated in various ways in the literature, including "injury", "injustice", "culpable action", "offence", and "wrong" (Vitoria, 2010, p. 303; Reichberg, Syse, and Begby, 2006, pp. 288–332).

6 "*Ab ore leonis libera me, Domine*".

Works Cited

Aspe, Virginia. 2021. The Influence of the School of Salamanca in Alonso de la Vera Cruz's *De dominio infidelium en iusto bello*. In Thomas Duve, José Luis Egío, and Christiane Birr. eds. *The School of Salamanca: A Case of Global Knowledge Production*. Leiden: Brill, pp. 294–334.

Ávila Sandoval, Santiago. 2003. Una reflexión sobre la historia de la economía prehispánica. *Análisis Económico* 18(39), pp. 325–340.

Becker Lorca, Arnulf. 2016. *Mestizo International Law*. Cambridge: Cambridge University Press.

Bellamy, Alex. 2018. Francisco de Vitoria (1492–1546). In Daniel Brunstetter and Cian O'Driscoll. eds. *Just War Thinkers: From Cicero to the Twenty-First Century*. New York: Routledge, pp. 77–91.

Braun, Christian Nikolaus. 2023. *Limited Force and the Fight for the Just War Tradition*. Washington, DC: Georgetown University Press.

Brunstetter, Daniel. 2018. Las Casas and the Concept of Just War. In David T. Orique and Rady Roldan-Figueroa. eds. *Bartolomé de Las Casas, O.P.: History, Philosophy, and Theology in the Age of European Expansion*. Leiden: Brill, pp. 218–242.

Brunstetter, Daniel and Francisco Lobo. 2024. R2P, the Imperial Critique, and Self-Determination: Recovering the Narrative of the Tlaxcaltecas. *Global Responsibility to Protect* 16(3), pp. 213–238.

Brunstetter, Daniel and Dana Zartner. 2011. Just War against Barbarians: Revisiting the Valladolid Debates between Sepúlveda and Las Casas. *Political Studies* 59(3), pp. 733–752.

Burrus, Ernest J. 1963. Alonso de la Veracruz's Defence of the American Indians (1553–54). *Heythrop Journal* 4(3), pp. 225–253.

Burrus, Ernest J. 1968. *The Writings of Alonso de la Vera Cruz.* St. Louis: St. Louis University.

Carballo, David. 2022. Governance Strategies in Precolonial Central Mexico. *Frontiers in Political Science* 4, pp. 1–13.

Carillo Cázares, Alberto. 2000. *El debate sobre la guerra chichimeca, 1531–1585: Derecho y política en Nueva España.* 2 vols. Zamora: El Colegio de Michoacán.

Carmagnani, Marcello. 2011. *The Other West: Latin America from Invasion to Globalization.* London: University of California Press.

Cerezo de Diego, Prometeo. 2015. *Alonso de Veracruz (1507–1584) y el derecho de gentes.* Madrid: Universidad Complutense de Madrid.

Coker, Christopher. 2010. *Barbarous Philosophers.* London: C. Hurst & Co.

Godinas, Laurette. 2007. *Alonso de la Veracruz, fray. De dominio infidelium et iusto bello. Sobre el dominio de los infieles y la guerra justa.* Ed. crítica, introd., notas y trad. Roberto Heredia Correa and colab. Olga Valdés García. México: unam-iifl, p. lxxxviii, p. 200. (Ediciones Especiales, 44) *Boletín del IIB* 12(1–2), pp. 267–270.

Gómez Robledo, Antonio. 1984. *El magisterio filosófico y jurídico de Alonso de la Veracruz.* Mexico, DF: Porrúa.

Gómez Robledo, Antonio. 1989. *Fundadores del derecho internacional. Vitoria, Gentili, Suárez, Grocio.* Mexico, DF: UNAM.

Heredia, Roberto. 2004a. *Fray Alonso de la Vera Cruz. Sobre el dominio de los indios y la guerra justa.* Mexico, DF: UNAM.

Heredia, Roberto. 2004b. Fray Alonso de la Veracruz, 'de dominio infidelium et justo bello'. Reseña bibliográfica (1958–2003). *Tzintzun: Revista de Estudios Históricos* 39, pp. 59–92.

Ignatieff, Michael. 2003. *Empire Lite: Nation-Building in Bosnia, Kosovo and Afghanistan.* London: Vintage.

Koskenniemi, Martti. 2009. Miserable Comforters: International Relations as New Natural Law. *European Journal of International Relations* 15(3), pp. 395–422.

Koskenniemi, Martti. 2021. *To the Uttermost Parts of the Earth: Legal Imagination and International Power, 1300–1870.* Cambridge: Cambridge University Press.

Lantigua, David. 2011. *Fray Bartolomé's Advice to Armchair Theologians.* Available at: https://sites.nd.edu/schoolofsalamanca/2011/02/25/fray-bartolomes-advice-to-armchair-theologians/.

Lazcano, Rafael. 2007. *Fray Alonso de Veracruz (1507–1584): misionero del saber y protector de indios.* Madrid: Editorial Revista Agustiniana.

Lima, María de la Luz. 2015. *El control social en el México prehispánico y colonial.* México, DF: INACIPE.

Lobo, Francisco and Felipe R. Silvestre. 2020. The International Law on the Use of Force in the Americas: Understanding the Dynamics of Subjects and Obligations. In Maria Luisa Duarte, Rui Tavares Lanceiro, and Francisco de Abreu Duarte. Coord. *Ordem Jurídica Global do Século XXI.* Lisbon: AAFDL Editora.

Lupher, David. 2009. *Romans in a New World: Classical Models in Sixteenth-Century Spanish America.* Ann Arbor: The University of Michigan Press.

Matthew, Laure E. and Michel R. Oudijk. eds. 2007. *Indian Conquistadors: Indigenous Allies in the Conquest of Mesoamerica.* Norman: University of Oklahoma Press.

McKenna, Miriam Bak. 2023. *Reckoning with Empire: Self-Determination in International Law*. Leiden: Brill Nijhoff.

Navarrete, Federico. 2019. Las historias tlaxcaltecas de la conquista y la construcción de una memoria cultural. *Iberoamericana* 19(71), pp. 35–50.

Pereña, Luciano. 1998. La Escuela de Salamanca. Notas de Identidad. In Francisco Gómez Camacho and Ricardo Robledo. eds. *El Pensamiento Económico de la Escuela de Salamanca*. Salamanca: Ediciones Universidad de Salamanca, pp. 43–64.

Reichberg, Gregory, Henrik Syse, and Endre Begby. 2006. Francisco de Vitoria (ca. 1492–1546). Just War in the Age of Discovery. In Gregory Reichberg, Henrik Syse, and Endre Begby. eds. *The Ethics of War: Classic and Contemporary Readings*. Oxford: Blackwell Publishing, pp. 288–332.

Sands, Philippe. 2022. *The Last Colony*. London: Weidenfeld & Nicholson.

Scott, James C. 2020. *Seeing Like a State*. New Haven: Yale University Press.

Trindade, Antônio Cançado. 2014. Prefacio. In Pedro Calafate and Ramón Mandado Gutiérrez. eds. *Escuela Ibérica de la Paz*. Santander: Ediciones Universidad Cantabria, pp. 41–109.

Velasco, Ambrosio. 2019. La filosofía crítica de Alonso de la Veracruz a 500 años de la conquista. *Revista Portuguesa de Filosofia* 75(2), pp. 1023–1046.

Vitoria, Francisco de. 2010. *Political Writings*. Cambridge: Cambridge University Press.

5

MARTIN LUTHER (1483–1546)

Valerie Morkevičius

Contexts

Martin Luther (1483–1546) definitively shaped Protestantism in his lifetime and beyond. Born in Eisleben (in present-day Germany) to a prosperous peasant farmer, his father intended him to study law, but a famously close encounter with lightning prompted Luther to abruptly enter an Augustinian monastery instead. After his ordination as a priest, he was selected to study theology, receiving his doctorate from the University of Wittenberg in 1512. There, Luther studied with Johannes von Staupitz, Vicar General of the Augustinian friars in Germany, whose curriculum centered on Scripture and emphasized the study of Augustine over the Scholastics (Leppin, 2017, p. 141). Echoes of Augustine can be seen in Luther's distinction between spiritual and temporal authority, as well as in his framing of temporal authorities' use of the sword as a form of neighbor-love.

Luther visited Rome in 1510 but found his counterparts there disappointingly frivolous and intellectually unimpressive (Beutel, 2003, p. 8). This frustration followed him home. Pope Leo X's grand new basilica in Rome was an expensive undertaking. When his parishioners in Wittenberg began to purchase the papal indulgences sold to finance St. Peter's construction, Luther railed from the pulpit against the idea that cash could compensate for sin (Beutel, 2003, p. 8).

Luther sent his theological critique of indulgences—along with his *Ninety-Five Theses*—to the Archbishop of Mainz in 1517. His text instantly struck a nerve: why should Germans pay to construct a church in far-off Rome? Luther's friends excitedly distributed his letter, translating it from Latin into German. Printed copies were soon widely distributed. In Germany, this

DOI: 10.4324/9781003428688-6

touched off a theological debate with political implications regarding the appropriate relationship between religious and secular authorities.

Luther's critique also struck a nerve in Rome. By the summer of 1518, Luther was on trial for heresy. Thomas Cajetan (known to his contemporaries as the premier expert on Thomas Aquinas) questioned Luther (Wicks, 1983, p. 531). But the Dominican could not convince Luther to recant. Frederick the wise blocked Cajetan's plans to extradite Luther to Rome. Because the Pope hoped Frederick would support Rome's preferred candidate in the upcoming election for the new Holy Roman emperor, the matter was temporarily dropped.

Nonetheless, Luther was excommunicated in 1520. And in 1521, the new Holy Roman Emperor, Charles V, put him under the imperial ban. Rendering Luther legally dead, the ban effectively permitted anyone to kill him without legal consequences. Luckily for Luther, Frederick had him kidnapped and safely hidden away in Wartburg castle. The year Luther spent at Wartburg was productive: he translated the New Testament into German. The next year, he returned to Wittenberg where he continued writing and preaching about reform. (He also began a translation of the Old Testament, which would take another decade to complete.) By 1526, Luther was busy organizing a new church: establishing a supervisory body, laying out a new form of worship, and summarizing the faith's tenets in two catechisms. The Protestant Reformation—which would catalyze many wars, including those that engulfed the France of Montaigne's time (also discussed in this volume)—was born.

Text and Tenets

Despite his growing responsibilities for his new church and family, Luther continued to write prolifically. The definitive collection of Luther's works includes over 100 volumes. Here, we will consider just seven texts, written over the course of a politically tumultuous decade. In these, he explores temporal authorities' responsibility for maintaining the polity and the appropriate use of force.

In three early texts (Treatise on Good Works, 1520; To the Christian Nobility, 1520; and On Temporal Authority, 1523), Luther echoes the Augustinian distinction between the heavenly and earthly kingdoms. Church authorities are responsible for people's souls, a spiritual role imparting no temporal powers. Thus, individual believers may use their conscience and reason to judge the teachings of others. They are not required to simply accept any doctrines from church authorities (Luther, 2003a). Temporal authorities, on the other hand, have the duty to "protect [their] subjects and to punish theft, robbery, and adultery," crimes seriously undermining social order (Luther, 2007a, p. 174). God gave secular sovereigns the right to use force, whether embodied as "hangmen, constables, judges, lords, or princes," lest governmental authority become "despised" or "enfeebled" (Luther, 2007c, p. 285).

How can the use of force be reconciled with Christ's commandment to love one's neighbor? True Christians would have no need for kings or judges. No one would act unjustly. But most people, Luther finds, are not true Christians. Hence, they "need the law to instruct, constrain, and compel them to do good" (2007c, p. 279). Human sinfulness means that "it is out of the question that there should be a common Christian government over the whole world, or indeed over a single country . . . for the wicked always outnumber the good" (Luther, 2007c, p. 281). Without secular government, "men would devour one another. . . . No one could support wife and child, feed himself, and serve God. The world would be reduced to chaos" (Luther, 2007c, p. 281).

Because earthly order is a good, Luther, like Augustine, understood the work of temporal authority—including wielding the sword—as a form of neighbor-love. Love tolerates "no injustice" toward one's fellow (Luther, 2007c, p. 286). For one's neighbor, one should "seek vengeance, justice, protection, and help, and do as much as [one] can to achieve it" (Luther, 2007c, p. 291). However, although defending one's neighbor is clearly "a matter of necessity," one ought to suffer injustice rather than defending one's own self by lawsuit or by force (Luther, 2007c, pp. 288, 291).

Just as religious authorities must not exercise temporal power, temporal authorities must not interfere in matters of faith (Luther, 2007c, p. 274). However, temporal authorities may discipline religious authorities who violate temporal laws (Luther, 2007b, p. 267). Ideally, there should be a balance between the two spheres, as "neither one is sufficient in the world without the other" (Luther, 2007c, p. 282). Where "spiritual government alone prevails over land and people, there wickedness is given free rein and the door is open for all manner of rascality" and "where temporal government or law alone prevails, there sheer hypocrisy is inevitable, even though the commandments be God's very own" (Luther, 2007c, p. 282).

Because earthly power governs only what is effectively a temporary stopover on the way to eternity, Christians should not be overly concerned with how political life is organized. In Luther's words, "temporal power is but a very small matter in the sight of God and too slightly regarded by him for us to resist, disobey, or become quarrelsome on its account" (2007a, p. 175). Unjust earthly authorities cause physical harm but not spiritual harm. Hence, "it is less disastrous when the temporal power goes wrong than when the spiritual power does" (Luther, 2007a, p. 174).

Regrettably, temporal authorities do sometimes abuse their power. A good ruler should "pay more regard to [his subjects'] needs and necessities than to his own will and pleasure," but Christians must submit even to foolish or unjust rulers (Luther, 2007a, p. 176). Citing Paul (Romans 13 and Titus 3) and Peter (I Peter), Luther asserts a Christian obligation to obey secular authorities and laws. Sin thus arises when people "lie to the government, betray it, or are disloyal to it, neither obeying it nor doing as it orders and commands" (Luther, 2007a, p. 174). It is even sin simply to "speak evil of the government and curse it . . . in public or in private" (Luther, 2007a, p. 174). Secular authorities may only be disobeyed if they overstep their authority and make demands that would "compel us to do wrong against God or men" (Luther, 2007a, p. 174). Even then, one should accept the consequences disobedience to temporal authority engenders: "to suffer wrong destroys no man's soul, in fact, it improves the soul although it does inflict hurt to our body and our possessions" (Luther, 2007a, p. 174).

Luther drafted "On Whether Soldiers, Too, Can Be Saved," addressing the legitimacy of military service, in December 1526. Earlier that year, Holy Roman Emperor Charles V had urged Catholic rulers to forcefully suppress Protestants. Meanwhile, the Ottoman Empire threatened Christian Europe from the east. In this tense atmosphere, Luther eschews pacifism and urges his followers to actively engage in political life—including warfare in the service of their sovereigns.

Luther begins by reaffirming the Augustinian idea that the use of force is tragically necessary for social order:

> What men write about war, saying that it is a great plague, is all true. . . .
> But they should also consider how great the plague is that war prevents.
> If people were good and wanted to keep peace, war would be the greatest
> plague on earth. But what are you going to do about the fact that people
> will not keep the peace, but rob, steal, kill, outrage women and children,
> and take away property and honor? The small lack of peace called war or
> the sword must be a limit to this universal, worldwide lack of peace which
> would destroy everyone.
>
> *(2003e, p. 103)*

Thus, although "slaying and robbing do not seem to be works of love," Luther argues they are "precious and godly" because they protect the good (2003e, p. 103).

War may legitimately be waged by sovereigns against their subjects (suppressing rebellion) or by sovereigns against other sovereigns. Subjects, however, cannot legitimately make war against their sovereigns. Even if the rulers forbade "the gospel to be preached" or "robbed the poor," the obligation to obey temporal authority means that "it would not therefore be just or right to do wrong in return" (Luther, 2003e, p. 108). If the persecution is intolerable, they should flee rather than take up arms at home (Luther, 2003e, p. 108).

Unlike Thomas Aquinas, Luther does not make an exception for rebellion against a tyrant. Although an insane sovereign could be deposed ("since his reason is gone," he is no longer a man), but a tyrant "still has a conscience and his faculties" (Luther, 2003e, p. 108). A tyrant may eventually mend his ways, particularly if he has wise advisors. Furthermore, Luther values order, arguing that:

> [I]f it is considered right to murder or depose tyrants, the practice spreads and it becomes a commonplace thing arbitrarily to call men tyrants who are not tyrants, and even to kill them if the mob takes a notion to do so.
>
> (2003e, p. 108)

Fearing mob rule, Luther opines that "if injustice is to be suffered, then it is better for subjects to suffer it from their rulers than for rulers to suffer it from their subjects" (2003e, p. 108).

Luther also emphasizes nonresistance to tyranny in an "Admonition to Peace: A Reply to the Twelve Articles of the Peasants in Swabia," written on the eve of the Peasant Rebellion in February or March of 1525. Although Luther clearly empathizes with the peasants (who were partially motivated by Reformation ideas), he firmly denies their right to rebel:

> You say the rulers are wicked and intolerable, for they will not allow us to have the gospel; they oppress us too hard with the burdens they lay on our property, and they are ruining us in body and soul. I answer: The fact that the rulers are wicked and unjust does not excuse disorder and rebellion, for the punishing of wickedness is not the responsibility of everyone, but of the worldly rulers who bear the sword.
>
> *(2003b, p. 72)*

Permitting rebellion would permit any man to judge any other, and "then authority, government, law, and order would disappear from the world; there would be nothing but murder and bloodshed" (Luther, 2003b, p. 76).

Instead of using force to defend temporal goods, the peasants ought to render these things unto their sovereigns, for "the Christian law tells us not to strive against injustice, not to grasp the sword, not to protect ourselves, not to avenge ourselves, but to give up life and property, and let whoever takes it have it" (Luther, 2003b, p. 78). Christians should trust that God will punish their unjust leaders. In support, Luther cites numerous examples of tyrants meeting their just deserts drawn from history and the Scriptures.

After the uprising began, Luther forcefully condemned it in "An Open Letter on the Harsh Book Against the Peasants." Here, Luther distinguishes rebels from ordinary criminals. Thieves and murderers fear the ruler, harming only private people and property. Hence, "no one ought to attack such a murderer" himself, since only the ruler has authority to punish (Luther, 2003d, p. 97). But a rebel "attacks the head himself and interferes with the exercise of his word and his office" (Luther, 2003d, p. 97). This undermines order, thus:

> [R]ebellion is a crime that deserves neither a court trial nor mercy. . . . The rebel has already been tried, judged, condemned, and sentenced to death and everyone is authorized to execute him. No murderer does so much evil and none deserves so much evil. For a murderer commits a punishable offense, and lets the penalty stand; but a rebel tries to make wickedness free and unpunishable, and attacks the punishment itself.
>
> *(Luther, 2003d, p. 98)*

Because of their responsibility to uphold order, sovereigns should seek to avoid the chaos of war. Nonetheless, some wars between sovereigns are justifiable but never as the aggressor. "Whoever starts a war is in the wrong," Luther asserts, "and it is only right and proper that he who first draws his sword is defeated, or even punished in the end" (2003e, p. 113). Wars of "lawful self-defense," fought out of necessity, are "human disasters" but nevertheless permissible (Luther, 2003e, p. 114). By contrast, "wars of desire" are simply evil.

Luther also instructs his readers that a good cause does not necessarily license unrestrained violence, emphasizing the importance of fighting with the right motivations. Subjects may fight for their sovereigns out of obedience, accepting wages for doing so. But Luther decries those who fight only for the sake of pay (whether they be mercenaries or subjects), because such people are "not happy when there is peace and not war" (2003e, p. 117). Their desires are thus disordered: greed motivates them, not duty. Luther also criticizes soldiers who "strike and kill people needlessly simply because they want to," as their violence is also wrongly motivated (Luther 2003e, p. 117). But fighting with the right motivation does not necessarily imply the minimal use of force. Discussing the peasant revolt in "Against the Murderous, Thieving Hordes of Peasants," Luther writes:

> Let no one have mercy on the obstinate, hardened, blinded peasants who refuse to listen to reason; but let everyone, as he is able, strike, hew, stab and slay, as though among mad dogs. . . . It is better to cut off one member without mercy than to have the whole body perish by fire, or by disease.
>
> *(Luther, 2003c, p. 95)*

Somewhat surprisingly, given his call for obedience and opposition to rebellion, Luther permits conscientious objection. If one knows "for sure" that one's sovereign's cause is wrong, one must not fight (2003e, p. 117). Sovereigns may punish subjects who object, but Luther counsels that it is better to lose one's wages or land or reputation than to fight unjustly. "You must take that risk and, with God's help, let whatever happens, happen" (Luther, 2003e, p. 117). However, if one isn't certain about the justness of the war, one should obey one's sovereign. Christian soldiers, after all, do not "fight as individuals, or for their own benefits, but as obedient servants of the authorities under whom they live" (Luther, 2003e, p. 104).

Controversies

Luther exudes controversy, with the *Ninety-Five Thesis* serving as the catalyst for a schism in the Church that would engulf Europe in religious civil war between Catholics and Protestants through the sixteenth century and beyond. He even stirred controversy amongst those who shared his Reformist views. Luther and Andreas von Karlstadt fought bitterly over divine law, the use of images, and the nature of the Eucharist; Luther likewise clashed with Ulrich Zwingli over whether Christ's body and blood were literally present in the Communion elements (Edwards, 2003, pp. 196, 199). Luther's dispute with Thomas Müntzer, however, had political implications. Müntzer, a former Lutheran, led the German peasant movement that Luther decried in *Admonition to Peace*, rejecting the doctrine of obedience to earthly princes. Müntzer saw rebellion as necessary for achieving the social and political mission of the Gospel of Luke (Mjaaland, 2018, p. 447). Luther viewed Müntzer as an "arch-devil" leading the peasants astray (Edwards, 2003, p. 198).

Luther's primary Catholic opponent was Cardinal Tomasso de Vio Cajetan, the Dominican theologian charged with investigating his alleged heresy. While Luther was inspired by Augustine, Cajetan was well known for his detailed commentary on Aquinas' *Summa Theologiae*. Cajetan wrote prolifically countering Luther's claims regarding faith in sacraments, including penance and the Eucharist, as well as the priesthood of all believers. Decades after his 1518 encounter with Luther, Cajetan recommended concessions that Rome might make to avoid a schism with the German church. A close reading of these suggests that Cajetan appreciated Luther's Biblical exegesis. But he underestimated the unwillingness of both Rome and Luther to tolerate a diversity of outward practices in cases where humility might require acknowledging some uncertainty as to what specifically is required (Cajetan, 1978, pp. 41, 46).

Another Dominican, the Spaniard Francisco de Vitoria, considered the implications of Luther's work for political theology. Vitoria excoriated Luther for denying that ecclesiastical laws are binding (Vitoria, 1991, p. 175). Against Luther's understanding of sovereignty as resulting from God's grace, Vitoria argued that human relations naturally produced sovereignty (1991, p. 255). And in *On the Law of War*, Vitoria misinterpreted Luther's condemnation of aggressive war, labeling him a pacifist: "Martin Luther, who has left no nook untainted with his heresies, denies that Christians may lawfully take up arms, even against the Turks (1991, p. 296).

Luther's stance on the Turks, whom his contemporaries like Vitoria, Las Casas, and Sepúlveda (covered in this volume) considered "barbarians," proved especially controversial. Luther's conviction that God had instituted *all* temporal authorities and that *only* temporal authorities may make the decision to use force led Pope Leo X to condemn him for claiming that "to fight against the Turk is the same as resisting God, who visits our sin upon us with this rod" (Luther, 2003e, p. 121).

Luther responded with *On War Against the Turk* (1529), arguing that the Pope was wrong to drive Christian princes to "[attack] the Turk . . . making war on him, before they amended their own ways and lived as true Christians" (2003f, p. 123), because as a religious authority, the Pope ought to care more for men's souls than for foreign policy. Additionally, Luther asserted that because violence cannot be used on behalf of the sacred, it was sinful to champion war against the Turks "as though our people were an army against the . . . enemies of Christ" (2003f, p. 123). Likewise, because temporal authorities could only fight for temporal goods, Luther criticizes Emperor Charles for asserting his role as "head of Christendom and as protector of the church" as a justification for fighting the Ottomans (Luther, 2003f, p. 130). The separation between sacred and temporal authority thus leads Luther to a pragmatic, earthly tolerance:

> Let the Turk believe and live as he will, just as one lets the papacy and other false Christians live. The emperor's sword has nothing to do with the faith; it belongs to physical, worldly things, if God is not to become angry with us. If we pervert his order and throw it into confusion, he too becomes perverse and throws us into confusion and all kinds of misfortune.
>
> *(2003f, p. 130)*

Luther's secularization of temporal power leads him to reiterate his condemnation of aggressive war: "I shall never advise a heathen or a Turk, let alone a Christian, to attack another or begin a war" (2003f, p. 123).

Legacy

Luther's legacy has shaped religious thought down to our own time. Those seeking to reconcile obedience to the state and to God have drawn inspiration from his ideas, although not always in consistent ways. Three mid-twentieth-century Lutheran theologians—Paul Tillich, Dietrich Bonhoeffer, and Reinhold Niebuhr—are especially interesting to note for the diverse ways in which they engaged Luther to address the pressing challenges across the two world wars of the twentieth century. All three were ordained in the

Evangelical Union Church. All drew explicitly on Luther's ideas but none-theless sometimes departed from his teachings as they renegotiated his main tenets in light of their own contexts, particularly in response to the aggressive rise of Nazi Germany.

Paul Tillich (1886–1965), born in Germany, earned a doctorate in Philoso-phy before his ordination. The outbreak of the First World War prompted Tillich to volunteer as a chaplain in the German army, with which he served at Verdun (Schüßler, 2009, p. 5). When the Nazis assumed power in 1933, Tillich was suspended from his university position. Luckily, he was invited to teach at Union Theological Seminary in New York, where he relocated dur-ing the tumultuous 1930s.

Like Luther, Tillich separated religious and political authority. He believed the Church had to balance two competing goods: an appreciation for the state as an earthly good and a belief in the ultimate unity of mankind. Because order is a good, churches should support any institutions that pro-duce it, including the nation-state. Lutheranism, he argued in a 1934 essay, "thus can subordinate itself to any social order or political structure without attempting to exercise any direct critical influence" (Tillich, 2014, p. 65). However, the Church can cede too much to the secular sphere. Tillich criti-cizes churches for "[surrendering] many and often decisive aspects of the claims which they are in duty bound to assert, above all in reference to social justice and the unity of mankind" (2014, p. 67). Tillich particularly high-lights Christians' obligation to resist the Nazi's "new myth"—a nationalism championing the community of blood over the community of sacrament and replacing the Christian values of "humility, love and hope" with "the pagan virtues of courage, power and enthusiasm" (2014, p. 71).

Tillich's vocal opposition to Nazism led the American Office for War Infor-mation to hire him to produce radio sermons transmitted into Germany. His wartime radio addresses explore how Christians should respond to the vio-lence of the Nazi state, even if this meant disobedience. In a series of speeches in the autumn of 1942, he both follows and departs from Luther.

While Nazis regularly believe "we are fighting them because we hate them, thoroughly, with our whole heart, until their complete destruction," Tillich follows Luther to propose that his listeners resist hating the enemy: you can be "superior to the National Socialists to the degree to which you keep your-selves free of hatred toward them! You are identical to them to the degree to which you permit yourselves to hate them" (1998, p. 63). More boldly, he departs from Luther by questioning blind obedience to an unjust state:

> What should actually be defended, and is it worth the defense? There are things that are worth being defended, and there have been defensive wars that have finally led to victory. But is that which National Socialism is defending also worth being defended by the German people? It is clear

what National Socialism wants to defend: the power that it has seized, first in Germany, then in Europe. But is this an object worthy of defense by the German people, a defense that, whether successful or not, will transform Germany into a heap of rubble?

(1998, p. 67)

Although Tillich believed the Nazis must be defeated, he cautioned that even a justified war must have limits:

Permit me now to leap from the ethical-religious to the political and to say: a victory over National Socialism will be attained only if the victors gain it without hatred. If it is won with hatred, then the conquered have, in reality, been victorious, and a new period of hatred will drag the world further into the abyss.

(1998, p. 64)

He then invokes, as a lesson, the Treaty of Versailles, which ended the First World War but so severely punished the Germans that it set the stage for the next war.

The end of the Second World War came with a nuclear bang, but the further development of nuclear weapons during the Cold War arms race truly shook Tillich (Craig, 1992, p. 689). An atomic war would "[produce] destruction without the possibility of a creative new beginning. It annihilates what it is supposed to defend" (Tillich, 1990, pp. 160–161). Tillich thus calls on "everyone who is aware of the possibility of mankind's self-destruction" to engage in "resistance against the suicidal instincts of the human race" across the political, religious, and moral spheres (1990, p. 158). Reprising the theme of the Church's responsibility to offer practical guidance to believers, Tillich warns that Christians should avoid falling prey to propaganda about the enemy, remembering that faith transcends national boundaries and particular historical moments (1990, pp. 158–159). To avoid sleepwalking into atomic war, Christians must resist the state's demands for unquestioning submission.

As he warned in *Systemic Theology*, penned across the Cold-War-infused 1950s, contemporary Christianity has been rendered impotent by its willing subjection to nation-states. Churches have failed to maintain their "radical otherness" (1967, p. 216). If they accept the liberal approach that relegates faith to the private sphere, they will become nothing more than a "benevolent social club," no longer speaking truth to power (1967, p. 216).

Reinhold Niebuhr (1892–1972), the American-born son of a German immigrant minister, earned a master's degree at Yale Divinity School in 1915. As the pastor of a small German congregation in Detroit, he condemned the rise of the Ku Klux Klan and supported auto union organizers, before accepting an invitation to teach at Union Theological Seminary in New York in 1928. He co-founded the Union for Democratic Action, which advocated for an interventionist and internationalist foreign policy, in 1941.

Niebuhr's pastoral work on the home front during the First World War forced him to square his conviction that the violence of war is often in vain (and his doubts about the morality of modern methods of warfare) with his very Lutheran understanding of his congregants' duty to loyally serve their new state (Lathangue, 1989, p. 8). Niebuhr upheld the Church's obligation to be "unmistakably patriotic," while always remaining "unmistakably Christian" (Chrystal, 1977, p. 293). Ultimately, Niebuhr broke with his youthful pacifism, disappointed that the American desire for peace failed to keep it out of the war. Niebuhr still respected true Christian pacifism, which he thought served as a valuable "reminder . . . that the relative norms of social justice, which justify both coercion and resistance to coercion, are not final norms," but cautioned that pacifists risk falling into a heresy if they discarded "the Christian doctrine of original sin as an outmoded bit of pessimism," in favor of "the Renaissance faith in the goodness of man" (1987b, p. 104). Assuming that "love is guaranteed a simple victory over the world" is as politically naïve as it is theologically dubious. Nothing in human history suggests this precept is true, and ignoring reality is a recipe for failure (Niebuhr, 1987b, p. 104). The rise of Nazism would prove Niebuhr right.

Like Tillich, Niebuhr criticized the contemporary Church for avoiding "involvement in the ambiguities of politics," thus leaving Christians without guidance in the face of modern complexity (1987a, p. 96). This quietism stems from "insufferable sentimentality" and a mistaken emphasis on neutrality: "the neutral Church is usually an ally of the established social forces" (Niebuhr, 1987a, p. 96). Although the Church has no authority to make rules for the temporal sphere, it should offer moral direction to those involved in politics.

For Niebuhr, justifying the use of force hinges on its intended results: "a political policy cannot be intrinsically evil if it can be proved to be an efficacious instrument for the achievement of a morally approved end" (2001, p. 171). Beyond self-defense, Niebuhr believed force could be used to accomplish justice in a world lacking effective institutions. States are too selfish "to make the attainment of international justice without the use of force possible" (Niebuhr, 2001, p. 110). As discussed in the chapter on Niebuhr in this volume, war was a lesser evil compared to combating greater evils such as Nazism. Legitimizing war in pursuit of a more just international order takes Niebuhr beyond Luther and closer to Augustine.

Nonetheless, Niebuhr does not embrace the civilized–barbarian dichotomy discussed across the Aristotle–Sepúlveda–Mills chapters, arguably a product of Catholic just war thinking's incorporation of Aristotle as an authority. Lest ill-considered uses of force undermine the good of order, Niebuhr counsels humility in the pursuit of justice. We should abandon the self-righteousness that fuels conflict, instead recognizing that "political controversies are always conflicts between sinners and not between righteous men and sinners" (Niebuhr, 1987b, p. 114). To this end, we should "restrain the impulse of justice," rather than seeking a vindictive peace (Niebuhr, 1987b, p. 114). Victory is no proof of virtue. Nor is power: states must avoid giving in to "an idolatrous idea of their own importance," reflecting instead on what justice demands they should cede (Niebuhr, 1987a, p. 98).

Dietrich Bonhoeffer (1906–1945) navigated being a doctor of theology and an ordained pastor during the rise of Nazi Germany. In 1933, Bonhoeffer joined the Confessing Church, decrying the Evangelical Union Church's willing submission to the Nazi party. In his view, the state's order to defrock pastors of Jewish descent overstepped its temporal authority. Bonhoeffer became one of the new Church's leaders, although he grew frustrated with his colleagues' emphasis on keeping the church separate from politics rather than taking practical action, such as coordinating help for the Jews. A staunch critic of the Nazis from the very beginning—and persecuted for it across the 1930s and 40s—Bonhoeffer was ultimately executed for his involvement in the German resistance.

Bonhoeffer was, from his youth, deeply patriotic. Describing German soldiers' sacrifices in terms Tillich would recognize, Bonhoeffer writes that they fought in the First World War with an "imperturbable consciousness of their duty, with an inexorable self-discipline and with a glowing love for the fatherland" (2013c, p. 129). But Bonhoeffer was also conscious that patriotism could be blinding:

When the war broke out . . . we thought it to be our duty to stand for our country and we believed of course in our essential guiltlessness. You cannot expect in such a moment of excitement an objective and detached valuation of the present conditions.

(2013c, p. 131)

Precisely such sober evaluation could have prevented the war, Bonhoeffer suggests, if all sides had considered their own responsibility for the tensions. With this lesson in mind, Bonhoeffer cautions against pursuing peace vindictively, reflecting on the heavy burden placed on ordinary people by the Versailles Treaty's reparations. He thus closes his lecture in an idealistic tone—a marked contrast to Tillich and Niebuhr—calling for the defeat of "meanness, selfishness, slander, hatred, prejudice among the nations" for the sake of everlasting peace (Bonhoeffer, 2013c, p. 134).

Bonhoeffer contrasts nationalism—which elevates the state as a symbol of hope—with Christian faith, which puts its trust in the cross (2012, p. 16). Nationalism, Bonhoeffer suggests, wishes to convince us that "Christ has not conquered; *we conquer*" (Bonhoeffer, 2012, p. 18). A parochial devotion to one's own country leads away from the international ecumenicism of the Church. Faced with conflicting loyalties, the Christian must choose the heavenly kingdom. This choice earned Bonhoeffer the label of pacifist enemy of the state.

The backlash he received in Nazi Germany for his critical stance against war and patriotism led Bonhoeffer to accept a position at Union Theological Seminary. Feeling guilty for abandoning his countrymen in need, however, he returned weeks later. As he explains in a 1939 letter to Reinhold Niebuhr, who had helped him secure the post:

> I have made a mistake in coming to America. I must live through this difficult period in our national history with the Christian people of Germany. . . . [They] will face the terrible alternative of either willing the defeat of their nation in order that Christian civilization may survive, or willing the victory of their nation and thereby destroying our civilization. I know which of these alternatives I must choose.
>
> *(Bethge, 2000, p. 655)*

As Germany threatened world war, Bonhoeffer chose to resist the patriotic and militarist demands Nazi Germany made on Christians. States are power-hungry, proud, and greedy, but individuals must not be blindly obedient: "what is sinful for an individual person, however, can never be a virtue for a *nation*" (Bonhoeffer, 2013a, pp. 354–355).

Bonhoeffer acknowledges three typical appeals to justify war: claiming war "works for the maintenance of the State and future peace," that war "is an irresistible event, over which no one has any power," or that "war reveals a heroic world of sacrifice (2013b, p. 394). But he finds these customary justifications unsatisfactory in the face of Christ's command to love one another. Neighbor-love "gives up its life for a brother, whether he is on this side or on the other side. Pure love quite simply cannot lift up a sword against a Christian, because that would mean to lift it against Christ" (Bonhoeffer, 2013a, pp. 355–356). And yet Bonhoeffer, like Niebuhr, finds secular pacifists' arguments unsatisfying. Their assertion that a rational organization could suppress war, "so as to reveal the world as a good world," reveals a failure to perceive mankind's fall from grace (2013b, p. 394). Thus, both the traditional justifications of war and secular pacifism are "equally unchristian" (Bonhoeffer, 2013b, p. 394).

Instead, Bonhoeffer places the commandment "Thou shalt not kill" at the center of his ethics (2013b, p. 394). The Church must oppose the presumption that violence serves a productive end:

> To the objection: War creates peace: the Church answers: This is not true, war creates destruction. To the objection: The nation must defend itself:

the Church answers: Have you dared to entrust God, in full faith, with your protection in obedience to His commandment? To the objection: Love for my neighbor compels me: the Church answers: The one who loves God keeps His commandments.

(Bonhoeffer, 2013b, pp. 394–395)

But pacifism, for Bonhoeffer, was not synonymous with passivity. Active in the German resistance, he was even involved in a plot to kill Hitler. How could he justify this? The answer lies in his belief that God is still speaking. Ethics are inherently limited and limiting: limited in the sense that they are "inseparably linked with particular persons, times, and places" and limiting in the sense that they serve as formal boundaries to right action (Bonhoeffer, 1995, pp. 267, 280). However, Christ's self-sacrifice has set Christians free (Bonhoeffer, 1995, p. 280). The only limit is God's commandment, which "both in its contents and in its form . . . is concrete speech to the concrete man" (Bonhoeffer, 1995, p. 273). While ethics may present one with conflicting duties, God's commandment "leaves room only for obedience or disobedience" (Bonhoeffer, 1995, p. 273).

Such radical freedom is a grave responsibility. Action without the support of ethics is performed "in the surrender to God of the deed which has become necessary and which is nevertheless . . . free; for it is God who sees the heart, who weighs up the deed, and who directs the course of history" (Bonhoeffer, 1995, p. 245). This freedom makes it possible to fulfill the law by suspending it: "In war, for example, there is killing, lying, and expropriation solely in order that the authority of life, truth and property may be restored" (Bonhoeffer, 1995, p. 257). Such acts may respond to God's call but must still be recognized as bearing the "guilt of the violation of the law" (Bonhoeffer, 1995, p. 258). In short, they could be justified to defy domestic powers aiming at global evil.

Conclusion

As just war debates today navigate the schism between the legalist (Walzerian) and reductive individualist (revisionists like Jeff McMahan) moral approaches, exploring Luther's reflections on war and obedience—and their legacy—shows that an individual's moral conscience can direct both the performance of just war duties and resistance to transgressions the state seeks to impose.

Luther's belief in the value of earthly order has inspired the Lutheran community's continued engagement with political life, and inevitably, with navigating the call to war. This has present-day implications. The separation of sacred and temporal authority in Lutheran thought means that such debates inherently involve questions of just what obedience to the state demands.

Certainly not blind patriotism. Conscientious objection? Outright resistance? The answer lies in adjudicating the relationship between obedience to God and obedience to temporal authorities and understanding where, in a given context, one's loyalty lies.

Works Cited

Bethge, Eberhard. 2000. *Dietrich Bonhoeffer: A Biography*. Minneapolis, MN: Fortress Press.

Beutel, Albrecht. 2003. Luther's Life. In Donald K. McKim. ed. *The Cambridge Companion to Martin Luther*. Cambridge: Cambridge University Press, pp. 3–19.

Bonhoeffer, Dietrich. 1995. *Ethics*. New York: Touchstone.

Bonhoeffer, Dietrich. 2012. National Memorial Day: Berlin, Reminiscere (Memorial Day), February, 21, 1932. In Isabel Best. ed. *The Collected Sermons of Dietrich Bonhoeffer*. Minneapolis, MN: Fortress Press, pp. 13–21.

Bonhoeffer, Dietrich. 2013a. Christ and Peace. In Clifford J. Green and Michael P. DeJonge. eds. *The Bonhoeffer Reader*. Minneapolis, MN: Fortress Press, pp. 352–356.

Bonhoeffer, Dietrich. 2013b. Fanø Theses Paper and Address: The Church and the Peoples of the World. In Green and DeJonge, *The Bonhoeffer Reader*, pp. 393–397.

Bonhoeffer, Dietrich. 2013c. Lecture on 'War'. In Greene and DeJonge, *The Bonhoeffer Reader*, pp. 128–134.

Cajetan, Tommaso de Vio. 1978. *Cajetan Responds: A Reader in Reformation Controversy*. Edited by Jared Wicks. Eugene, Oregon: Wipf and Stock.

Chrystal, William G. 1977. Reinhold Niebuhr and the First World War. *Journal of Presbyterian History* 55(3), pp. 285–298.

Craig, Campbell. 1992. The New Meaning of Modern War in the Thought of Reinhold Niebuhr. *Journal of the History of Ideas* 53(4), pp. 687–701.

Edwards, Mark U., Jr. 2003. Luther's Polemical Controversies. In Donald McKim. ed. *The Cambridge Companion to Martin Luther*. Cambridge: Cambridge University Press, pp. 192–206.

Lathangue, Robin N. J. 1989. *Reinhold Niebuhr and Liberal Pacifism, 1914–1940*. MA thesis, McMaster University.

Leppin, Volker. 2017. Luther: A Mystic. *Dialog: A Journal of Theology* 56(2), pp. 140–144.

Luther, Martin. 2003a. The Freedom of a Christian (1520). In J. M. Porter. ed. *Luther: Selected Political Writings*. Eugene, OR: Wipf and Stock Publishers, pp. 25–36.

Luther, Martin. 2003b. Admonition to Peace: A Reply to the Twelve Articles of the Peasants in Swabia (1525). In J. M. Porter. ed. *Luther: Selected Political Writings*. Eugene, OR: Wipf and Stock Publishers, pp. 71–84.

Luther, Martin. 2003c. Against the Robbing and Murdering Hordes of Peasants (1525). In J. M. Porter. ed. *Luther: Selected Political Writings*. Eugene, OR: Wipf and Stock Publishers, pp. 85–88.

Luther, Martin. 2003d. An Open Letter on the Harsh Book Against the Peasants (1525). In J. M. Porter. ed. *Luther: Selected Political Writings*. Eugene, OR: Wipf and Stock Publishers, pp. 89–100.

Luther, Martin. 2003e. Whether Soldiers, Too, Can Be Saved (1526). In J. M. Porter. ed. *Luther: Selected Political Writings*. Eugene, OR: Wipf and Stock Publishers, pp. 101–120.

Luther, Martin. 2003f. On War against the Turk (1529). In J. M. Porter. ed. *Luther: Selected Political Writings*. Eugene, OR: Wipf and Stock Publishers, pp. 121–132.

Luther, Martin. 2007a. Treatise on Good Works (1520). In Theodore G. Tappert. ed. *Selected Writings of Martin Luther Vol. 1*. Minneapolis, MN: Fortress Press, pp. 97–196.

Luther, Martin. 2007b. To the Christian Nobility of the German Nation Concerning Reform (1520). In Tappert. ed. *Selected Writings of Martin Luther Vol. 1*. Minneapolis, MN: Fortress Press, pp. 251–354.

Luther, Martin. 2007c. Temporal Authority: To What Extent It Should be Obeyed (1523). In Theodore G. Tappert. ed. *Selected Writings of Martin Luther Vol. 2*. Minneapolis, MN: Fortress Press, pp. 265–320.

Mjaaland, Marius Timmann. 2018. Sovereignty and Submission: Luther's Political Theology and the Violence of Christian Metaphysics. *Studies in Christian Ethics* 31(4), pp. 435–451.

Niebuhr, Reinhold. 1987a. The Christian Witness in the Social and National Order. In Robert McApee Brown. ed. *The Essential Reinhold Niebuhr: Selected Essays and Addresses*. New Haven, CT: Yale University Press, pp. 93–101.

Niebuhr, Reinhold. 1987b. Why the Christian Church is Not Pacifist. In Robert McAfee Brown. ed. *The Essential Reinhold Niebuhr: Selected Essays and Addresses*. New haven, CT: Yale University Press, pp. 102–122.

Niebuhr, Reinhold. 2001. *Moral Man and Immoral Society: A Study in Ethics and Politics*. Louisville, KY: Westminster John Knox Press.

Schüßler, Werner. 2009. Tillich's Life and Works. In Russel R. Manning. ed. *The Cambridge Companion to Paul Tillich*. Cambridge: Cambridge University Press.

Tillich, Paul. 1967. *Systemic Theology: Three Volumes in One*. Chicago: University of Chicago Press.

Tillich, Paul. 1990. *Theology of Peace*. Edited by Ronald H. Stone. Louisville, KY: Westminster John Knox Press.

Tillich, Paul. 1998. *Against the Third Reich: Paul Tillich's Wartime Addresses to Nazi Germany*. Edited by Ronald H. Stone and Matthew Lon Weaver. Louisville, KY: Westminster John Knox Press.

Tillich, Paul. 2014. The Totalitarian State and the Claims of the Church. *Social Research: An International Quarterly* 82(1), pp. 49–77.

Vitoria, Francisco de. 1991. *Political Writings*. Edited by Anthony Pagden and Jeremy Lawrance. Cambridge: Cambridge University Press.

Wicks, Jared. 1983. Roman Reactions to Luther: The First Year. *The Catholic Historical Review* 69(4), pp. 521–562.

6

MICHEL DE MONTAIGNE (1533–1592)

Daniel R. Brunstetter

Introduction

The subject of war is ubiquitous in Michel de Montaigne's famous sixteenth-century book, the *Essays*. Its contents were penned across a lifetime of violent upheaval. No less than eight monstrous civil wars pitting Catholics against Protestants, wars of a ferocious nature that showed no limits, engulfed Montaigne's France. These were wars where excess, as opposed to restraint, was the norm. Especially against civilians. From the first passages in which he explores themes of vengeance, mercy, and the merits of parley, penned in the 1570s, to the last pages written toward the end of his life in the late 1580s, tinged with regret, the *Essays* reveal how Montaigne was fascinated with, and tormented by, the appeal with which mankind takes up arms. But Montaigne was not just a writer; he was a political actor who tried to forge a bridge of moderation between warring parties—a role which carried great personal risk (Desan, 2017). That he wrote amid this precarity, when violent death lurked just around the corner, gives the pages of the *Essays* where he broaches the subject of war a unique tenor.

Montaigne was a humanist, profiting from the Renaissance revival of the classics to forge his views on politics and war at a time when divine authority was biased by schism and violent rivalry. Plutarch and the skeptics, including Seneca and Epictetus, were his favorites. He rejected inherited authority, especially Aristotle and Christian theology, broadly speaking, because he found them too rigid. He was interested in books about the conquest of the Americas and had things to say about the cruelty of the Spaniards, though he never read canonical just war thinkers such as Francisco de Vitoria or Francisco Suarez. And probably not Bartolomé de las Casas either. It is safe to say that he knew nothing of the just

DOI: 10.4324/9781003428688-7

war tradition as such—he never speaks in terms of the standard *jus ad bellum*, *jus in bello*, *jus post bellum* categories—and yet for those interested in the ethics of war, Montaigne's turn to the classics offers profound insights drawn from a set of sources often neglected by traditional accounts of just war, but which were essential to the humanist education (O'Driscoll, 2018). These open an alternative window into the ethics of war—one that focuses on human character by peering into the souls of warriors past (Duff, 1999).

The *Essays*, despite its cliched reputation of being a skeptic's treatise constructed on shifting moral sands, is a book filled with political and moral prescriptions and proscriptions (Schaefer, 1990; Thompson, 2018). It is no wonder that the first English translation, published at the aurore of the seventeenth century and said to have inspired Shakespeare, carries the title: *The Essayes, or Morall, Politike, and Millitarie Discourses*. Contemporary scholars point to Montaigne's arguments for religious toleration and individual liberty as hallmarks of Modernity (Levine, 2001; Hartle, 2013). These arguments have much to do with the way Montaigne reacted to the politics of his lifetime, which was shaped and driven by war. Indeed, the *Essays* showcase how personal ethics, military ideals (of the noble class), and the literary ideals of the Humanist are intertwined (Supple, 1984). When it comes to grappling with war, Montaigne was not so much concerned with when and how to make war, as he was with the emotions, virtues, and vices that shape human intuitions about war. One scholar suggests that his goal in writing was "taming the early modern aristocratic culture of violence and cruelty" and replacing it with one of restraint and clemency (Quint, 1998, p. ix). Given Montaigne's renown and the place war holds in the *Essays*, it is surprising there has been little interest in Montaigne from those interested in the ethics of war—a shortcoming that this essay seeks to correct (Brunstetter, 2022).

Contexts

Montaigne was born in 1533 in the south-west of France. Europe was beginning the age of exploration and discovery of unknown lands, as well as the rediscovery of classical texts. He had a humanist's education in his youth, trained and served as a lawyer in the Bordeaux parliament during his formative years, before retiring to his chateau to dedicate himself to writing in 1570. Montaigne was called out of retirement in 1581 to serve two terms as the mayor of Bordeaux. His time in politics traversed a tumultuous period in French history. Growing tensions between Protestants and Catholics erupted into a civil war in 1562, launching a period known as the Wars of Religion: violence, religious hatred, political intrigue, and personal precarity shaped Montaigne's activities and writing. Marching armies, marauding bands of fanatics, sieges, pitched battles, civilian massacres, and pillaging ravaged the landscape for more than three decades.

Montaigne did not sit by idly during the tumult. He actively confronted the challenges of morally navigating the politics that fueled war's eruption and quelled its flames. He served as an envoy for kings (Henri III), queens (Catherine de Medici), and royal pretendants (Henri de Navarre, the future Henri IV), seeking to build bridges of tolerance amid the bloodshed (Desan, 2017). The simmering violence persisted until 1598, when Henri IV proclaimed the Edict of Nantes establishing the freedom of religion in France, but not before, according to some estimates, four million people perished. Montaigne would not live to see the lasting peace. He died in 1592 of illness, at the age of 59.

Texts and Tenets

Montaigne published a singular book: the *Essays*. Begun in 1572, it went through various published iterations in his lifetime, the first being in 1580. He continued working on it until his death, after which multiple posthumous editions that incorporated additions that he had planned to insert at the time of his passing were published. The *Essays* is an undulating work, divided into three books comprising some 107 different essays. The essay was a novel form that Montaigne is credited with inventing. Not a treatise, not a dialogue in the Platonic style, and not a scholastic disputation or philosophical discourse, the essay—which comes from the French word *essayer* or to try—was the author's free-flowing attempt to explore a particular subject as his thoughts on it enter his mind (and sometimes come to a learned conclusion). The *Essays* cover a host of topics, but the book's originality lies in Montaigne's decision to pen a self-portrait that takes shapes across the essays themselves:

> Painting myself for others, I have painted my inward self with colors clearer than my original ones. I have no more made my book than my book has made me—a book consubstantial with its author, concerned with my own self, an integral part of my life.
>
> *(II.18, pp. 504–5)*[1]

This exploration of the self has a universal feel insofar as Montaigne claims that each of us "bears the entire form" of the human condition (III.2, p. 611). Hence, the reader—as many who have read the *Essays* can attest—sees much of his or her own self in Montaigne's meandering prose.

For our purposes, it is important to note that many of the essays deal with war and politics. The civil wars that consumed France during Montaigne's lifetime thrust war upon him. As he openly notes, his own life was at times at the mercy of a Machiavellian enemy intent on indiscriminate massacre, while his discrete political activities involved building bridges between warring factions to set the stage for peace. The *Essays* thus showcase Montaigne's own inquiries into why and how men make war. Their political and moral bent, coupled with the pedagogical side of teaching mankind how to live, suggests that the *Essays* might be read in the genre of the mirror of princes, like Xeno-phon's *The Life of Cyrus*, Erasmus' *The Education of a Christian Prince*, or Machiavelli's *The Prince*; indeed, at places in the *Essays,* Montaigne indi-rectly rebuked the in-vogue tenets of Machiavellian statecraft (Engster, 1998).

Montaigne had his sources of predilection to guide his inquires, though his disdain for intellectual authority shaped which sources he admired. At the dawn of the Renaissance, the Christian just war tradition was not his frame of reference. The canonical line from Augustine via Gratian to an Aristote-lian-inspired Aquinas did not shape his reflections on war, while contempo-raries such as Vitoria and the Spanish scholastics were not part of his library, even though he was interested in the injustice of the New World conquests (Losse, 2013, pp. 61–83). Montaigne cast his gaze on Greek and Roman clas-sics, interlaced with accounts of non-Western war customs that piqued his curiosity. He turned away from the Christian sources of authority because they had lost their credibility in the tumult of his century. To quote one of his more famous quips: "the laws of conscience, which we say are born of nature, are born of custom" (I.23, p. 83).

The opening essays set the tone. Epaminondas, Alexander the Great, and Lucious Marcius; Thebes, Athens, and Rome—these are the people and places that populate his imagination. Polybius and Plutarch are his sources of predilection. Yet, studying them leads to no universal rules regarding war-fare. "Truly man is a marvelously vain, diverse, and undulating object," he concludes (I.1, p. 5). Warfare in the world of the ancients was, of course, structured by rules (O'Driscoll, 2015). Montaigne's musings are peppered with anecdotes about them: Alexander slaughtering the men and enslaving the women and children of Thebes once victory was achieved, the Athenians forgoing military advantage to retrieve the bodies of their dead, and nostalgic Romans' disdain for *la ruse dans la guerre* (treachery in war) as antithetical to fighting fair as well as their requirement for starting wars only after hav-ing announced them and designating the time and place of battle. Montaigne entertains the notion that these nostalgic Romans hit upon universal morals governing war: "Deceit," he ponders, "may serve for the moment; but only that man considers himself overcome who knows he was downed neither by trick not by luck but by valiance, man to man, in a fair and just war." Only to reject the idea with a quote from Virgil: "It clearly appears from the language of these good men that they had not yet accepted the fine saying: courage or ruse against an enemy, who cares" (I.5, p. 16)?

Montaigne's early musings on war are typical of his general line of inquiry. Instead of bowing to universal laws that govern human affairs, which some-times necessitates weeding out examples that do not fit the mold, Montaigne sought out examples of customs that showcase human diversity. The Greeks, he observed, embraced *la ruse*, as did newer generations of Romans. But not the kingdom of Ternate:

> [A]mong those nations that we so smugly call barbarians, custom has it that the never start a war without having first announced it, adding to this an amply declaration of the means that they have to employ it: what sort of men and how many, what supplies, what offensive and defensive weapons. But also, that being done, if their enemies do not yield and come to an agreement, they give themselves the right to their worst, and do not think they can be reproached for treason, ruse, and any other means whatever that serves to conquer.
>
> *(I.5, p. 17)*

The war ethics of Ternate, a Sultanate geographically located in present-day Indonesia, has its intrinsic value, but Montaigne references it for a different reason. Across the *Essays*, other "barbarians" surface as counterexamples to European war: the Persians, the Turks, the Aztecs, the Tupinamba cannibals of Brazil, and more. Montaigne was not looking for commonalities and moral overlap. Rather, he saw the diversity of human customs as a challenge to the so-called universal truths, which some authoritative figures declared ought to govern humanity: "So many humors, sects, judgments, opinions, laws, and customs teach us to judge sanely our own, and teach our judgment to recognize its own imperfections and natural weakness" (I.26, p. 116).

Montaigne is famously known as a skeptic. He rejects Aristotle, the "god of scholastic knowledge" as being historically contingent: "Before the principles which Aristotle introduced were in credit, other principles satisfied human reason, as his satisfy us at the moment. . . . They are no more exempt from being thrown out than their predecessors" (II.12, p. 429). "Reason," he says, is "an instrument of lead and of wax, stretchable, pliable, and adaptable to all biases and measures; all that is needed is the ability to mold it" (II.12, p. 425). Instead, he embraces the skepticism of the Stoics: "It seems in truth that nature . . . has given us our share only presumption. This is what Epictetus says, that man has nothing properly his own but the use of his opinions. We have nothing but wind and smoke for our portion" (II.12, p. 360). Much has been written about the multiple layers of Montaigne's skepticism, but the takeaway point is this: doubt what others thrust upon you as philosophical truths (Hartle, 2005). They are the fruit of a particular culture's presumption, that is, the unfounded belief in its own moral superiority, which Montaigne calls "our natural and original malady" (II.12, p. 331). This would include the universal principles regarding the morality of war proffered by thinkers of the just war tradition. Had Montaigne been aware of the tradition as such—he did not have the same frame of reference as his scholastic contemporaries and was thus not educated to draw from the canon—he probably would have enjoyed listing the diversity of claims among canonical thinkers to cast doubt on the tradition's authority.

The skeptical turn does not lead to living an amoral, or immoral, life. Nor does it mean there are no moral intuitions that might govern warfare. In absence of a reliable authority, Montaigne takes the inward turn to peer into the soul to unlock the human condition and explore where emotions and morals collide. Montaigne's examination of himself, his flaws and virtues, his truths and contradictions, is where the novelty of the *Essays* lies (Bakewell, 2010). It was a very different enterprise compared to the popular *Confessions* penned by Augustine, which tracked the Saint's conversion from paganism toward Christian virtue. Montaigne's introspection cut against the grain of inherited morality by questioning the virtues of Aristotelian Scholasticism and Christianity, more broadly speaking. Where others decried vice, Montaigne indulged human imperfections. One scholar argues that Montaigne is rebuking both Ciceronian and Christian forms of virtue, which he sees as dangerous guides to political action (Thompson, 2013, p. 197). Another scholar goes as far to argue that Montaigne targeted the pedagogical failure of Aristotle to form the good ruler, replacing the Philosopher with his own philosophical teaching designed to moderate the prince (Hartle, 2013, p. 116).

The key to Montaigne's pedagogical project is the emulation of great souls. If Montaigne himself is not a worthy source, as he tells his readers time and time again, then who was? Enter Plutarch. As one scholar notes, Montaigne "turned to the Greeks as models to give more weight to the moral enterprise of the *Essays.* . . . [Montaigne's] study of the laws and philosophy of war were part of his mediation on mankind" (Christodoulou, 1992, p. 89). As the world around him sank in savagery, one can imagine Montaigne tucked away in his tower library trying to escape those who peddled dubious claims about justice:

> [W]hoever boasts, in a sick age like this, that he employs a pure and sincere virtue in the service of the world either does not know what virtue is, since our ideas grow corrupt with our conduct (indeed, hear them portray it, hear most of them glorying in their behavior and making their rules; instead of portraying virtue, they portray injustice pure and simple, and vice, and present it thus falsified for the education of princes.
>
> *(III.9, p. 759)*

There he sits perusing the thick leather-bound volumes of Plutarch's *Les vies des hommes illustres Grecs et Romans* (or *Illustrious Lives of the Greeks and Romans*) and the aptly titled *Oeuvres Morales et Meslees* (*Moral and Mixed Works*). Such reading was not a frivolous pastime but an exercise in character building (Edelman, 2019). Plutarch in Montaigne's opinion "is a philosopher who teaches us virtue" (II.32, p. 549). The stakes could not be higher, given that among his readers were kings and pretendants to the throne. Montaigne gave a personal copy to both Henri III and Henry of Navarre. One can speculate on the conversations they had when they met, or what the king and future king learned from the *Essays* about war, ethics, and the human condition.

In the essay "On the Education of Children," Montaigne explicitly tells would-be readers to crack open Plutarch's *Lives*. He warns that nothing should be accepted on

> [M]ere authority or trust. Let not Aristotle's principles be principles to him any more than those of the Stoics or Epicureans. Let this variety of ideas be set before him; he will choose if he can; if not he will remain in doubt. Only the fools are certain and assured.
>
> *(I.26, p. 111)*

By reading the *Lives*, the student will associate "with those great souls of the best ages." In doing so, they should not be overly interested in trivial facts: "Let [the student] be taught not so much the histories as how to judge them" (I.26, p. 115). The attitudes, dispositions, motivations, and quality of judgement of the great souls interest Montaigne more than the actual events because they provide insight into human character.

It should come as no surprise that the lives of the Greek and Roman great souls were shaped by war. And their actions in war were shaped by virtue and vice. Instead of assuming that reason can lead men to act justly in war, Montaigne observes that humans have a natural penchant for excess and cruelty. "Our structure, both public and private, is full of imperfections . . . our being is cemented with sickly qualities: ambition, jealousy, envy, vengeance, superstition, despair dwell in us." (III.1, p. 599). Indeed, "Nature herself," says Montaigne, "attaches some instinct for inhumanity" (II.11, p. 316). This penchant for inhumanity can be found in the greatest of souls in the *Lives* but so too can the keys to controlling them. Montaigne wants his student to know

> [W]hat springs move us, and the cause of such different impulses in us. For it seems to me that the first lessons in which we should steep his mind must be those that regulate his behavior and his sense, that will teach him to know himself and to die well and live well.
>
> *(I.26, p. 117)*

Among the most noble souls from the works of Plutarch, two stand out. Alexander the Great: "so many outstanding virtues were in him—justice temperance, liberality, fidelity to his word, love for his people, liberality towards the vanquished" (II.36, p. 571). He also has specific "military virtues—diligence, foresight, patience, discipline, subtlety, magnanimity, resolution, good fortune." His overall character is impeccable, even though Montaigne does reproach specific actions, namely the razing of Thebes and the slaughter of its valiant defenders undertaken when his anger and thirst for revenge overwhelmed his virtuous disposition (I.1, p. 5). "It is impossible," Montaigne remarks, "to conduct such great movements according to the rules of justice; such men require to be judged in gross, by the master purpose of their actions" (II.35, p. 571). There is something unsettling in Montaigne giving Alexander's outbursts a free pass, as if he was excusing the injustice committed in war. But the example of Alexander points to the natural place that anger and vengeance hold in men's souls. These human frailties are parts of the human condition, just like valor and moral integrity. And Montaigne ultimately wants his pupil to know about them to better control them.

Hence his admiration for Epaminondas. Although Plutarch's version of Epaminondas' life was lost, Montaigne read a version cobbled together from surviving ancient texts mimicking Plutarch's style that was commonly added to sixteenth-century editions of the *Lives* as a sort of appendix (Hanson, 1999). Montaigne admires not so much his valor and glory but his "wisdom and reason." He was the "first man of Greece," and with regard to his character and conscience,

[H]e very far surpassed all those who have ever taken to manage affairs. For in this respect, which must principally be considered, which alone truly marks what we are, and which I weigh alone against all others together, he yields to no philosopher, not even Socrates.

(II.36, p. 573)

Montaigne was not the first to offer such accolades; Cicero before him did the same, which should be a sign for those interested in the just war tradition to inquire further.

Epaminondas was a Theban general and statesman from the fourth century BC who liberated the city-state from Spartan subjugation and propelled it to a position of power and primacy in ancient Greece. Following Montaigne's advice, the reader of the *Lives* need not get lost in the historical details but focus instead on Epaminondas' character traits, which were put on full display in a battle when his own freedom was at stake. Consider the following passage:

Truly that man was in command of war itself, who made it endure the curb of benignity at the point of its greatest heat, all inflamed as it was and foaming with frenzy and slaughter. It is a miracle to be able to mingle some semblance of justice with such actions; but it belongs only to the strength of Epaminondas to be able to mingle with them the sweetness and ease of the gentlest ways, and pure innocence.

(II, 36, p. 609)

Alexander, we have seen, was prone to excess in war. Ann Hartle argues that in placing Epaminondas above Alexander, Montaigne was rebuking the authority of Aristotle as a teacher of human virtue. As demonstrated in the opening chapter of this volume, Aristotle was not only a teacher of Alexander, but also his virtue ethics has been chosen as a model source by some contemporaries involved in military ethics, not to mention foundational to Alisdair MacIntyre's body of thought. Montaigne saw things through a different lens. "It would seem," Hartle writes,

> [T]hat the two stories of Alexander's cruelty are intended to manifest, in particular, the failure of Aristotle's teachings to moderate the infinite desires and passions of Alexander and, in general, the ineffectiveness of Aristotle's restraints on the power of the natural master.
>
> *(2013, p. 116)*

But Aristotle does not share the entire burden of blame. Montaigne tells us also that the otherwise heralded Caesar proclaimed that "the times for justice and for war were two," while Gaious Marius, also a Roman general of some renown but known for his endless ambition and self-seeking opportunism, explained that "the noise of weapons kept him from hearing the voice of the laws." These were flaws Epaminondas did not have: "even for the inestimable good of restoring liberty to his country [he] scrupled to kill a tyrant or his accomplices without due forms of justice" (III.1, p. 609). Instead, he valued clemency over vengeance and virtue over excess valor. This would have been a stark observation for sixteenth-century readers of the *Essays*, who would no doubt have been making mental comparisons to French men of war on both sides of the tumult.

What would a sixteenth-century reader in France get from a Montaignian education? The reader—maybe a future king—is put face to face with the great souls of the past and implicitly led to ask: how do I compare? As a form of ethical instruction, it is worth noting the importance of being shown examples of how to live and act that differed from what the reader witnessed in sixteenth-century France, where religious zeal and hatred were the predominant attitudes, and excess violence was the norm of warfare against an irreconcilable enemy. Montaigne draws the contemporary lessons from the Epaminondas example in a later essay entitled "Of the useful and the honorable":

> Let us not fear, after so great a preceptor, to consider that there are some things illicit even against the enemy; and that not all things are permissible

for an honorable man in the service of his king, or the common cause, or of the laws. For our country does not come before all other duties. . . . It is a lesson proper for the times.

Montaigne knows that any education in virtuous warfare will have to peel back the indoctrination of authoritative and divine sources. To his readers tempted by such misguided appropriations of justice, Montaigne asks them to call a spade a spade: "Let us take away from wicked, bloody, and treacherous natures this pretext of reason. Let us abandon this monstrous and deranged justice and stick to more human imitations" (III.1, p. 610).

Epaminondas, who was in a position of power but refused to bend power to commit injustice, illustrates a moral intuition that those dallying in politics and war could imitate. Despite "justifications" offered by so-called authorities that would legitimize immoral acts in the name of the state, the king, or one's religion, Montaigne suggests a different course drawn from reading Plutarch: "I want [my student] to refrain from doing evil, not for a lack of power or knowledge, but for a lack of will" (I.26, p. 123).

Controversies

Montaigne courted controversy politically and philosophically. While many hold the image of the sage Montaigne holed up in his tower library reading and writing while France consumed itself in pitiless violence, this is a false image. Montaigne's life story revolves around his political activities and reveals the challenges of navigating the religious tensions in France: as a Catholic mayor of a mostly Protestant Bordeaux, and as a secret emissary to a Catholic King in negotiation with a Protestant potential heir to the throne (Desan, 2017), Montaigne followed a path of tolerance and pursued the reconciliation of feuding sides instead of taking sides. This was a dangerous stance. The history books are dotted with accounts of Protestant lynch mobs in the Bordeaux region and the tragic fate of those caught up in the frenzy. The most infamous display of calculated violence was the St. Bartholomay's day massacre in 1572, where Catholics killed over 3,000 Protestants who had assembled in Paris for the wedding of Henry of Navarre to the king's sister, Margaret of Valois. This sparked further massacres of Protestants across France, as well as revenge massacres enacted by Protestants against Catholics. Montaigne might have been safer had he chosen sides, but then again, he seems to think that his earnest belief in tolerance saved him on several occasions when Protestant ruffians could have killed him but spared him because of his reputation.

The astute reader finds the philosophical roots of Montaigne's political stance sprinkled across the essays: his skepticism of religious doctrine, his arguments for toleration of the Other, his abhorrence of cruelty, his critique of pedantry and presumption. In hindsight, it is easy to think this was a clear-cut choice, that any reasonable person would ply the middle path. However, war, especially civil war, has the quality of polarizing people and creating entrenched factions. There were a few who followed in his stead. Montaigne praises the Protestant General François de la Noue, who showed "conscientious affability . . . amid such injustice of armed factions" (II.18, p. 502). The curious among us could read his *Discourse politiques et militaires* published in 1587, which explores the moral virtues of the military profession and is sometimes compared to the *Essays*. But one example does not resolve the issue at hand, which is quelling the appetite of the multitude who, in times of tumult, all too easily flock to the side of the fanatics and exult in killing their enemy, including civilians. For Montaigne, the key to dissuading the masses lies in persuading those who court power to think differently. The flaws of Alexander and the virtues of Epaminondas being the mirror in which to see one's own self.

It is hard to gauge Montaigne's influence amidst the tumult of his century. But in the generations that followed—after a hiatus of unpopularity that coincided with the *Essays* being placed on the Catholic Church's *Index of Prohibited Books* in 1676 because of his anti-theological arguments—Montaigne garnered the reputation of being *the* moralist born from an era of butchery. Voltaire, Diderot, and Benjamin Franklin were among the Enlightenment thinkers who read the *Essays* through a moral lens (Desan, 2018, p. 10). Closer to our time, a phrase penned in the diary of Jean Guéhenno, a French professor living through the Nazi Occupation of France in the Second World War, captures the lure of Montaigne:

> The most admirable thing about these "Essays"—so reasonable, so moderate—is that they were written during the Wars of Religion, when excess was the rule. . . . I think of what a Montaigne would be like today, caught between the various "leagues" that are drenching the world in blood, but trying to define the order of human thought, as Montaigne defined the order of French thought four centuries ago.
>
> *(2016, p. 9)*

What solidified Montaigne's philosophical reputation is this: when opposing sides were digging up—or indeed making up—theological and philosophical authorities to justify killing the other side, Montaigne had the moral wherewithal to steer clear of the frenzy while proffering a different take on the ethos of war. The *Essays* is a source, introspective at its core, for readers to probe the human condition in times of tumult (Brunstetter, 2022).

Legacy

My reading of Montaigne has been a parallel endeavor to a career spent writing about the just war tradition and making sense of the contemporary wars that animated my research. A tattered copy of the *Essays* sits on my shelves, its pages covered in marginalia. From time to time, I pluck it off and peruse it. Parallel endeavor is not quite right. Reading Montaigne helps me to make sense of war in a way that the just war tradition does not. Hence the first legacy: Montaigne reminds his readers to be skeptical of authority and philosophy, to doubt that Truth jumps from the pen of Aquinas or Jeff McMahan (to mention two canonical figures who come to very different conclusions), to recognize that just war theorizing might not provide all the answers. One need not read Montaigne to have similar doubts, but the introspective element to the *Essays* obliges the reader to question one's own assumptions about war and ethics to probe the extent to which one holds blind faith in this or that authority. The point of the exercise is not to be a total skeptic and doubt everything but, rather, to perennially interrogate the frames that inform how *I* think about war and ethics. This leads to a second legacy, which deals with sources.

Every age has its wars, its moral authorities to judge the permissions and restraints, as well as the exceptions to the rules. Those who navigate the wars ravaging their lifetime can turn to the existing authority, or seek out alternative sources readily available, to make sense of war. The narrative of the just war tradition holds special authority over how we understand the do's and don'ts of modern war—a story to which I have contributed (Brunstetter, 2021). The tradition is a well to draw from, or a river that flows through history to use James Turner Johnson's image (2009, p. 252). But there are other wells, other rivers. Montaigne's turn to Plutarch showcases one source which turns the focus away from universal rules, whether derived from religious authority or human reason, toward human intuition. The inward turn places emotion squarely into the mix, above reason and thus intimately essential to morality. Ambition, fear, rage, hatred, anger, extasy, presumption, spite, sadness, honor, compassion, mercy, arrogance, self-assurance, jealousy, sacrifice, glory, zest, and more emotions populate the soul. And these are normal; they are what makes us human. Acting ethically may not be so much about knowing the rules but about controlling the mind, which can be learned by mingling with warrior souls of the past. This has implications for how we teach the ethics of war to those who may one day do the fighting and to the leaders who send them to war. Teaching the principles and rules that some authority figure has devised is insufficient at best and maybe even the wrong approach if we accept some interpretations of Montaigne. Reasoned principles that become moral checklists divorce ethics from its ultimate subject: understanding and regulating human intuitions and actions. Moreover, each culture has its own ethics of war, and while there may be similarities, the differences can be crucially divisive (Bryant, 2021). For Montaigne, probing in earnest the best souls caught up in the throes of war was a means to return our humanity, with all its imperfections and contradictions, to the core of morality.

Montaigne chose Epaminondas as his ideal: who would you choose? Answering this question is a personal choice, and we can debate the criteria for deciding who makes the list, but in choosing to read Montaigne, I have made mine.

Note

1 All quotations are from Donald Frame's translation of the *Essays* (Montaigne, 1943). It is standard practice in Montaigne studies to quote the book number, essay number, and page number.

Works Cited

Bakewell, Sarah. 2010. *How to Live: A Life of Montaigne in One Question and Twenty Attempts at an Answer*. London: Chatto & Windus.

Brunstetter, Daniel R. 2021. *Just and Unjust Uses of Limited Force: A Moral Argument with Contemporary Illustrations*. Oxford: Oxford University Press.

Brunstetter, Daniel R. 2022. What Reading Montaigne during the Second World War Can Teach Us About Just War. *Journal of International Political Theory* 18(3), pp. 355–374.

Bryant, Michael. 2021. *A World History of War Crimes: From Antiquity to the Present*. 2nd Edition. London: Bloomsbury Academic.

Christodoulou, Kyriaki. 1992. L'image du guerrier grec dans les *Essais* de Montaigne. In Gabriel-André Pérouse, André Thierry, and André Tournon. eds. *L'homme de guerre au xvie siècle*. Saint-Étienne: Presses de l'Université de Saint Étienne.

Desan, Philippe. 2017. *Montaigne: A Life*. Princeton: Princeton University Press.

Desan, Phillipe (ed.). 2018. *Les usages philosophiques de Montaigne*. Paris: Hermann.

Duff, Tim. 1999. *Plutarch's Lives: Exploring Virtue and Vice*. Oxford: Oxford University Press.

Edelman, Christopher. 2019. Plutarch and Montaigne. In Katerina Oikonomopolou and Sophia A. Xenophontos. eds. *Brill's Companion to the Reception of Plutarch*. Leiden: Brill, pp. 479–492.

Engster, Dan. 1998. The Montaignian Moment. *Journal of the History of Ideas* 59(4), pp. 625–650.

Guéhenno, Jean. 2016. *Diary of the Dark Years, 1940–1944*. Translated by David Ball. Kindle Edition. Oxford: Oxford University Press.

Hanson, Victor Davis. 1999. *The Soul of Battle: From Ancient Times to the Present Day, How Three Great Liberators Vanquished Tyranny*. New York: Free.

Hartle, Ann. 2005. Montaigne and Skepticism. In Ullrich Langer. ed. *The Cambridge Companion to Montaigne*. Cambridge: Cambridge University Press, pp. 183–206.

Hartle, Ann. 2013. *Montaigne and the Origins of Modern Philosophy*. Evanston: Northwestern University Press.

Johnson, James Turner. 2009. Thinking Historically about Just War. *Journal of Military Ethics* 8(3), pp. 246–259.

Levine, Alan. 2001. *Sensual Philosophy: Toleration, Skepticism, and Montaigne's Politics of the Self*. Lanham: Lexington Books.

Losse, Deborah N. 2013. *Montaigne and Brief Narrative Form*. London: Palgrave Macmillan.

Montaigne, Michel de. 1943. *The Complete Essays of Montaigne*. Translated by Donald M. Frame. Stanford: Stanford University Press.

O'Driscoll, Cian. 2015. Rewriting the Just War Tradition: Just War in Classical Greek Political Thought and Practice. *International Studies Quarterly* 59(1), pp. 1–10.

O'Driscoll, Cian. 2018. Keeping Tradition Alive: Just War and Historical Imagination. *Journal of Global Security Studies* 3(2), pp. 234–247.

Quint, David. 1998. *Montaigne and the Quality of Mercy: Ethical and Political Themes in the Essays*. Princeton: Princeton University Press.

Schaefer, David L. 1990. *The Political Philosophy of Montaigne*. Ithaca: Cornell University Press.

Supple, James. 1984. *Arms Versus Letters: The Military and Literary Ideals in the Essais of Montaigne*. Oxford: Clarendon.

Thompson, Doug. 2013. Montaigne's Political Education: *Raison d'etat* in the *Essais*. *History of Political Thought* 34(2), pp. 195–224.

Thompson, Douglas I. 2018. *Montaigne and the Tolerance of Politics*. Oxford: Oxford University Press.

7

JOHN BROWN (1800–1859)

John Kelsay

Introduction

On Sunday, 16 October 1859, a small band of men began their march toward Harpers Ferry, Virginia. Armed with rifles and homemade pikes, they had been training for weeks on a small farm in southwestern Maryland. Their intention? To carry out plans crafted by their leader, John Brown. Specifically, they hoped to (1) take possession of the Federal Armory located at the Ferry; (2) canvas several nearby plantations, capturing the owners and recruiting slaves to join John Brown's fighting force; and (3) assume success, after which they would depart quickly, disappearing with their new members into the heavily wooded mountains nearby, where they would be joined by others bringing more rifles and pikes. The ultimate goal would be to conduct raids further south, recruiting more slaves to join in a guerrilla campaign designed to bring down the southern economy, thus forcing an end to the south's "peculiar institution." After initial success with respect to (1) and (2), the plan fell apart. For reasons that remain unclear, Brown and his newly enlarged force remained at the Armory where, confronted by local militia, they engaged in a shoot-out with casualties on both sides. And when a band of Marines commanded by Robert E. Lee arrived, defeat became certain. A number of Brown's men, including two of his sons, were killed. Badly wounded, Brown himself was arrested. Taken to a nearby jail, he would stand trial for treason against the Commonwealth of Virginia. He would be convicted, with the sentence—death by hanging—carried out on December 2.

We know that John Brown is not part of the canon of just war tradition. As I hope to show, however, the democratic and egalitarian commitments

DOI: 10.4324/9781003428688-8

that led him to Harpers Ferry also shaped his thinking about the justification and conduct of armed force. John Brown and others like him provide a good example of the way that the concerns expressed in the vocabulary of *jus ad bellum* and *jus in bello* resonate whenever people think seriously about the relationship between morality and power. First, though, let us ask the question: who was John Brown, and how did he become the kind of person ready to plan and carry out the Harpers Ferry raid? In order to answer, we should begin with his family, then trace the development of his career in the movement to abolish slavery.

Context

Born on 9 May 1800, in Torrington, Connecticut, John was the oldest surviving son of Owen and Ruth (nee Mills) Brown. (Born several years earlier, two other sons died in infancy.) From the start, John's parents endeavored to pass on their faith, which joined the Christian faith to the cause of abolition. Owen came from the Congregationalist and Presbyterian churches in which the family took part, while Ruth was fostered by reading a rather long meditation on the Golden Rule ("do unto others as you would have them do unto you") by Jonathan Edwards the Younger. Following Ruth's death in 1808, Owen's influence loomed particularly large in John's life. Primarily a businessman who attained success as a tanner, Owen would remarry and gradually move the family west, to upstate New York and then to Ohio (at the time known as the "Western Reserve"). There he served as an officer in the church in Hudson, as well as participating in the Underground Railroad—Hudson being one of the final U.S. stops for fugitive slaves on the way to freedom in Canada.

John thus imbibed the abolitionist understanding of Christianity from an early age. Echoing the sermon by Edwards the Younger, he would later declare that he believed in two things: the Golden Rule and the Declaration of Independence, saying "I believe they mean the same."[1] (Brown, 1859, p. 56, note 46) He followed his father into the tanning business, became a devoted reader of abolitionist publications, and a regular churchgoer. Once married, he sought out other business opportunities to support his growing family (Brown's first wife died, and he remarried; between these two marriages, he fathered 20 children, several of whom died in childhood, while others would die as participants in their father's military adventures). Most of Brown's business ventures failed, so that he would eventually declare bankruptcy to deal with debts. Nevertheless, his faith and devotion to the abolitionist cause sustained him, and he came to be well known, especially among opponents of slavery in New England, and over time, among black abolitionists like Frederick Douglass and Henry Highland Garnet. Garnet was an early promoter of the idea that abolitionists needed to move from having an emphasis on moral suasion to a judgment that political and even physical force would prove necessary to advance the cause of freedom. In a famous speech delivered to an 1843 convention of free blacks, Garnet argued that resistance on the part of enslaved persons was not only permitted; it should rather be considered a duty. Brown's own move in this direction seems to coincide with the development of Garnet's ideas; both men would influence Douglass' somewhat later acceptance of the utility of armed force. For such activists, developments from the late 1840s and throughout the 1850s suggested that advocates of slavery possessed increasing power. When Texas entered the Union in 1845, it did so as a state friendly to the institution; in 1850, the Congress reaffirmed earlier legislation requiring the return of fugitive slaves; and the Kansas–Nebraska Act of 1854 left open the possibility that one or both territories might be incorporated as slave states. For men like Brown, such developments required a forceful response. As we shall see, he took it upon himself to organize for action.

Texts

Unlike some others covered in this volume, John Brown did not write a formal treatise on the justice of armed force. Never a scholar (though for a brief period in his younger years, he studied for the ministry), Brown developed his thoughts in shorter pieces and in correspondence. Both may usefully be related to actions. Thus, in response to the Fugitive Slave Act, John Brown helped with the organization of the League of Gileadites, a kind of civilian defense organization made up mostly of free blacks living in Springfield, Massachusetts. The League took its name from a biblical story in which the descendants of Gilead unite under a charismatic leader in order to defend themselves against aggression. Even so, the Springfield "branch" of the League signaled its union on 15 January 1851 by adopting the "Words of Advice" written by Brown:

> Nothing so charms the American people as personal bravery. . . . *No jury can be found in the Northern States that would convict a man for defending his rights to the last extremity.* . . . Should one of your number be arrested, you must collect together as quickly as possible. . . . Let no ablebodied man appear on the ground unequipped. . . . Your enemies will be slow to attack you after you have once done up the work nicely. . . . *Stand by one another, and by your friends, while a drop of blood remains; and be hanged, if you must, but tell no tales out of school. Make no confession.* . . . Union is strength.
>
> *(Brown, 1851, pp. 7–9)*

The document concludes with a formal agreement and a set of resolutions pointing to action consistent with the "Words of Advice." Overall, the logic of the text points to the notion that defense of oneself, one's property, family, and close colleagues is an established right, recognized by all fair-minded people.

For Brown, that same right justified the use of force on the part of "free soil" settlers in Kansas territory. The 1854 act mentioned earlier established terms by which a majority of those living in the area would eventually hold elections, and the state legislature would decide whether to enter the Union as a slave or a free state. Wealthy southerners sponsored poorer white people willing to move to Kansas; ironically, most of these did not in fact own slaves, but they were in favor of the institution. With help from "ruffians" crossing the border from Missouri—where slavery was established—these settlers made life difficult for those, like John Brown's sons, who moved to Kansas in hopes of securing farmland and, in the end, ensuring a state where slavery would not have legal status. When his sons requested their father come to their aid, he responded, using his abolitionist connections to secure money and weapons for their defense. Having delivered these, the "old man" (as people began to refer to him) decided to join the fighting as commander of a unit within a larger force led by one of sons.

In a letter addressed to his wife and others now living in North Elba, New York, Brown recounts some of his adventures.

> We were called to the relief of Lawrence, May 22. . . . Next day our little company left, and during the day we stopped and searched three men. Lawrence was destroyed in this way: Their leading men had (as I think) decided, in a very *cowardly* manner, not to resist any Government official to serve it, notwithstanding the process might be wholly a bogus affair. The consequence was that a man called a United States marshal came on with a horde of ruffians which he called his posse, and after arresting a few persons turned the ruffians loose on the defenceless people. They robbed the inhabitants of their money and other property, and even women of their ornaments, and burned considerable of the town.
>
> *(Brown, 1856, pp. 12–13)*

Brown's description of the sack of Lawrence is crafted so that a reader will understand that his own actions constitute a defensive response. His own unit broke off from the main force, after which:

> [W]e encountered quite a number of proslavery men, and took quite a number prisoners. Our prisoners we let go; but we kept some four or five horses. We were immediately after this accused of murdering five men at Pottawatomie, and great efforts have since been made by the Missourians and their ruffian allies to capture us.
>
> *(Brown, 1856, p. 13)*

We shall return to the Pottawatomie affair; for now, it will suffice to note again Brown's intention to represent his actions as a matter of defensive force, as well as the fact that he asked his wife to make sure that one of his most prominent supporters, Gerrit Smith of New York, read his letter. The text proceeds with descriptions of further encounters with more or less organized pro-slavery forces. In some of these, Brown's men and their allies are successful, while others do not go so well. Finally, they met a regular army force. Charged with bringing some order to the chaos brought by competing militia, a Colonel Sumner "compelled us to disband; and we, being only a handful, we obliged to submit."

As the next passage suggests, however, this did not mean that Brown and those with him were giving up the fight. Invoking the example of King David, a portion of whose early career involved fleeing to the "wilderness," where he led a band of guerrilla warriors (I Kings 18–31), Brown explains:

> Since then we have, like David of old, had our dwelling with the serpents of the rocks and the wild beasts of the wilderness; being obliged to hide away from our enemies. We are not disheartened, though nearly destitute of food, clothing, and money. God, who has not given us over to the will of our enemies, but has moreover delivered them into our hand, will, we humbly trust, still keep and deliver us. We feel assured that He who sees not as men see, does not lay the guilt of innocent blood to our charge.
>
> *(Brown, 1856, p. 15)*

As with the League of Gileadites, so with Brown's actions in Kansas—these he justified as matters of self-defense. Once it became clear that the free soil or antislavery settlers would prevail, however, Brown turned his attention to a different sort of venture. Moving his men to Iowa, he began to train them for an invasion into the "heart of Africa"—his way of referring to the slave states in the southern United States. And for that, he felt the need for a somewhat different justification. So, in January 1858, John Brown visited the home of Frederick Douglass in Rochester, New York. There he composed a Provisional Constitution for the People of the United States; in April, the text was presented to the community of free blacks and escaped slaves in Chatham, just south of Toronto. Those gathered for this constitutional convention discussed the document article by article, approving each and then adopting the text as a whole. At the same time, they voted to appoint Brown to the office of commander in chief of an armed force. He took copies with him to Harpers Ferry, hoping to distribute them following the success of the raid. Instead, they provided evidence useful to the state's case against him. For our purposes, however, they point to Brown's version of just war thinking, in particular by filling in his understanding of right authority, just cause, and right intention.

First, with respect to authority—the old man seemed to understand intuitively that his venture into "Africa" required authorization. Self-defense, assuming an imminent threat, did not. But an action like the raid on Harpers Ferry, along with the ongoing guerrilla campaign Brown and his men envisioned, did. Since he claimed to fight in the service of justice for the enslaved descendants of Africans, members of the Chatham community stood as representatives for their disenfranchised brothers and sisters. Following his arrest, Brown responded to questions from Virginia Senator James M. Mason and others. Referring to the Provisional Constitution, the Senator asked whether Brown considered his force a military organization; the response "I did in some measure. I wish you would give that [Provisional Constitution] your close attention" drew a further query, as to whether the old man considered himself the "commander-in-chief of this provisional force." Again, the response raised further questions: "I was chosen, agreeably to the ordinance of [the Provisional Constitution] commander-in-chief of that force" (Brown, 1859, p. 47). Brown knew this placed him outside the standard command and control measures outlined in the Constitution of the United States of America. Since he considered the elected officials of his time as hopelessly corrupt with respect to the issue of slavery, he—like others before and since—sought for a substitute and found it in the Chatham community. During the same interview, he also appealed to God and to standards of justice; even with those ideas, however, the legal idea of the consent of the governed loomed large for Brown.

Second, with respect to just cause, we may begin with the preamble to the Provisional Constitution.

Whereas slavery throughout its entire existence in the United States is none other than a most barbarous unprovoked and unjustifiable War of one portion of its citizens upon another portion, the only conditions of which are perpetual imprisonment and hopeless servitude or absolute extermination in utter disregard and violation of those eternal and self-evident truths set forth in our Declaration of Independence. Therefore, We, Citizens of the United States, and the oppressed people who by a recent decision of the Supreme Court are declared to have no rights which the White Man is bound to respect, together with all other people degraded by the laws thereof, Do for the time being Ordain and establish for ourselves the following Provisional Constitution and Ordinances the better to protect our Persons, Property, Lives, and Liberties, and to govern our actions.

(Brown, 1858, pp. 26–27)

Forty-eight articles follow; of these, 44 deal with the organization of the military force envisioned by John Brown. Only one article proved controversial at the Chatham convention—article 46, which indicates that those adopting the Constitution are not trying to overthrow the existing federal or state governments. They instead intend to "amend and repeal" laws supportive of slavery, fighting under the same flag "that our Fathers fought under in the Revolution" of 1776.

Several items in the preamble are worthy of note. First, the text identifies enslaved persons as citizens of the United States, who thus possess the rights outlined in the Declaration of Independence. And second, despite the Supreme Court's 1857 ruling in the case of Dred Scott, Brown's Constitution declares that those enslaved may, together with others committed to the cause of freedom, organize themselves to resist aggression.

In one sense, then, the Provisional Constitution continues the theme of Brown's earlier writings, that is, that self-defense is an established right. The kind of forceful action and especially the organization of an army with its own command and control structure suggest a somewhat different argument, however. Remembering the speech of Henry Highland Garnet mentioned earlier, with its stipulation that enslaved persons not only have a right to resist oppression but also a duty, we may understand the Provisional Constitution as an extension of that argument. Slaves should resist, yes; but others should come to their aid. A letter composed by Theodore Parker, a prominent Unitarian minister and abolitionist, shortly after Brown's raid, makes the idea clear.

> There are men in all the Northern States who feel the obligation which citizenship imposes on them—the duty to help those slaves. Hence arose the ANTI-SLAVERY SOCIETY. . . . Hence comes CAPT. BROWN'S EXPEDITION—an attempt to help his countrymen enjoy their natural right to life, liberty, and the pursuit of happiness. He sought by violence what the Anti-Slavery Society works for with other weapons. The two agree in the end, and differ only in the means. Men like Capt. Brown will be continually rising up among the white people of the Free States, attempting to do their *natural duty* to their black countrymen—that is, help them to freedom.[2]
>
> *(Parker, 1859, pp. 135–136)*

Brown's Constitution thus builds on old notions regarding the duty to come to the aid of those in need—what today we would call a duty to rescue or a responsibility to protect victims of aggression. His idea of just cause is thus tied to a number of judgments currently recognized by scholars of the just war tradition and of domestic and international law.

Finally, what of right intention? As a number of interpreters of just war tradition put it, this criterion may be understood as the link between right authority, just cause, and those criteria requiring just conduct: aim of peace, proportionality of ends, reasonable hope of success, and last resort, as well as the *jus in bello* requirements of discrimination and proportionality of means. With respect to the aim of peace and last resort, the Provisional Constitution seems very clear. As the passages quoted before indicate, the goal is a more inclusive and just America. As Brown and other abolitionists often put it, the success of their movement would be a fulfillment of the promises of the Declaration of Independence, a kind of Second American Revolution. And armed force is justified, other means of action having failed.

For the proportionality of ends and reasonable hope of success, as well as the *in bello* criteria, we must look beyond the Constitution, to reports of Brown's conversations with Frederick Douglass and other allies, as well as to the orders he gave those joining him at Harpers Ferry. From the former, we know that Brown's friends tried to persuade him that the raid was a doomed venture, which would in the short run make things worse for the people he wanted to help. In response, Brown liked to quote Romans 8: 31: "If God is for us, who is against us?" When Douglass refused to join in the raid, the old man expressed sadness; but he added that if in the end the venture failed and he died, he might be worth more to the slave's cause in death than in life.

With respect to the *jus in bello* criteria, we know that Brown ordered his men to avoid killing or unnecessarily causing injury to noncombatants. When he learned that two of those with him had killed a black porter working on the train coming through the area, he immediately moved to disassociate his efforts from such activity. As well, we have the testimony of plantation owners brought to the Armory as prisoners that he treated them well; indeed, when Brown later spoke about why the plans for a swift departure and march into the mountains went awry, he indicated that he dallied because of concern for the welfare of these people.

The lack of a major treatise on the justice of war, then, need not prevent our understanding of John Brown's reasoning. In shorter pieces, correspondence, and above all the Provisional Constitution and Ordinances, the old man's ideas seem very clear. As well, his actions in planning and carrying out the raid at Harpers Ferry speak loudly about his notions of justice—though as we shall see, some of his contemporaries expressed reservations or raised questions regarding Brown's activities, not only at the Ferry but also earlier, in Kansas.

Controversies

The great controversy for Brown had to do with the institution of slavery itself, and the raid on Harpers Ferry contributed to the regional polarization that culminated in the Civil War. Along the way, though, Brown dealt with other controversies, many of them reflecting divisions in the abolitionist movement. As Sinha (2016) and others demonstrate, the majority of white abolitionists began with an emphasis on moral suasion. For many, among them the well-known journalist and publisher of *The Liberator*, William Lloyd Garrison, Christian faith traveled with an emphasis on nonviolence of the type envisioned in the Sermon on the Mount (Matthew 5–7). Faced with questions regarding the 1837 death of Elijah Lovejoy, a Presbyterian minister, journalist, and ardent abolitionist, for example, Garrison and others in the movement spoke of his "martyrdom," while abstaining from the militant language characteristic of Brown and others. The reason? Lovejoy died trying to prevent a mob from destroying his printing press; in so doing, he used a gun. For Garrison and for the many Quakers devoted to the slave's cause, such a use of force went against the deeply held convictions.

Thus, the question of *whether* armed or "physical" force might be justified loomed large, particularly as Brown and others began to develop their ideas beginning in the 1840s. In so doing, white abolitionists like Brown followed the example of many black abolitionists. Drawing on a number of sources, including the example of slave revolts like Nat Turner's or earlier of Touissant l'Ouverture's successful rising in Haiti, black leaders proved more ready to embrace the notion of justified force than Garrison and his allies. David Walker's *Appeal*, with its apocalyptic vision of divine judgment as a response to the sin of slavery, was published shortly after Turner's revolt and circulated widely. Increasingly getting impatient with the lack of progress toward abolition, Brown, Henry Highland Garnet, and others began to make the case for armed force. As they did so, they, like their opponents among the abolitionists, appealed to the Bible. The Sermon on the Mount was not their favorite text, however. Instead, Brown in particular drew from the Presbyterian and Congregationalist churches in which he participated; these stressed the importance of taking "the whole counsel" of Scripture. Thus, the Old Testament, with its emphasis on God as a warrior, proved important, as did the Apocalypse or Book of Revelation, with its depiction of Christ as leader of the armies of Heaven. Never finally resolved between abolitionists, this controversy nonetheless seemed to disappear with the advent of the Civil War; friends and allies of John Brown judged that the war proved his case.

A second controversy, also never finally resolved—even to this day—had to do with the killings at Pottawatomie. Recalling Brown's letter recounting his first adventures in Kansas, we know that shortly following the sack of Lawrence, he and others stood accused of murdering five men. We also know that in that letter, he neither confirmed nor denied the charge; from other sources, we find him saying that he did not carry out any of the killings, though he approved those who did.

The entire affair confused Brown's supporters at the time and continued to do so until the presentation of new evidence some years later. The facts are as follows: after the disaster at Lawrence, Brown and his men determined to make their way to Pottawatomie Creek, where a number of pro-slavery settlers were in residence. There they took prisoners—one or two at a time. Following a brief interrogation by Brown, who was clearly in command, his companions attacked these men. Using short swords, they hacked the prisoners to death, leaving the mutilated bodies to be discovered the next day. In total, five settlers died. Returning to his camp, Brown reportedly told his force that there would no further work of this type.

As noted, the question of why Brown and his men did this, as well as regarding his rather coy presentation of his own role, proved controversial in his own time and subsequently. Particularly for his admirers, the manner of the killings seems troubling. As more than one opponent of slavery remarked, the deaths of pro-slavery settlers seemed just; however, mutilation did not. Tracing various pieces of evidence, historians provide a number of plausible interpretations of the event. Perhaps the most convincing proposal is that of Stephen B. Oates (1984), who notes that in the days prior to the Pottawatomie killings, five free-state settlers died at the hands of pro-slavery forces, with the bodies found in a condition similar to Brown's victims. In that sense, Pottawatomie seems to be the sort of thing just war thinkers refer to as a reprisal—a response "in kind" to a previous wrong, with the intention of deterring similar action in the future.

Legacy

As previously noted, the Harpers Ferry raid failed, and John Brown, along with several of his men, was taken to the jail in nearby Charlestown. The interview with Senator Mason and others took place there; published shortly after in a prominent New York newspaper, this would be the first in a series of letters and speeches by which the old man made his case to America and also to admirers in Britain, France, and elsewhere.

The raid secured John Brown's status as the most famous among the militant opponents of slavery. The aftermath made him something more, however. As his correspondence and declarations went out through the media of his day, along with reports of his sober, humble demeanor at the jail and eventually the courtroom, even those with reservations regarding the use of force found themselves moved. For these, as for those already supportive of Brown, his behavior after the raid was reminiscent of numerous biblical characters, so that recognition as a martyr for the slave's cause followed. One might say that he became larger than life: for some, as a martyr for the cause of freedom; for others, particularly in the southern states, a symbol of a wider conspiracy to bring an end to that region's way of life.

In terms of American history, then, it seems appropriate to say that John Brown's legacy is ambiguous. What of his just war thinking?

One of the issues contemporary scholars debate has to do with the ethics of irregular war—armed resistance, revolution, and other uses of force typically associated with groups that operate outside the established command and control procedures of a state. Responding to such conflicts, some argue that the criterion of right or legitimate authority is no longer necessary to our evaluation of the justice of a particular case. Others see things differently, however. Finlay (2015), Gross (2015), and Scheid (2015) all argue that armed groups involved in irregular campaigns do need to win support of a broader population, if they are to associate their cause with justice. Gross' comment is representative. Just cause is:

> [A]nchored narrowly in the right to self-determination and the right to live a dignified life. Legitimate authority requires a peoples' consent or trust, which guerrillas gain through a range of formal and informal political and social institutions.
>
> *(2015, p. 23)*

Such a comment reflects John Brown's thinking as well. Believing firmly that a just society requires that authority rest on the consent of those governed, his consultation with the Chatham community sought to ground a claim to right authority, as well as to establish a just cause for the raid at Harpers Ferry and the broader campaign he envisioned.

Then, too, John Brown has a lot to offer to discussions of racial justice. No less a figure than W.E.B. Du Bois composed an admiring biography of Brown, praising the old man's commitment to freedom and equality for all people. Decrying the continuing failures of America to honor the rights of all its citizens, regardless of skin color, Du Bois asked:

> Has John Brown no message—no legacy, then, to the twentieth century? He has and it is this great word: the cost of liberty is less than the price of repression. The price of repressing the world's darker races is shown in a moral retrogression and an economic waste unparalleled since the age of the African slave trade.
>
> *(1909, p. 287)*

Continuing in this vein, Du Bois argues that the truth John Brown embodied—that a society cannot afford to put off dealing with the claims of justice—is valid for all places and times.

> This, then, is the truth: the cost of liberty is less than the price of repression, even though the cost be blood. . . . This calls for the abolition of hard and fast lines between races, just as it called for the breaking down of barriers between classes. Only in this way can the best in humanity be discovered and conserved
>
> *(1909, p. 295)*

What, then, is John Brown's legacy? When it comes to addressing issues dealing with the justice of armed force, I have argued that his scattered writings, correspondence, and speeches reflect concerns that mirror standard criteria of the just war tradition. The claim is not that he was familiar with the kind of historically or philosophically oriented conversations most contemporary scholars envision when thinking about that tradition. What Brown does reflect, though, is the way the tradition's concerns are difficult to avoid whenever people want to think seriously about the relationship between power and morality. As such, we do well to consider his example.

Notes

1 Here and elsewhere, all quotations from Brown's speeches, writings, and other items are taken from Stauffer and Trodd (2012). Where italics are used, those are present in their original text.
2 Also included in Stauffer and Trodd.

Works Cited

Brown, John. 1851. Words of Advice (in Stauffer and Trodd, below).
Brown, John. 1856. Dear Wife and Children, Everyone (in Stauffer and Trodd).
Brown, John. 1858. Provisional Constitution and Ordinances for the People of the United States (in Stauffer and Trodd).
Brown, John. 1859. Interview with Senator Mason and Others (in Stauffer and Trodd).
Du Bois, W. E. Burghardt. 1909. *John Brown*. New York: International Publishers.
Finlay, Christopher J. 2015. *Terrorism and the Right to Resist*. Cambridge: Cambridge University Press.
Gross, Michael J. 2015. *The Ethics of Insurgency*. Cambridge: Cambridge University Press.
Oates, Stephen B. 1984. *To Purge This Land With Blood: A Biography of John Brown*. Amherst: The University of Massachusetts Press.
Parker, Theodore. 1859. To Francis Jackson (in Stauffer and Trodd).
Scheid, Anna Floerke. 2015. *Just Revolution*. Lanham, MD: Lexington Books.
Sinha, Manisha. 2016. *The Slave's Cause: A History of Abolition*. New Haven: Yale University Press.
Stauffer, John and Zoe Trodd (eds.). 2012. *The Tribunal: Responses to John Brown and the Harpers Ferry Raid*. Cambridge, MA and London: The Belknap Press of Harvard University Press.

8

PIERRE-JOSEPH PROUDHON (1809–1865)

Alex Prichard

Introduction

Proudhon's *War and Peace: On the Principle and Constitution of the Rights of Peoples* (2022 [1861]), is arguably the first extended reflection on the ethics of war in the industrial age, remains one of the only anarchist approaches to this subject, and has been almost completely ignored by mainstream academia, by those who have engaged his thinking, and by anarchists who have not (e.g., Castleton, 2017a; Jun, 2020).[1] Proudhon's approach to the ethics of war, which spans seven published and unpublished books, engages and develops out of the neo-Roman discourse of *jus gentium*. This branch of Roman law is often translated, somewhat anachronistically, as *the rights of nations* but, given nations are distinctly modern inventions, is more accurately understood as that form of right that emerges from the common practices of peoples, collectives, or any political groups. Unlike the state of nature, which is presumed to be populated by individuals, jus gentium emerges when individuals coalesce into groups. Modern states sought to transcend, to tame, and to represent this pluralism by rendering the people a single constituent power, represented in and through popular sovereignty.

For *jus* to be *jus*, it was assumed by most philosophers that it had to be grounded in something that transcended this chaotic, seemingly brutish empirical, anarchic world. Appeals to reason, nature, God, or state sovereignty were appeals to right that did not rely on the fickle whims of the people. *Jus gentium* was a fallback position at best and certainly could not justify the right of war.

But Proudhon disagreed. He argued that the jurists are wrong to assume that right, justice, morality, history, progress, or the ethics of war come from

DOI: 10.4324/9781003428688-9

anywhere else. Unlike the pacifists studied in this volume (Rosa Luxemburg, Martin Luther King Jr., and maybe G.E.M. Anscombe, for example), for Proudhon the right of war is immanent to the development of peoples. Whether war should be fought or not, and how it is fought, reflects the will of the people. For better or worse, *jus ad bellum* and *jus in bello* cannot transcend or supersede *jus gentium*. *War and Peace* and the writings that followed are not therefore explicitly works of just war theory. The focus on *jus gentium* provides the historical and sociological under-laboring for such a theory.

War and Peace is arguably one of Proudhon's most anarchistic books, precisely because it seeks to recover *jus gentium* as the positive, moral, and historical grounding for politics as such. It is an explicit development of the epistemological and moral anti-foundationalism of Proudhon's wider theory of justice, an account of the economic causes of war, and a defence of the science of political economy as the necessary precondition of the pursuit of a just peace. It had significant impact after his death on the ideas of Tolstoy in particular. This chapter's final aim is to convince you that Proudhon's theory of the ethics of war rewards and demands further research.

Contexts

Pierre-Joseph Proudhon (1809–1865) is arguably the first self-professed anarchist in the history of political thought. Born into a peasant family in Besançon in France during the Napoleonic wars, he remembered the poverty and famines that blighted his family and how this forced him out of school and into work before he could finish his Baccalaureate. Fortunately, his father managed to convince the local press to take on this precocious child as an apprentice typesetter, and through the publication of the Latin Vulgate Bible and cutting-edge political and social theory, Proudhon taught himself Hebrew and Latin and became intimately acquainted with the works of Charles Fourier, one of the foremost socialist theorists of that time and a compatriot from Besançon. Then, in 1832, his brother Jean-Etienne died in suspicious circumstances during military service, an event which, he said, made him an "implacable enemy of the established order" (cited in Vincent, 1984, p. 48).

Texts and Tenets

Proudhon's most famous text for English readers is undoubtedly his first full monograph, *What is Property? Or an Investigation into the Principle of Right and of Government* (1840). It was considered by Marx to be the first scientific treatment of the topic of private property, and its publication and subsequent notoriety ensured Proudhon became one of foremost public intellectuals of his time. Following the 1848 revolution, he was elected to the revolutionary French Assembly. His politics and temperament led to a confrontation and then a duel with his fellow Assembly member, Felix Pyatt. Thankfully, no one was hurt, but the duel taught him a lot about the ethics of violence, as I will unpack later in the chapter. Though this likely attempt to assassinate him had failed, within a year, he was thrown into jail for criticizing the direction of the second republic and Louis Napoleon Bonaparte's "Caesarism". He was still incarcerated when Louis Napoleon crowned himself Emperor Napoleon III, following his *coup d'état* in 1851.

While in jail, he penned three more books, got married, fathered his first child, and began his journey towards the theory of decentralized federalism, after striking up a close friendship with the Italian republican Giuseppe Ferrari. On his release, he pursued numerous publishing initiatives, including books and newspapers, and continued a voluminous correspondence. He also watched his attempt to establish Bank of the People fail and incurred substantial personal debt as a result. Publishing was his only source of income, but, thankfully, he was prolific and his works very popular indeed. By the time of his death, he had penned over 50 volumes, with still more manuscripts being transcribed and published to this day.

A cursory glance at this published output demonstrates a striking consistency in his core concerns. For example, the subtitles of his first book *What is Property?* (1840) and *War and Peace* (1861) both show that *the philosophy of right* was Proudhon's primary focus across his career. At first it was refracted through political economy, then war and finally federal constitutionalism. His *magnum opus*, for which he was exiled in 1858, was a four-volume, million-word exploration of Enlightenment moral philosophy entitled *De la Justice dans la Révolution et dans l'Église. Études de philosophie pratique* (for a discussion, see Prichard, 2013). Published in Brussels in 1858, this comprehensive, encyclopedic, neo-Kantian and neo-Comtean exploration of the foundations of modern philosophy and practical ethics consisted of 12 books covering topics like labor, love and marriage, the state, ideas, and many, many more. A second edition was published in Paris later that year, while multiple further printings were halted by state censorship and repression. Leo Tolstoy recounts smuggling a copy back to Russia with him in 1861.

Proudhon turned to international politics in the final five years of his life and was practically alone among European socialist revolutionaries to do so. Furthermore, while most other European revolutionaries fled to London and Manchester to escape state repression, and there witnessed the smog of industrialization, Proudhon went to Belgium, the lynchpin of European international security. As he put it, it was his Belgian "exile that brought me the inspiration for this book" (Proudhon, 2022, p. 110).

Proudhon engaged international politics through a neo-Kantian lens, which was fairly typical, but which would have been more difficult had

Joseph Tissot, Kant's primary translator in France, not been a close friend of Proudhon's. Proudhon develops an original realist, sociological reading of Kant to this end. This involved arguing that reason and right emerge from or are "immanent" to society. *Jus gentium*, or the rights of peoples, disprove the argument that there are transcendent universal and abstract principles that can shape society. The challenge for social science is to derive right from social evolution rather than impose it upon it, he argued.

War and Peace was only the first of seven books on this topic. The others include *The Principle of Federation* (Proudhon 1863), a book on the 1815 Treaties of Vienna (Proudhon 1864), two studies of the errors of the Risorgimento and Mazzini's nationalism (Proudhon 1862, 1865), a posthumous study of Europe's "natural borders" (Proudhon 1868), and an unpublished history of the Polish partitions (see the transcribed chapters in Ferretti and Castleton, 2016). Unique in their own right, these could also be read as appendices to the moral philosophy of *De la Justice*, case studies in the global history of ethics.

Proudhon consulted all the texts we would expect of a scholar of the just war tradition in preparing *War and Peace*, except perhaps, for Saint Augustine and Thomas Aquinas. There is an extensive discussion of the ancients, including Cicero, Homer, Virgil, the Justinian Digest, and the Latin Vulgate Bible, which Proudhon himself translated as a typesetter in Besançon some 25 years earlier. There is extensive reflection on the literature on the philosophy of right, including extensive commentary on the works of Hobbes, Grotius, Wolff, Vattel, Pufendorf, Kant, and Hegel, among others. The book includes an extensive engagement with the writings of well-known military strategists, including Baron General Jomini, Friederich Ancillon, and Adolphe Thiers. The latter oversaw Proudhon's incarceration and later ordered the guns to be turned on the Paris Communards. He was also one of the most celebrated historians of the French Revolutionary Wars in his time.

Controversies

Proudhon makes four demonstrable but controversial contributions to our understanding of the right of war, each one refracted through the prism of *jus gentium*. The first is an exploration of the historical and social construction of *jus gentium*, which he develops to theorize the anarchy of the rights of peoples. Politics is predicated on the whims of the people, he argues. While he doesn't put it quite like this, we might assume he wouldn't be averse to the claim that all politics, not just international politics, take place in anarchy.

Second, history demonstrates, he argues, that the *right of force* is central to politics as such and is only scaled differently at the national/international level. Fundamentally, all politics is the same: an expression of the "right of force". Inertia and reason are both forces too, he argues, and war is only the most cataclysmic use of force in history. Because of the rituals and rationalizations of it, war is also demonstrably a moral and an intimately social pursuit. This claim is decidedly idiosyncratic, but empirically (as opposed to ideally) it is incontrovertible. Those who fight in wars do so collectively, often hand to hand, with a conception of the good and of right firmly in mind. They draw on society's material and moral support and derive their conceptions of valor and their military technology and skill from society too. Conceptions of the good, sustained by the use or threat of force, *make* societies, which is why they are also used to defend them.

Third, Proudhon argues that despite this social element, because war is unpredictable and fundamentally destructive, chaotic, and unpredictable, its moral claims are routinely nullified in practice by atrocities. Nevertheless, our systems of justice and right are the product of the victorious in war: the outcomes of wars are right making, but that does not make them right. Indeed, the salutary/educational effects of war are derived from these inevitable failings or shortcomings. People may rebel against their new dominating power, and new conceptions of justice emerge from this process of societal critical self-reflection.

Finally, Proudhon argues that only by better understanding war and the right of force, can we hope to develop a lasting peace. A better, more realistic account of the right of force must be based, he argued, on the science of political economy. *War and Peace* is also therefore, a history of the transformation of warfare, from piracy, to plunder, to territorial conquest, and, as he put it in *What is Property?*, an appeal for "justice in equality" and "order in anarchy" (Proudhon 1994, p. 209). In *War and Peace*, he argues that the struggle for equality in anarchy extends beyond the polity to the macrohistorical process between peoples. For as long as the principles of political economy are kept out of theories of just war, it is likely that war will remain with us. Only "agro-industrial federalism" could lead to peace, he argued.

Let me unpack each of these claims in a little more detail. Proudhon argued that justice or right is "immanent" to personhood and society, rather than transcendent or universal, or the outcome of rational or theological divination. In other words, common law, precedent, history, and what he calls "the universal testimony of humanity" (for a full discussion, see Prichard 2022a), all point to the fact that both sides of all disputes consider themselves to be in the right, and nothing can adjudicate between them but force: laws do not enforce themselves; right emerges out of this conflict of reciprocal rights claims and shapes them in turn. The right of war, however, is protean, he argues, "divine", even, because it remains beyond reason. It is the premise and empirical fact on which all other rights and political institutions are founded, and it cannot be abolished since no force is overwhelming enough to accomplish that task once and for all.

Proudhon identifies this right of war in what he calls the "moral phenomenology of war". Book one of *War and Peace* sets this out in detail and shows how the moral justifications of war are constitutive of war and sew war into the moral fabric of society. Texts in the history of political thought, including the Bible, the epics, the histories of Rome, the Napoleonic Wars, and the pagan myths (Proudhon's focus was Eurocentric), are taken to be windows on their times and also an empirical record of the right of war in history. For example, the Bible is fundamentally structured around an epic war between good and evil, is replete with the justifications for the most egregious wrongs by our standards, and like the pagan mythology it incorporated, sublimated, and transcended, shapes conceptions of war and right to this day. In short, the "moral phenomenology of war" is ubiquitous in society.

This turn to history, or what we might now call discourse analysis, was as much shaped by Proudhon's reading of Vico as anyone else during this period. Where Proudhon departs from his contemporaries, however, is his reluctance to see in history some transcendent *telos* towards a necessary present or future. Proudhon is quite clear that his "genealogy of right" (Proudhon, 2022, p. 449) does not result in the transcendence of the common or voluntary law of the peoples. This common right of war underpins all right and justice, and the latter are fundamentally unstable, relative, or contingent, as a result.

Peace is only possible because of the implied threat of violence, he argues, or because violence has been sublimated or institutionalized. On the other side of the coin, war is a "a cry for peace", he says (Proudhon, 2022, p. 100), an attempt to address the social conflicts that have emerged over time: through the force of arms, insurrection, riots, rebellion, revolution, and the economic inequalities they engender or reflect. For example, Proudhon discusses Napoleon's clamoring for "peace at all costs" and his dream of universal sovereign peace that was predicated on a monopoly of violence first and right second. The reason Napoleon failed in his bid for European supremacy, Proudhon argues, is because he simply didn't have the force to back up his republican ambition (Proudhon 2022, pp. 257–268). This does not denude revolutionary right of its virtue, but it does qualify it by administering a heavy dose of realism. But, against the conservatives, Proudhon argued that the 1815 settlement stopped nothing. As he put it, "The genie of war was [only] nailed to a rock by the Holy Alliance" (Proudhon 2022, p. 101).

Constitutions and their corresponding penal, civil, and legal codes might imply or specify the terms of an agreement to reject the use of direct violence, but they also presuppose the unequal structural effects of settled distributions of power *post bellum*. The violence that was halted through negotiated compromise is a pact, a bargain, the balancing of power. In other words, society is founded upon sublimated violence; it is pacified by the imbrication of violence into the everyday workings of society, from law and incarceration to city walls and fortifications, to the exploitation of one class for the sustenance of the other, or through public debt and military expenditure, etc. Society is structured by these resulting relations of force between the "*aristoï* or *optimates*" over the "plebs [. . . and] *ignavi*", proclaimed as the sovereign will. The anointed, appointed, or lauded are simply more powerful, and society develops complex, contested, and ultimately contingent moral rules to justify and drive these evolving balances of power (Proudhon, 2022, p. 126).

The second claim Proudhon makes is that *jus gentium* cannot be superseded or transcended by reason or the sovereign will of either a king or any other group. No such group has been successful because the idea of domination through universal sovereignty is repugnant to all self-respecting peoples. It is their right to rebel against such domination. It is not reason, God, or any other such "providential" force underpinning right that drives the evolution of societies but rather this "immanent" struggle for justice against domination. Conflict, disagreements, or agonism, within and between peoples, is the motor, the dynamism of the evolution of justice. Justice, Proudhon argues:

> [C]onsists of each member of the family, or of the city, or of the species, even as he asserts his freedom and dignity, acknowledging these things within others and showing them as much honour, consideration, power and enjoyment as he seeks for himself. That respect for humanity in ourselves and in our fellows is the most basic and most consistent of our affections.
>
> *(Proudhon 2022, p. 162)*

Ironically, then, the defense of dignity, equality, and mutual aid also leads to conflict (Proudhon 2022, pp. 139–142). This right of force is the least understood law of nature, he argues. Without a science of force or power, how can we hope for justice or peace?

So, what is the right of force? For Proudhon, everything exerts competing or countervailing forces, like positive and negative currents in a battery; their dynamic equilibrium is what generates a charge. This explicitly relational social ontology was derived from his reading of Auguste Comte, who "made the relation the basis of his positivism, and has excluded metaphysics and theology in its name" (Proudhon, 1990, p. 1140). Balances of forces emerge in nature through this complex social interaction, he argues, but because society changes and evolves in its moral, scientific, and geological context, relations of forces are always contingent, related to, and effected by other forces and our evolving understanding of them, and so cannot come to a transcendent equilibrium. It is the rationalization of this contingent balance through religion, poetry, politics, and science that distinguishes humans from the rest of nature, because we are able to imbue our lives with meaning, expressed through language and wider discourse, and spans generations.

Force is not simply gauged by "vigour" and "brawn", or "accumulated wealth and productivity". Like Clausewitz, his contemporary Proudhon argued that it "also includes faculties of spirit, courage, virtue, discipline" (Proudhon, 2022, p. 251). Force becomes power when it is collectivized. Power is a collective noun, plural, and relational. Proudhon subdivides power between "*puissance*", which is a function of brute force, and "*pouvoir*" which is the conscious collective coordination, administration, of whatever division of labor it takes to achieve a common or group goal (what he calls "social power"). Both are necessary, of course, and the most successful societies are those that best combine the two. But to do so effectively requires science. As we come to better understand our society and its needs, we find new and better ways to organize, rebalance our divisions (of labor), reinterpret our history, and envisage new futures. This causes us to re-constitute our societies internally and in relation to our neighbors. But given we are often members of more than one group at once, this process is also radically complex, results in split loyalties, and makes disagreement inevitable. Finally, there is no natural border to this process. In fact, as Proudhon observed, the most similar communities in France and Germany live on either side of its so-called natural border, the Rhine, and bear little relation to their respective metropolitan centers (Proudhon, 1868).

Proudhon argues that because there is no transcendent way to adjudicate, and because the right of force is inherent in all things, the "right of war" must be considered "equally just on both sides". To argue otherwise is not only to ignore what he calls the "testimony of the human race" and other synonyms but also to denude war of its inherent social and moral dimension, to overlook the ethos that drives it, and to conflate pillage and plunder, or "banditry", with war.

What distinguishes war from piracy, Proudhon argues, is the collective social and moral content of the struggle. And what is permitted in war is a function of the moral and technological development of the societies fighting it. Nevertheless, the rules of war, just like the relations of force on which they are grounded, are reciprocal; "what is permitted for one is permitted for the other" (Proudhon, 2022, p. 138).[2] In other words, since there is nothing to stop it, anything goes. And thus the practice of war "shames" it. Since war cannot rid itself of these "infractions", it cannot decide questions of right once and for all. As Proudhon puts it:

> Thought of as the verdict of force, war is sublime [. . .] But there is no iniquity with which it is not tainted to its roots: it may have a face like an archangel and have the name of God inscribed on its shield; but it has a dragon's feet and tail.

(Proudhon 2022, pp. 450–451)

Proudhon's reading of Thucydides' "Melian Dialogue" is instructive here (Proudhon, 2022, pp. 175–178). Proudhon argues that the Melians derived their right to resist, and the "righteousness" of their position, from the symmetry and reciprocity of the right of force and right of war; that is, from *jus gentium* as dictated by the Gods. The Melians claimed that this symmetry of the right of force and war was divine: they had a divine *right* to refuse to submit to Athens, regardless of the latter's power. If might *is* right, if "it were merely a matter of giving force its due, compromise would be easy" (Proudhon, 2022, p. 231), Proudhon says. But as the Melians demonstrated: it isn't. The right to resist is a function of the sanctity of collective dignity, not of reason. The Athenians' appealed to the Gods to justify the right of force on their side too. Mutually lofty (and exclusive) appeals to justice (by the Athenians and the Melians) ensured the battle became an atrocity. The Melians were wiped out, and the Athenian Empire lasted barely another ten years.

As should be clear by now, Proudhon's third potential contribution to the just war tradition would be to see the laws of war as immanent to and emergent from war itself. How armies are structured, their chains of command, how they're organized, and how they fight reflect the moral conscience of the people from which the army is drawn, the circumstances it faces, the material context in which they are fighting, and their available weaponry. An army's level of skill, cohesion, or depravity are all relative to (but not reducible to) their sociological and historical origin or context. Wars are thus only rarely chaotic, and they are certainly not anti-social: in practice, war is an expression of society's underlying normative architecture, and norms and rules are followed even on the battlefield. But, notwithstanding tales of camaraderie between enemies during "lulls in the fighting" (Proudhon, 2022, p. 321), "it has not been possible to rid the dueling between States of the horrors that besmirch it" (Proudhon, 2022, p. 256).

Proudhon's use of an analogy between warfare and dueling (one he was well positioned to comment on) is instructive here (Proudhon, 2022, book III, chapter 3). Proudhon argues that dueling has, over the course of history, transformed from a public battle of jousting and gladiatorial violence, often through proxies, into the nineteenth-century conception of high chivalry, pistols at dawn, or gunslinging. The violence became more immediate and up-close, less spectacle, more solemnity.

War, by contrast, has moved in the opposite direction. From close quarter combat, the military revolutions in the eighteenth and nineteenth century pushed armies apart. Canons, machine guns, and artillery became more precise, with death being a near certainty, and this led to the development of ever thicker armor plating to protect soldiers and iron-hulled ships to protect navies. The immediacy, proximity, and valor that characterize dueling cannot be retained in an industrial age. Indiscriminate killing becomes slowly the norm. Likewise, the horrors of the Crimean War led to the invention of field ambulances, field hospitals, and the Red Cross. The Crimea was also the site of the fatal Charge of the Light Brigade and the beginning of the modern disenchantment with war. Pacifism was an invention of the nineteenth century, he says, but most have taken the wrong lesson from this:

> Others, so-called friends of humanity, congratulate themselves at the sight of weapons and machinery of war keeping abreast of the progress of industry and becoming more and more murderous. War, they say, will peter out on account of the excesses of its destructive capability. They fail to see that ending war that way leads to political and social disorganization [from pluralism to monopoly]. Once weapons have reached the point where numbers and discipline, as well as courage, no longer mean anything in warfare, it is farewell to majority rule, farewell to universal suffrage, farewell to the empire, farewell to the republic, farewell to government of any form. It will be power to the most villainous.
>
> *(Proudhon, 2022, pp. 282–283)*

Proudhon is of course anticipating the emergence of total war and totalitarianism in the twentieth century, which would shape the thought of Luxemburg, Sturzo, Anscombe, and Niebuhr, all studied later in this volume. But is peace the answer, he asks? Exploring this brings us to the fourth and final contribution Proudhon makes to the just war tradition. While peace is perhaps only tangentially the object of just war theory, summarizing Proudhon's account of peace and how to achieve it is useful in so far as it illustrates why he thought war couldn't be constrained, mollified, or tamed. Proudhon does not think war can be abolished by fiat or law, nor is he a pacifist. But he is of the view that unless we transform the right of force into something less violent, life in the industrial age will be a very bloody hell.

Proudhon was an advocate of what we today call transformative justice, the notion that given crime is a social construct, then transforming the individual and society is key to solving the problem of crime. Without the rehabilitation of society and the individual, the social causes of crime are untouched. By analogy, he says that if "universal pacification" is the end, then political economy is the means (Proudhon, 2022, p. 448). You cannot abolish war or incarcerate war criminals and leave the conditions which caused the violence untouched. "Transforming" war, he argues, involves transplanting the warrior spirit into industry, where the urge to production and destruction can be given less deadly outlets. Similar processes are the hallmark of social or civilizational changes. The violence of chattel slavery was sublimated or transformed into wage slavery, national animosities transformed into national sport, and the transformation of autocratic and absolutist violence into the (Weberian) philosophy of state sovereignty, for example.

But none of these will bring a lasting peace unless the problems of material inequality are addressed. Political economy is key, he argues. Proudhon makes a distinction between "poverty" and "pauperism", where the former is an ascetic state of contentment with the basics of a good life, and the latter is a condition of destitution caused by structures of social domination and expropriation. Fix the latter, while aiming for the former, and society will be pacified, he argues. Today we call this "de-growth". The route to that end point goes through empowering people to work, to labor, and to produce, to unleash their creativity, and to provide the basics of life to one another, communally.

The absence of this harmonious state of affairs is caused by what Proudhon called a "RUPTURE OF THE ECONOMIC EQUILIBRIUM" (Proudhon, 2022, p. 347 All caps in original). This rupture is caused by the exploitation of one group for the ease and comfort of another; it is unsustainable and ultimately ends in conflict, whether that is the American Civil War or European Imperial conflict in the Crimea. In antiquity, war was the plundering of one people for the benefit of another. Alexander the Great was the first to transform this plundering into conquering and ruling from afar or by proxy. With the emergence of territorial conquest and direct rule, the doctrines of state sovereignty and private property, standing armies, conscription, and the professionalization of the military, it became necessary to defend gains from the outside and from internal threats. In these new garrison states, violence is monopolized by the elite, and society is pacified for fear the privileges of the few will be seized by the many. Proudhon coined the term "*militarisme*" for this conjoining of the military and the state and the term "*governmentalisme*" to reflect the expansion of the modern ideology of government as a means of pacifying society (for more, see Prichard, 2022a, pp. 34–36).

But each of these phases was ultimately a battle for post-war reparations or what Proudhon calls "indemnities" (2022, p. 451). As is well known, war was central to the creation of money, mainly through the production of public debt and bonds. Likewise, Proudhon was quite clear that colonialism was driven by a parasitic imperative to address domestic economic contradictions, though he was disappointingly silent over the fate of the Algerians, unlike, for example, Fanon, who is studied later in this volume, for whom violence served a different tool (see Zouache, 2016).

Legacy

The reception of *War and Peace* was muted in its time, but its arguments have echoed quietly through European social theory to this day, not only on the left but regrettably (but perhaps unsurprisingly) on the far right too. Thankfully, right-wing uptake of Proudhon's ideas passed away with the Vichy regime, literally with the passing of Henry Moysett in 1948. Moysett was the editor of the 1927, collected works edition of *War and Peace*. He taught at the French military and naval academies and then served in Marshal Pétain's government as minister for labor between 1941 and 1942. He was subsequently charged with treason.

Both E.H. Carr (1950) and Schapiro (1945) argued that Proudhon's *War and Peace* was a progenitor of fascism, but this is a highly anachronistic reading that ignored Proudhon's anti-statism. As Edward Castleton has shown (2017b, pp. 232–242), Proudhon was familiar with Darwin's arguments, and they both agreed that there was no necessary telos to the evolution of society or to the struggle for existence. For both Proudhon and Darwin, social evolution is "polygenetic": it has multiple starting points, with complex intermingling of species, meaning there is no grand design in nature and no superior people.

The more salubrious, pacifist, and prefigurative tones of *War and Peace* echoed through to Tolstoy, who spent some time with Proudhon in April 1861, just as he was finalizing the drafts of the serialized version of *War and Peace* for publication. As Boris Ėĭkhenbaum (1982, pp. 175–194) has shown, Proudhon's impact on Tolstoy was significant. Tolstoy's arguments in the appendix to *War and Peace* about the relationship of war to the philosophy of history and his denunciation of the "Great Man" thesis of history are distinctly Proudhonian. As I have speculated elsewhere, it is also likely that Proudhon's knowledge of the Crimean campaigns came first hand from Tolstoy too (Prichard, 2022a, pp. 27–30). *War and Peace* also shaped Raymond Aron's ideas about the ethics of war (Aron 1966, pp. 600–606) and French sociology of law (Gurvitch, 1947), and the most recent French edition of *War and Peace* is pitched as a contribution to Bergsonian phenomenology (Panero, 2012). Nevertheless, *War and Peace* remains unjustifiably ignored by Anglo-American scholarship, especially students of just war theory. Much more work is needed to bring this book in from the cold.

Conclusion

If we were to ask Proudhon which wars are justified and how we should fight them, it is likely he would answer by saying they are all just in the combatants' eyes. Ethical discourse reflects and only marginally constrains the exercise of military violence. The study of the ethics of war ought to lead, he suggests, to the study of political economy. Until society is pacified on the basis of a structural balance of power that gives the people their due, structural inequalities will be challenged through the exercise of the right of force, within and between peoples. It is through social science and the theory of rights that durable and persuasive rules to govern the use of force can be brought within reach.

Notes

1 Anarchism is an anti-capitalist and anti-statist political philosophy that puts mutual aid at its heart. Anarchy, on the other hand, can be defined as the absence of a final point of authority (Prichard, 2022b).
2 This is the sort of moral truism common to Chomsky's (2003) criticisms of the just war tradition.

Works Cited

Aron, Raymond. 1966. *Peace and War: A Theory of International Relations*. Translated by Richard Howard. Edited by Annette Baker Fox. New York: Doubleday.
Carr, Edward. H. 1950. Proudhon: The Robinson Crusoe of Socialism. In E. H. Carr. ed. *Studies in Revolution*. London: Macmillan, pp. 38–55.
Castleton, Edward. 2017a. Pierre-Joseph Proudhon's War and Peace: The Right of Force Revisited. In Béla Kapossy, Isaac Nakhimovsky, and Richard Whatmore. eds. *Commerce and Peace in the Enlightenment*. Cambridge: Cambridge University Press, pp. 272–299.
Castleton, Edward. 2017b. Une Anthropologie téléologique: Fins et origines des peuples et des hommes selon Pierre-Joseph Proudhon. In Vincent Bourdeau and Arnaud Macé. eds. *La Nature du socialisme: Pensée sociale et conceptions de la nature au xixe siècle*. Besançon: Presses Universitaires de Franche-Comté, pp. 197–242.
Chomsky, Noam. 2003. Commentary: Moral Truisms, Empirical Evidence, and Foreign Policy. *Review of International Studies* 29(4), pp. 605–620.
Eïkhenbaum, Boris. 1982. *Tolstoi in the Sixties*. Translated by Duffield White. Ann Arbor: Ardis, pp. 175–194.
Ferretti, Federico and Edward Castleton. 2016. Fédéralisme, identités nationales et critique des frontières naturelles: Pierre-Joseph Proudhon (1809–1865) géographe des 'États-Unis d'Europe'. *Cybergeo: European Journal of Geography*. Available at: https://doi.org/10.4000/cybergeo.27639.
Gurvitch, Georges. 1947. *Sociology of Law*. London: Routledge and Kegan Paul Ltd.
Jun, Nathan. 2020. Anarchism and Just War Theory. In Luis Cordeiro-Rodrigues and Danny Singh. eds. *Comparative Just War Theory: An Introduction to International Perspectives*. London: Rowman and Littlefield, pp. 11–30.
Panero, Alain. 2012. Avant-Propos: D'une phénoménologie inédite de la guerre à une nouvelle éthique du regard. In Pierre Joseph Proudhon. ed. *La Guerre et la Paix*. Édition critique d'Alain Panero. Paris: Nuvis, pp. 5–74.
Prichard, Alex. 2013. *Justice, Order and Anarchy: The International Political Theory of Pierre-Joseph Proudhon*. Abingdon: Routledge.
Prichard, Alex. 2022a. Introduction. In Pierre-Joseph Proudhon. ed. *War and Peace*. Oakland: AK Press, pp. 1–36.
Prichard, Alex. 2022b. *Anarchism: A Very Short Introduction*. Oxford: Oxford University Press.
Proudhon, Pierre-Joseph. 1862. *La Fédération et l'Unité en Italie*. Paris: E. Dentu.
Proudhon, Pierre-Joseph. 1863. *Du Principe Fédératif et de la Nécessité de reconstituer le parti de la révolution*. Paris: Ernest Flammarion.
Proudhon, Pierre-Joseph. 1864. *Si Les Traités de 1815 ont Cessé d'Exister*. Paris: E. Dentu.
Proudhon, Pierre-Joseph. 1865. *Nouvelles Observations sur l'Unité Italienne*. Paris: E. Dentu.
Proudhon, Pierre-Joseph. 1868. *France et Rhin, Oeuvres Posthumes de P. J. Proudhon*. Paris: A. Lacroix, Verboeckhoven et Companie.
Proudhon, Pierre-Joseph. 1990 [1988]. *De la Justice dans la Révolution et dans l'Église: Études de philosophie pratique*. 4 vols. Paris: Fayard.

Proudhon, Pierre-Joseph. 2022 [1861]. *War and Peace: On the Principle and Constitution of the Rights of Peoples*. Edited and Introduced by Alex Prichard. Translated by Paul Sharkey. Oakland: AK Press.

Proudhon, Pierre-Joseph. 1994. *What is Property?, Or, An Inquiry into the Principle of Right and of Government*. Cambridge: Cambridge University Press.

Schapiro, J. Salwyn. 1945. Pierre-Joseph Proudhon, Harbinger of Fascism. *The American Historical Review* 50(4), pp. 714–737.

Vincent, K. Steven. 1984. *Pierre-Joseph Proudhon and the Rise of French Republican Socialism*. Oxford: Oxford University Press.

Zouache, A. 2016. Proudhon et la Question coloniale algérienne. *Revue économique* 67(6), pp. 1231–1244.

9

CARLOS CALVO (1824–1906)

Pablo Kalmanovitz

Introduction

Carlos Calvo was by far the most influential Latin American international jurist of the nineteenth century. The "Calvo doctrine" and "Calvo clause" are arguably the best-known and most significant Latin American contributions to public international law in the late nineteenth and early twentieth centuries. In a nutshell, the doctrine holds that private claims of wrong by foreign citizens should normally be adjudicated in national courts, not through diplomatic means nor by military intervention or threats thereof. The Calvo clause is an application of the doctrine to contractual practice. It was often included in contracts between states and private actors to the effect that domestic remedies must be exhausted before invoking diplomatic protection for alleged breaches of contract. Many national constitutions in Latin America have mandated that such clauses be included in contracts (Shea, 1955; Tamburini, 2002; Greenman, 2018).

While usually remembered as a prominent defender of jurisdictional nationalism and sovereign equality, Calvo also made key contributions to the legal doctrines of *jus ad bellum* and state sovereignty, as this chapter endeavors to show. As a publicist of international renown in the late nineteenth century and follower of canonical intellectual figures in the history of the law of nations, Calvo belongs to the "regular war tradition" within the broad history of just war thinking (Kalmanovitz, 2015, 2020, pp. 69–126). Most importantly, he articulated an account of *jus ad bellum* more restrictive than European and U.S. practice towards Latin America at the time. To that effect, he pushed creatively and strongly, both intellectually and diplomatically, to outlaw resort to state force to collect private debts abroad or to claim compensation for damages suffered by subjects during social unrests.

DOI: 10.4324/9781003428688-10

Calvo interrogated critically the conditions under which alleged wrongs to foreigners could be treated as injuries to their states—even serious enough to constitute a *casus belli*—and in the process theorized the nature of state responsibility and of acts and wrongs attributable to states. Furthermore, he famously contributed to build the notion that Latin American states had legitimacy and authority equal to their European counterparts, as a consequence of which conflicts over private claims by foreigners should be adjudicated in national courts. For Calvo, as for other thinkers in the regular war tradition, when a political association had sovereign standing—that special form of legitimate authority—it had special rights and obligations vis-à-vis other sovereign states. In the situation that most interested Calvo, sovereign states were not liable to attack for alleged wrongs caused to foreigners in their territory; they had the sovereign right to adjudicate such disputes in national courts. It was precisely because the young Latin American republics had sovereign standing that other states must defer to their jurisdiction, instead of themselves adjudicating unilaterally and enforcing militarily.

In the 1860s, when Calvo began publishing his writings and treaty compilations, the legality of military intervention to do justice to subjects abroad was contested and politically acute. For the law of nations of the eighteenth century, it had been settled doctrine that states could intervene when another state failed to protect or do justice to their subjects. The obligation to protect subjects inside and outside state territory was—and still is—a characteristic feature of sovereignty. If the wrong was sufficiently serious, it could amount to a lawful cause for war. At least since Hugo Grotius, furthermore, effective protection and the proper administration of justice were understood to be central responsibilities of sovereign states, indeed what differentiated them from "bands of brigands" (Kalmanovitz, 2020, pp. 62–65).

This understanding of state responsibility came to be revised after the French Revolution and subsequent revolutionary upheavals in nineteenth-century Europe. As Calvo set out to write in Paris in the 1860s, European state practice was moving towards *not* recognizing the responsibility of states for private losses suffered during social unrests or revolutions. By contrast, through the nineteenth and early twentieth centuries, it was common practice in Latin America that private claims of wrong raised by French, British, and North American nationals escalated into diplomatic crises. State military responses went from port blockades to ground invasions, to military occupation and regime change in the extreme case of France's 1861 intervention in Mexico.

Texts and Tenets

Carlos Calvo was born in Montevideo, Uruguay, in 1824, to a family of successful traders in the River Plate region. He moved to Argentina at an early age and studied law in Buenos Aires. At 28, he began his diplomatic career as Argentina's consul to Montevideo, thanks to the political connections of his wealthy family. In 1860, he moved to Europe to represent Paraguay as *chargé d'affairs* in London and Paris. The success of this diplomatic mission, along with Calvo's political skills and tact, gave him a level of access to European diplomatic and intellectual circles very rare for Latin Americans at the time. He settled in Europe for life, where he served in high-level diplomatic posts for Argentina, including minister in Paris, Berlin, and the Holy See, as well as mentor to young Latin American students and diplomats in Europe (Marichal, 2015; Tamburini, 2002; Davis, 2021).

In the late 1860s, Calvo became a prominent member of the circle of professional international lawyers coming together around Brussels and Paris. He was one of two non-Europeans founding members of the *Institut de Droit International* (the other was from the United States), which was the most important organization during the formative years of the international legal profession (Becker-Lorca, 2014, p. 2; Koskenniemi, 2002). The first number of the main publishing outlet of the *Institute*—the influential *Revue de droit international et de législation comparée*—first published in French Calvo's case against foreign interventions to collect private debts (Calvo, 1869).

Thanks to generous funding from the Paraguayan and Peruvian foreign offices, while in Paris, Calvo undertook two vast treaty compilation projects that sought to incorporate the Latin American region into the European scholarly and practical field of international law (Marichal, 2015, pp. 721–728). To show how embedded the Latin American region was in European legal and diplomatic practice, Calvo followed as model G.F. von Martens' historical compilations of European treaties, which by the mid-nineteenth century had become the most authoritative reference in international law (Koskenniemi, 2008). But while Martens had focused nearly exclusively on European legal relations, Calvo undertook an analogous task for Latin America, from the times of the Spanish and Portuguese conquests to the present, compiling relations among European empires before Latin American independence and among the independent republics, European powers, and the United States.

Calvo had a team of assistants search and copy relevant documents in the archives of the British Museum, the Imperial Library in Paris, and the Archive of the Indies in Sevilla. From 1862 to 1866, they produced Calvo's "complete collection of the treatises, conventions, capitulations, armistices, and other diplomatic acts of all the states of Latin America, from the Gulf of Mexico to the Cape of Horn, from 1493 to the present" (it actually ran until 1825) which was edited in 11 volumes and published in Spanish and French in Paris, Buenos Aires, and Madrid. The collection included, in addition to treaties and diplomatic memoranda, current Latin American statistics on population, international trade, geography, and more. The second compilation was the "historical annals of the revolution of Latin America, with supporting documents, from 1808 to the recognition of the independence of that vast continent," which was published in five volumes between 1864 and 1867, also in Paris in both Spanish and French.

In 1864, Calvo published his first monograph in Spanish, *La América del Sur ante la Ciencia del Derecho de Gentes Moderno* [South America before the science of the modern law of nations] (Calvo, 1864), which is an expanded memoir of his work as Paraguayan representative in London and Paris. It features prominently detailed analyses of spurious claims raised by North American, British, and French nationals in Argentina, Uruguay, and Paraguay, all of which escalated to diplomatic disputes and armed interventions. In 1868, Calvo published in Paris the first edition of his legal treatise and *magnum opus*, *Derecho Internacional Teórico y Práctico de Europa y América* (Calvo, 1868), the French version of which was published in 1870 and subsequently revised and vastly expanded in five editions until 1896. The book made Calvo's name in European and Latin American diplomatic, intellectual, and legal circles and contains the fullest articulation of the Calvo doctrine.

Contexts

Contemporary scholarship on Calvo has emphasized his outstanding role in the "semi-peripheral" appropriation and redeployment of the European law of nations (Becker-Lorca, 2014; Obregón, 2006a, 2006b). He was semi-peripheral in the sense of being a representative from the outskirts of the European world system that consolidated in the late nineteenth century. He simultaneously identified with European law, values, and standards and affirmed semi-peripheral differences. Like other international Latin American intellectuals of the nineteenth century, notably Andrés Bello and José María Torres Caicedo, Calvo sought to leverage European international law in order to defend the sovereign standing of the emerging Latin American republics, partly through an appeal to recognize their distinctive state-building processes (Mantilla Blanco, 2021). Their position towards the inherited corpus of the European law of nations was distinctively ambivalent.

As a descendant of Europeans, Calvo took himself to be part of the metropolitan centre and identified with European cultural norms, standards of civilization, and legal professionalism. But, on the other hand, he sought to challenge the metropolitan centre "with notions of [Latin America elites'] own regional uniqueness (as natives of America)" along with their own civilized and civilizing practices (Obregón, 2006a, p. 817). Like other nineteenth-century semi-peripherals, Calvo "did not perceive international law as coming from Europe, as a foreign and distant model, but rather as part of their own Spanish tradition, which connected them not only to modern European international law but also to Roman law" (Obregón, 2006a, p. 824). And yet, this self-perception was contested in Europe and North America. In the second half of the nineteenth century, European international law became a global hegemonic discourse that in principle excluded non-Europeans and the uncivilized from the society of states. Facing contestation about the status of Latin American republics, Calvo understood that mastery of European legal language and doctrine was necessary for defending the political standing of the young Latin American republics. But Calvo's mastering of the European legal canon was not a passive form of reception but rather a "reinterpretation of rules" that simultaneously sought to earn the inclusion of Latin America in the society of civilized states and be recognized as different (Becker-Lorca, 2014, pp. 42–44). In an intellectual jiu-jitsu of sorts, he sought to turn European norms around against European practices in Latin America while affirming the legitimacy, distinctiveness, and standing of the new states (Becker-Lorca, 2014, p. 66).

To make this argument, Calvo had to face a hard trade-off. On the one hand, European powers and the United States were often ready to abuse diplomatic protection rights and to disqualify *a priori* the ability of Latin American states to administer justice. As Calvo was keen to show in detail, unilateral action was often taken in pursuit of spurious claims by reckless adventurers. On the other hand, however, the emerging Latin American states were in fact often incapable of securing the investments, goods, integrity, and lives of foreigners in their territory. National courts were not always capable of delivering justice, particularly not during civil wars. In some cases, therefore, diplomatic protection could be justified. Moreover, flatly refusing to recognize the sovereign right of diplomatic protection would dissuade potential immigrants and investors from providing much-needed skills and capital. This was a fundamental consideration for Calvo, for whom the free flow of European migrants, international trade, and foreign investment were essential in the construction of viable Latin American republics (Davis, 2021). Therefore, both the private interests of foreign investors and the public interest and authority of the new republics had to be simultaneously promoted and advanced, even if they sometimes appeared to pull in opposite directions.

This tension between national jurisdiction and diplomatic protection set the stage for Calvo's reformulation of the doctrine of *jus ad bellum*. He argued and abundantly illustrated that coercive interventions undermined both private and public interests in the long run. Crucial in this intervention was Calvo's close familiarity and insight into the effects of military interventions by naval powers in Latin America. Such interventions not only were unjust and caused material losses to fragile Latin American states but also strengthened domestic factions, toppled governments, and undermined in multiple ways the project of state building in the region. And without functional states, Calvo insisted, the expansion of finance, European populations, and global trade would be impossible.

Controversies

As noted earlier in the chapter, the proposition that states could intervene coercively in order to protect their nationals abroad was well established in the European law of nations of the eighteenth century. When Calvo made his case for the exclusion of these interventions, he engaged extensively with the writings of Emer de Vattel, the most influential eighteenth-century European authority in the law of nations (Kalmanovitz, 2020, pp. 69–78). At the beginning of Calvo's chapter on "the reciprocal obligations of state" in his 1868 legal treatise, where the central case of the doctrine is first made, Vattel is cited to introduce the crucial distinction between perfect and imperfect state rights. While perfect rights are "accompanied by the right of compelling those who refuse to fulfil the correspondent obligation," imperfect rights are not.[1] In this language, the question for Calvo was whether damages caused to foreigners can create perfect obligations in states. In this regard, he wrote, Vattel was the one who "discussed at greatest length the extent to which states are responsible for the acts of their private citizens" (Calvo, 1868, §§293, p. 295).[2]

Singularly important for Calvo was Vattel's chapter on "the Concern a Nation may have in the Actions of her Citizens" in *Le Droit des Gens*, which discusses state responsibility for damages caused to foreigners. Vattel rightly took this to be part of the larger theme of "what share a state may have in the actions of her citizens, and what are the rights and obligations of sovereigns in this respect" (II.71, p.298). While Calvo duly claimed to be faithful to Vattel, he in fact omitted key elements in Vattel's discussion and creatively reinterpreted elements that were marginal or ambiguous.

Vattel begins his account of state responsibility for damages to aliens with the following categorical claim, which was hardly auspicious for Calvo:

> Whoever uses a citizen ill, indirectly offends the state, which is bound to protect this citizen; and the sovereign of the latter should avenge his wrongs, punish the aggressor, and, if possible, oblige him to make full reparation; since otherwise the citizen would not obtain the great end of the civil association, which is safety.
>
> *(II.71, p. 298)*

It is a fundamental responsibility of states to prevent their subjects from injuring foreign subjects. To deny this would be to undermine the foundations of international society, for: [I]f you let loose the reins to your subjects against foreign nations, these will behave in the same manner to you; and . . . we shall see nothing but one vast and dreadful scene of plunder between nation and nation. (II.72, p. 299)

If wrongs are committed against foreign nationals, states must ensure that justice is done and "compel the transgressor to make reparation for the damage or injury . . . or inflict on him an exemplary punishment" (II.76, p. 299).

Importantly for present purposes, according to Vattel, "the sovereign who refuses to cause reparation to be made for the damage done by his subject . . . renders himself in some measure an accomplice in the injury, and becomes responsible for it" (II.77, p. 300). In the extreme,

> [T]he princes whose subjects are robbed and massacred, and whose lands are infested by those robbers, may justly level their vengeance against the nation at large. . . . All nations have a right to enter into a league against such a people, to repress them, and to treat them as the common enemies of the human race.
>
> *(II.78, p. 301)*

Such measures, Vattel explained, could be permissibly taken when a state, "by its manners and by the maxims of its government . . . accustoms and authorises its citizens indiscriminately to plunder and maltreat foreigners, to make inroads into the neighbouring countries etc" (II.78, p.301). In the case of "savage nations," furthermore, the "nation at large," that is every member of the state, could be punished for it.

Calvo duly referenced and emphatically agreed with all these quotes from Vattel (Calvo, 1868, §296). But in reality, Calvo's deepest fear was precisely that post-independence Latin American states be de-ranked in such a dramatic way. For in Latin America, some European aliens did truthfully claim to have been "indiscriminately plundered and maltreated," and Argentina's dictator Rosas did send troops to Uruguay during its civil war, that is, "made inroads into neighbouring countries" among other wrongs.

Helpfully for Calvo, however, Vattel was characteristically ambivalent in introducing two caveats in his discussion. First, he wrote, "if a sovereign, *who might keep his subjects within the rules of justice and peace*, suffers them to injure a foreign nation . . . he does no less injury to that nation, than if he injured it himself" (II.72, p. 299, emphasis added). It arguably follows that, if a sovereign *cannot* keep his subjects within the rules of peace and justice, he may not be held responsible for the wrongs they commit. Put differently, states should not be held strictly liable for their subjects' wrongs but rather be subject to some sort of due diligence standard. Second, Vattel wrote, "as it is impossible for the best regulated state . . . to model at his pleasure all the actions of his subjects . . . it would be unjust to impute to the nation or the sovereign every fault committed by the citizens" (II.73, p.299). It would certainly be unreasonable to make every single fault of foreigners a cause for international conflict; injuries to aliens would have to be manifest, significant, and consequential enough to merit intervention.

Vattel did not discuss the relevance of state capacity for state responsibility, and the common-sense proposition that not every private claim of injury should be escalated into a diplomatic crisis does not tell us which private claims could be permissibly escalated. And yet Calvo quoted these qualifications in support of the claim that the question of whether states can be held responsible for injuries caused to foreigners during civil wars had been authoritatively settled in the negative. "Almost every publicist," Calvo concluded, "is in agreement with the basics of the doctrine" (Calvo, 1868, §293).

The reality, however, is that Vattel had by no means settled the question; his argument overall operates rather against Calvo's conclusion. There was no consensus on the matter neither at the time of Calvo's writing nor for decades afterwards (Shea, 1955; Tamburini, 2002; Greenman, 2021), although Calvo's prominent interventions in European legal debates were certainly *meant* to settle it.

In addition to his creative reconstructions of Vattel, Calvo produced his own lines of argument against foreign interventions. He began his discussion in *Derecho Internacional Teórico y Práctico* by highlighting the practical significance of the question whether states could be held accountable for damages to aliens caused by factions during civil wars. The answer to this question, he wrote, "affects not only the international rights of states, but also each people's own exclusive and particular legislation" (Calvo, 1868, §291, p. 387). The very foundations of state sovereignty were at stake.

Calvo produced three separate but complementary lines of argument. The first concerns the odious privileges that resulted if interventions based on claims from citizens abroad were accepted; the second is an argument from duress; and the third consists in citing an important number of precedents and pronouncements drawn from European state practice. In characteristic semi-peripheral fashion, Calvo's three-fold legal argument was meant to leverage the authority of the European law of nations and the standard of civilization as constraints on the use of European force.

In the first argument, Calvo denounced two kinds of privileges. First and foremost, if the right to intervene unilaterally is recognized in international law, powerful states would be favoured over weaker ones because they have far more compelling means of force, notably naval forces that could blockade ports essential to a nation's subsistence. Consequently,

> [T]he more powerful states would never listen to reclamations made by weak states, which by contrast would have to attend and comply with theirs, as has happened more than once with compensations demanded by some European governments to South American states.
>
> *(Calvo, 1868, §291, p. 388)*

A perfect right to secure justice for nationals abroad may be effective in the hands of powerful states, but it would be non-existent for weak states. Far from advancing justice, such right would only accentuate power differences among states. It results in the rule of armed force, not the rule of law, and should therefore be condemned and renounced by civilized nations. If we recognize, with Vattel, the rule of law and sovereign equality as foundational principles of international law, then the right of unilateral intervention must be rejected.

The kind of situation to which this argument refers was familiar to Calvo from his work as representative of Paraguay. His 1864 memoir made a point of discussing in detail cases of spurious claims raised by foreign nationals, which resulted in naval blockades and resort to force by Great Britain, France, and the United States (Calvo, 1864). The main case in the memoir, which Calvo called the "*cuestión Canstatt,*" involves a dual Uruguayan and British citizen who was accused of plotting to kill Paraguayan president Solano López. British diplomats confronted Paraguay for Canstatt's detention and demanded his release. Upon Paraguay's refusal, a diplomatic crisis escalated to the point of two British frigates firing against president López's official vessel when leaving Buenos Aires after a diplomatic visit.

Also paradigmatic in Calvo's memoir was the case of U.S. consul Eduard Hopkins, who in 1853 arrived in Asunción, Paraguay, to represent U.S. interests and carry on his own private business. Eventually facing financial failure, Hopkins alleged that his investments had been arbitrarily undermined by Paraguayan authorities and demanded compensation (Calvo, 1864, pp. 13–18). In reaction to Paraguay's refusal, together with an incident involving a U.S. steamboat venturing into the Paraná River being fired at by Paraguayan forces, U.S. president Buchanan sent a naval force of 20 frigates and 2,000 troops, along with a special negotiator in charge of "doing justice" to Hopkins and the owners of the damaged steamboat (Calvo, 1864, p. 24). Eventually, there was a peaceful settlement but of course under the shadow of the U.S. military threat and to Paraguay's disadvantage.

Not included in the 1864 memoir but important for Calvo and discussed at length in his 1868 legal treatise was the invasion of Mexico in 1861 by the triple alliance of Spain, France, and Great Britain. Calvo called it the "*cuestión Jecker,*" and its roots were on president Benito Juarez's temporary suspension of foreign debt payments after a financially disastrous civil war (Calvo, 1868, §§87–91). Among the standing claims was Mr. Jecker's, a Swiss national who loaned a significant sum to general Miramón, Juárez's predecessor and enemy during the civil war, who used the funds to unsuccessfully fight Juárez.

The intervention of the triple alliance began with an ultimatum and the following war manifesto:

Holy debts recognized by treaties have ceased to be honoured. . . . Such state of affairs must put the allied governments in the regrettable case of demanding, not just reparations for the past, but also guarantees for the future. . . . We must point Mexico to the path that leads to its happiness.[3]

In addition to doing justice to foreign investors, the Alliance sought to stabilize the political situation in Mexico after the civil war. Jointly with conservative Mexican elites, France propelled Maximilian of Habsburg to be crowned as emperor of the newly created Mexican Empire, an enterprise that did not last long nor end well for Maximilian. Of all the interventions in Latin America, this was for Calvo the most brazen and contrary to sovereignty. The allies' demand of guarantees for the safety of their nationals, "left to the free will and supreme decision of [the allied] powers, the kind of government that would be established in Mexico" (Calvo, 1868, §89, p. 174).[4]

The second form of odious privilege denounced by Calvo belonged not to states but to foreigners abroad in relation to their national counterparts (Calvo, 1869, p. 417). The practices Calvo denounced gave foreigners the power to ignore national jurisdictions and appeal directly to their foreign offices for diplomatic protection. Nationals did not have an equivalent opt out option, of course, and this for Calvo constituted a "profound attack against one of the constitutive elements of state independence, which is territorial jurisdiction" (Calvo, 1869, p. 417). The sort of private claim typically raised in such cases should be adjudicated in domestic courts; diplomatic protection could be activated only in cases of manifest denials of justice.

Third, Calvo made an argument from duress that turned on the logic of fairness in attributions of responsibility. As he put it,

> Far from us the idea of not recognizing the titles of claims based on the rule of common law [*règle de droit commun*] according to which every person must compensate for the damages she causes. This principle is applicable in normal times and under ordinary circumstances, but could one logically think to extend it to *serious situations of force majeure*, which overthrow a whole order of established things and often take a country to the brink of abyss? These situations appear to us essentially different.
>
> *(Calvo, 1869, p. 422 emphasis added)*

Note how careful Calvo is of not giving the impression of denying the existence of state responsibility or the sovereign right of diplomatic protection. He again quoted Vattel approvingly to the effect that "the sovereign who refuses to cause reparation to be made for the damage done by his subject becomes responsible for it" (II.77, p. 300). His point, rather, was that civil unrest should count as a form of duress which excluded the responsibility of an otherwise responsible agent (Calvo, 1868, §295).

The "serious situations of *force majeure*" to which Calvo referred in the passage quoted were drawn from European practice, which after the

Congress of Vienna often agreed with him. He discusses a large number of cases to illustrate how European states denounced interventionist practices as unfair and unduly intrusive on sovereignty. When Calvo quoted Vattel to the effect that not every wrong should be the cause for diplomatic complaint, he was also referencing a long commentary inserted by Paul Pradier-Fodéré to the 1863 edition of *Le Droit des Gens,* which concluded with the assertion that "we can consider settled in the law of nations the rule that legitimate governments are not liable to pay to foreigners compensation for damages and losses caused by rebels; foreigners must not have more rights than nationals" (Vattel and Pradier-Fodéré, 1863, p. 51). Calvo sought to leverage this understanding of mid-nineteenth century European practice and extend it to the Latin American region.

Legacy

Calvo's interventions in international law had multiple and variegated effects on the law and practice of resort to force in the late-nineteenth and early-twentieth centuries. Most famously, the Calvo doctrine—and Calvo's own personal involvement—propelled the Argentinean foreign minister Luis María Drago to denounce the 1902 British and German bombardment of ports and naval blockade of Venezuela. As in Mexico half a century earlier, the attack was triggered by general Cipriano Castro's failure to honour public debt following a destructive civil war. In reaction to the blockade, Drago published a diplomatic note addressed to the U.S. government which made an elaborate case for the illegality of coercive collections of debt. The argument incorporated the main tenets of the Calvo doctrine, on which basis Drago called for the activation of the defensive declaration in the Monroe Doctrine (Nettles, 1928).

The United States was unpersuaded, but Drago reintroduced the issue at the Pan-American Conference of 1906. This time strongly supported by U.S. diplomats, the Conference recommended that the 1907 Peace Conference at the Hague examine "the question of compulsory collection of public debts and the best means tending to diminish among nations conflicts of purely pecuniary origin," which representatives at the Hague duly did (Nettles, 1928, p. 207). Negotiations at the Hague led to article 1 of the Convention Respecting the Limitation of the Employment of Force for the Recovery of Contract Debts in which "the contracting Powers agree not to have recourse to armed force for the recovery of contract debts claimed from the Government of one country by the Government of another country as being due to its nationals." The agreement, however, excluded cases of refusal, neglect, or obstruction of arbitration procedures. It was a victory not for national jurisdiction but for international arbitration, but it resulted in a more restrictive norm of *jus ad bellum* that in effect barred unilateral foreign interventions in response to "pecuniary disputes."

Arguably the highest diplomatic point of Drago and Calvo's influence was the inclusion of Article 9 in the Montevideo Convention on the Rights and Duties of States (1933), according to which "nationals and foreigners are under the same protection of the law and the national authorities and the foreigners may not claim rights other or more extensive than those of the nationals." In virtue of this article, the standard of due diligence, which excludes state responsibility for damages to aliens, is the same level of protection received by nationals in domestic courts. Foreigners could not expect to be treated better or as in their own national courts, nor according to some objective standard of diligence.

The Seventh Pan-American Conference in Montevideo, which produced the Convention, has been hailed as an "immense accomplishment" for Latin American international lawyers and diplomats in its emphatic affirmation of the principle of non-intervention and the introduction of a formal conception of statehood that eventually replaced the substantive standard of civilization (Becker-Lorca, 2014, pp. 306–307).[5] Several articles in the Convention came to be widely recognized as expressions of customary international law, but not article 9. Contestation over its content was manifest during the first Codification Conference at the League of Nations (1930) which unsuccessfully attempted to codify state responsibility for wrongs to foreigners. No other codification attempt was ever made. Through the 1930s, the positions of the United States. and Latin American republics proved irreconcilable, with the United States advocating for an objective standard of due diligence and Latin Americans for the subjective standard of the Montevideo Convention (Greenman, 2021, pp. 157–173; Shea 1955, pp. 56–61).

In parallel to these international developments, and closely tied to them, the Calvo doctrine had a remarkable impact on Latin American constitutions. Between 1880 and 1930, most Latin American republics inserted in their constitutions articles to the effect that, before invoking diplomatic protection, foreigners must exhaust remedies in domestic courts, which entitled them to the same levels of justice and protection as nationals (Garcia-Mora, 1950).[6] In virtue of these "Calvo clauses," which were also inserted in contracts between private foreigners and public authorities, arbitration tribunals in the 1920s began declining cases in which petitioners had not exhausted local remedies (Shea, 1955, pp. 231–256).

If we consider, to conclude, the extent to which current international law aligns with Carlos Calvo's views, a mixed picture results. Regarding the *jus ad bellum* dimension of his doctrine, the principle of *jus contra bellum* in the UN Charter has effectively banned the use of armed force to collect private debt, a move that Calvo and Drago would certainly have celebrated. As both were keen to point out, penalties for failure to honour debts and contracts can be internalized in financial markets; the use of force to exact payment is unnecessary and contrary to economic interests. Regarding the jurisdictional dimension, by contrast, the nationalist principle of jurisdiction has been largely eviscerated in the practice of foreign investment. Objective standards of due diligence are now well established in international arbitration law, and arbitration has become the standard forum for adjudication of transnational investment disputes (Dumberry, 2021, pp. 246–281). If under the Calvo clause, foreigners had to renounce their right to invoke diplomatic protection without first exhausting domestic remedies, nowadays states often renounce their jurisdictional rights via treaty and willingly submit to arbitration panels, the rulings of which are underwritten not by the threat of force but of exclusion from financial power.

Notes

1 Vattel, *Law of Nations*, Prelim. 17, 75. All citations are to the Liberty Fund Edition, indicating book, paragraph number, and page number when quoting. Calvo cites this passage in Calvo (1868, §289).
2 Unless noted otherwise, all translations from French and Spanish are my own.
3 Available at https://www.memoriapoliticademexico.org/Efemerides/1/10011862.html
4 Needless to say, there was no single instance Calvo could refer to of Mexican or Paraguayan nationals mistreated in Great Britain or France followed by coercive measures from Mexican or Paraguayan authorities.
5 Article 1 of the Montevideo Convention famously states: "The State as a person of international law should possess the following qualifications: (a) a permanent population; (b) a defined territory; (c) government; and (d) capacity to enter into relations with the other States."
6 Including the constitutions of Bolivia, Ecuador, Honduras, Mexico, Peru, Venezuela, Costa Rica, Cuba, and El Salvador (Marichal, 2015, p. 730).

Works Cited

Becker-Lorca, Arnulf. 2014. *Mestizo International Law: A Global Intellectual History 1842–1933*. Cambridge: Cambridge University Press.
Calvo, Carlos. 1864. *La América del Sur ante la Ciencia del Derecho de Gentes Moderno*. Paris: A. Durand.
Calvo, Carlos. 1868. *Derecho Internacional Teórico y Práctico de Europa y América*. vol. Tomo Primero. Paris: Durand et Pedone-Lauriel.
Calvo, Carlos. 1869. De la Non-responsabilité des États a Raison des Dommages Soufferts par des Étrangers en Cas d'Émeute ou de Guerre Civiles. *Revue de Droit International et de Legislation Comparée* 1.
Davis, Teresa. 2021. The Ricardian State: Carlos Calvo and Latin America's Ambivalent Origin Story for the Age of Decolonization. *Journal of the History of International Law* 23.
Dumberry, Patrick. 2021. *Rebellions and Civil Wars: State Responsibility for the Conduct of Insurgents*. Cambridge: Cambridge University Press.
Garcia-Mora, Manuel. 1950. The Calvo Clause in Latin American Constitutions and International Law. *Marquette Law Review* 33(4), pp. 205–219.
Greenman, Kathryn. 2018. Aliens in Latin America: Intervention, Arbitration and State Responsibility for Rebels. *Leiden Journal of International Law* 31(3), pp. 617–639.

Greenman, Kathryn. 2021. *State Responsibility and Rebels: The History and Legacy of Protecting Investment against Revolution*. Cambridge: Cambridge University Press.

Kalmanovitz, Pablo. 2015. Early Modern Sources of the Regular War Tradition. In Seth Lazar and Helen Frowe. eds. *The Oxford Handbook of Ethics and War*. Oxford: Oxford University Press, pp. 145–164.

Kalmanovitz, Pablo. 2020. *The Laws of War in International Thought*. Oxford: Oxford University Press.

Koskenniemi, Martti. 2002. *The Gentle Civilizer of Nations: The Rise and Fall of International Law, 1870–1960*. Cambridge, UK; New York: Cambridge University Press.

Koskenniemi, Martti. 2008. Into Positivism: Georg Friedrich von Martens (1756–1821) and Modern International Law. *Constellations* 15(2), pp. 189–207.

Mantilla Blanco, Sebastián. 2021. José María Torres Caicedo and the Politics of International Law in Nineteenth-Century Latin America. *American Journal of Legal History* 61(2), pp. 177–210.

Marichal, Carlos. 2015. El nacimiento de los estudios internacionales sobre América Latina: Comentarios a las obras de José María Torres Caicedo y Carlos Calvo a mediados del siglo XIX. *Foro Internacional* 3(221), pp. 707–736.

Nettles, Edward. 1928. The Drago Doctrine in International Law and Politics. *The Hispanic American Historical Review* 8(2), pp. 204–223.

Obregón, Liliana. 2006a. Between Civilisation and Barbarism: Creole Interventions in International Law. *Third World Quarterly* 27(5), pp. 815–832.

Obregón, Liliana. 2006b. Completing Civilization: Creole Consciousness and International Law in Nineteenth-Century Latin America. In Anne Orford. ed. *International Law and its Others*. Cambridge: Cambridge University Press, pp. 247–264.

Shea, Donald. 1955. *The Calvo Clause: A Problem of Inter-American Law and Diplomacy*. Minneapolis: University of Minnesota Press.

Tamburini, Francesco. 2002. Historia y Destino de la Doctrina Calvo. *Revista de Estudios Histórico-Jurídicos* 24, pp. 81–101.

Vattel, Emer de and Paul Pradier-Fodéré. 1863. *Le Droit des Gens*. Nouvelle Édition. Paris: Librairie de Guillaumin.

10

ROSA LUXEMBURG (1871–1919)

Owen Worth

Introduction

This chapter will look at Rosa Luxemburg's life and work and pay specific attention to her outlook on war, conflict, and revolutionary change. It will examine her wider position on violence and of whether her understanding of confrontation provided any justification of 'just' violence in any form. Her position on pacifism was unique in the sense that she rejected any form of violence, conflict, and war in the revolutionary transformation of power. Yet, she died violently in the aftermath of numerous skirmishes between right-wing paramilitary Freikorps, while allegedly supporting violent insurrection during the chaos of the end in Berlin at the end of World War I (Jones, 2016). This chapter will argue that Luxemburg's understanding of, and approach to, politics was the key to her ethical position on war and violence.

Rosa Luxemburg's understanding of war was highly significant in the early part of the twentieth century where she maintained the socialist opposition to war against the growing support for militarism among working-class parties in the lead up to World War I. During the war itself, she positioned herself as an opponent of the use of military force. Yet, despite this, she often appears missing from literature on the ethics of war. As this chapter will show, Luxemburg's pacifism, rooted within her unique understanding of revolutionary socialism (and thus different from Martin Luther King Jr., studied in a later chapter), provides a perspective on war that fuses the economic alongside the ethical in a manner that is quite unique and largely overlooked. It is perhaps because Luxemburg is viewed as a revolutionary socialist and an economist, rather than as an ethicist, that her distinct approach to war has not received more attention. It is true, however, that her understanding of

DOI: 10.4324/9781003428688-11

war cannot be understood apart from her commitment to the pursuit of a revolutionary post-capitalism society. Underlying this commitment is a belief that a revolutionary spirit can be born out of the spontaneity of contestation, whereby new social norms and ideas can be generated (Chen, 2015). Any form of political strategy that relies upon practices common under capitalism would therefore limit the emancipatory potential of such a society (Luxemburg, 1971). It is within this 'critical' ontological mindset that Luxemburg's political life must be understood.

While Luxemburg's way of thinking about the ethics of war deserves recognition and attention on its own merits, it was also integral to her later and indeed wider understanding of capitalist expansion. In *The Accumulation of Capital*, perhaps her most noteworthy piece of work and certainly her *magnum opus* on political economy, she finishes by stating that 'capital increasingly employs militarism' to expand to non-capitalist societies so that it can seek new markets and further develop (Luxemburg, 1913, pp. 446–447). Here, war takes on a functional and an essential role for capitalism's survival. In the backdrop of World War I, this pivotal position of war saw her pacifism take on a more profound form that led to her imprisonment and to the publication of works such as the *Junius Pamphlet*. As I shall explain later in the chapter, these positions were also shaped by her lived experiences and from her various exchanges with contemporaries. As such, her thought both animated and illuminated the historical period through which she lived and vice versa.

Contexts

Rosa Luxemburg was born to a Jewish middle-class family in 1871 in Zamosc, Poland. Her family moved to Warsaw when Rosa was a small child, and she was brought up within the middle-class Jewish intelligentsia. Amidst an atmosphere of anti-Semitism and Russification, Luxemburg was raised in a multilingual environment that—in line with the radical Jewish tradition at the time—embodied a distinct transnational identity. She considered herself more German, even though she grew up in the Vistula Land of Poland, which was controlled by the Russian Empire, and later went to University in Zurich. It was during her time in Zurich that she first became involved in socialist politics. This involvement intensified over time. In 1893, she spoke at the Second International and co-founded the Social Democracy of the Kingdom of Poland (SDKP). The SDKP emerged as a counter organisation to the Polish Socialist Party which they believed was pursuing a distinctly national line of development. It was during this period that Luxemburg's own unique brand of internationalism was forged. Having completed a doctorate on the political economy of industrialisation in Poland in 1898, she moved to Germany.

She wedded Gustav Lübeck in a marriage of convenience, thus obtaining German citizenship and subsequently joining the Social Democratic Party (SPD) of Germany. Here, she took on a central role within the Second International and rose to significance as a leading theoretician within the Party. Her critique of Marxist revisionism generated significant attention within the wider socialist movement, and her focus on labour movements and on the significance of the strike challenged orthodox views on party organisation. In the aftermath of the 1905 Revolution in Russia, Luxemburg returned to Warsaw to write for the (now renamed to include Lithuania) SDKPiL's political publications. She sought in this role to stimulate oppositional forces against the Russian Empire. Her writings emphasised the importance of strike action as a means of attacking capitalism. She was arrested in 1906 in Poland but managed to smuggle herself back into Germany while awaiting trial. Back in Germany, she played an important part in the Second International, reiterating her stances on internationalism against Lenin's strategic form of national self-determination. It was the rejection of self-determination and of the nation within the context of proletarian politics that was to become highly significant within Luxemburg's brand of socialism, especially within the context of the twentieth century nation-state (Davis, 1976). Up to the outbreak of World War I, she would become a highly important figure within the SPD, representing them at international congresses and taking a leading role in the Party's education programme.

While the SPD voted to support the Kaiser and the war effort in 1914, Luxemburg maintained a passionate anti-war stance, arguing that militarism and socialism were incompatible with its values. She organised *Die Internationale* alongside Clara Zetkin, Karl Liebknecht, and Franz Mehring to contest the war. This was renamed the *Spartacus League* in 1916. As a consequence of her pacifism, she spent a large part of the war incarcerated in a variety of prisons. Nevertheless, by the end of the war, she had gained a large following. This led to the establishment of the Independent Social Democratic Party of Germany in 1917, a grouping predominantly made up of anti-war members of the SPD. Released from prison at the end of the war, she emerged to a Germany in a state of meltdown. With her former student Friedrich Ebert looking to head up a government in the aftermath of the Armistice, the *Spartacus League* was reformed in opposition. With the German revolution in full swing, Luxemburg and Liebknecht formed the German Communist Party (KPD) at the start of 1919. Within days of the formation of the new party, a set of uprisings instigated by factory stewards erupted. This unrest would subsequently become known as the *Spartacus Uprising*. Ebert and Gustav Noske, who was responsible for Military Affairs in the Council of the People's Deputies, which had assumed power during the revolution, enlisted the support of the far-right *Freikorps* to put down the uprising. The Freikorps went onto murder both Liebknecht and Luxemburg on 15 January.

Luxemburg was assaulted, beaten, and hit on the head with a rifle butt before being shot and her body thrown into a tributary of the river Spree (Evans, 2015; Gietinger, 2019).

In the aftermath of her death, Luxemburg became a singular figure within left-wing politics in the sense that she became a significant figure for different approaches (Ypi, 2022). Despite her criticisms of Lenin and of Bolshevism, she was well received in the Soviet Union. Though her positions received mixed responses from figures such as Stalin and Trotsky, she enjoyed acclaim in the Soviet Union, avoiding the charges of deviation that all other socialist variants brought (Nettl, 2019). She was particularly memorialised in East Germany, where she was regarded almost as the Godmother of the East German State (Weitz, 1994). Yet, the term Luxemburgism also came to be associated with a more libertarian and spontaneous form of Marxism at odds with the institutional party constructs found in both the parliamentary and Soviet-led forms of socialism. As such, Luxemburg was often proclaimed as the intellectual progenitor of a more emancipatory form of socialism that never quite materialised. Indeed, she has been touted as a key founder of the traditions of Marxist-Humanism and Marxist-Feminism (Dunayevskaya, 1991). At the same time, the SPD in Germany have also tried to claim her as their own and have argued that she made significant contributions to the later development of the Party. In doing so, many within the SPD have sought to distance the involvement of the Party in her murder (Gietinger, 2019).

As an effective socialist martyr, Luxemburg has thus come to occupy a position where everyone wishes to claim her. Leninists continue to maintain that she was one of their own and repudiate that the idea of a Luxemburgism alternative was a myth. Likewise, Social Democrats believe she was central to the building of post-war Social Democracy and Trade Unionism. For its part, the western Marxist tradition has heralded her as an independent thinker who inspired a wider critical tradition (Worth, 2012). As with many inspirational leaders, her work becomes littered with inconsistencies and open to multiple interpretations; yet, the tendency to declare Rosa as 'one of us' by respective (and often rival) traditions continues. Accordingly, while it is difficult to nail her philosophical and political views to any particular mast, this further affirms the originality of her thought—and contributes to the richness of her legacy.

Texts and Tenets

Rosa Luxemburg's work can be understood through her ethical approach to politics, her strategic thinking towards revolutionary change, and her interpretation of capitalism. These were often interlinked and informed by her understanding of dialectical materialism, which was so central to her worldview. As I outlined in the chapter introduction, Luxemburg's understanding of dialectical materialism led her to believe not only that a revolutionary politics would emerge from the struggle against capitalism but also that the form this revolutionary politics would take could not be comprehended or measured through the lens of bourgeois society (Luxemburg, 1971). This did not mean that all values were to be abandoned. There were still certain virtues that needed to be upheld. Her navigation of these tensions was perhaps most evident in her Bernstein critique entitled *Reform or Revolution* (1900). It was here that she first came to prominence within SPD circles, and here her independent radicalism shone though most vividly. Luxemburg rejected the parliamentary reformism of Bernstein, arguing that it was rooted within a vulgar form of opportunist determinism. By merely engaging within existing political structures and the reproduction of capitalism, Bernstein's reformism was merely extending the narrow ontology of bourgeoise liberal science, thus limiting the revolutionary potential of the working-class movement. It also placed down an understanding of dialectics from which the movement would emerge (Luxemburg, 1900 [1986]). From this departure point, Luxemburg understood this dialectical process to provide a wide variety of emancipatory features and conditions.

One of the conditions that Luxemburg was keen to stress was the importance of internationalism as a prerequisite to socialist emancipation. Her infamous piece on *The National Question* rejected the notion that nations had the right to self-determination in their development of socialism (1908). For Luxemburg, self-determinism was divisive, led to pernicious forms of national mythology, and ultimately legitimated (and generated) war between nations. In line with her arguments presented in *Reform or Revolution*, she suggested that the doctrine of national self-determinism merely reinforced the bourgeoise creation of nations and nationhood and was a key tool in dividing the working-class and limiting its potential. While the complexities of uneven development and equally the succession of national liberation movements that followed the collapse of European Colonialism would suggest a certain naivety in Luxemburg's analysis, it remains influential in the twenty-first century. There is still today a vibrant school of thought which supposes that any appeal to nationalism or self-determination, no matter how well-meaning or how strategic, necessarily creates divisions and limits potential.

Luxemburg's rejection of reformism and of national self-determination is paramount to understanding her approach to war. She viewed war as a necessary result of bourgeoise nation-building and—as I shall mention later—an entailment of capitalism. This view went hand in glove with her condemnation of militarism, her support of universal suffrage, and her condemnation of capital punishment. Her thoughts on these matters were expressed in a series of reflective articles published in the *Red Flag*, the socialist newspaper attached to the Spartacus League, upon her release from prison in 1918. Yet, her anti-war position was also strategic. Perhaps her most influential anti-war piece was the *Junius Pamphlet* written while incarcerated (1915). This showed how the Great War had left international socialism in tatters and had succeeded in pitting national working-class parties up against each other leading to the 'suicide' of any coherent international working-class movement. Luxemburg's response was to re-emphasise the importance of internationalism and to illustrate that strategic proletarian politics needed to reject any form of engagement with war and reclaim a class struggle through labour agitation and contestation.

The *Junius Pamphlet* not only reinforced her conviction that working-class opposition to capitalism had to be levied at an international level, but it also confirmed her belief in spontaneity. Rather than look to form strict organisational bodies to precipitate change, she affirmed her stance that real change would emerge from the practices of contestation themselves, which would generate a new form of revolutionary politics. In this way, she rejected the mechanical determinism of revolutionary change (Lowy, 1981). The *Junius Pamphlet* also provided another clue to her understanding of war, which is perhaps lost on wider commentators. As war is tied up with class politics, as Luxemburg avers, it becomes incompatible with any form of socialist politics. War also significantly developed under capitalism where it became more sophisticated and more institutionally embedded within society. As such, any attempt to strategically engage with its practices merely reinforces the capitalist system that it is enmeshed within. War is thus understood not merely as an unjustifiable form of violence in the manner that most pacifists in the liberal tradition fundamentally argue but also as a key social component and indeed bulwark of capitalism.

War also featured significantly in her (1913) classic, the *Accumulation of Capital*, which is regarded as her central contribution to political economy. Written just before the start of World War I and so preceding the *Junius Pamphlet*, it showed how war and conflict became essential for the capitalist system to develop. Yet, it should also be stressed that two thirds of the text is dedicated towards first providing a critical overview of the arguments within Volume II of *Capital*, where Marx outlines how the circulation of capital allows for its reduction and then second giving a more critical overview of how the classical economists understood reproduction. The third section, which provides the crux of her arguments, is thus just over 100 pages. Here, she argues that for capitalism to develop, it requires for its continued expansion to new markets that are non-capitalist. The imperialism and militarism of the latter part of the nineteenth century and early twentieth century should thus be understood within this context. Effectively new demand was artificially required by capital due to under-consumption in capitalist system, thus leading to the invasion and exploitation of non-capitalist societies.

The *Accumulation of Capital* has been both heralded as a classic and criticised as being flawed. In the latter, the lack of vigour in her understanding of under-consumption has long since been criticised, led initially from Austro-Marxist Otto Bauer at the time of publishing. This in turn saw her reply with a detailed response entitled 'anti critique' (1915). However, as these largely rested on the formulation of economic laws regarding the trajectory of capitalism, the very fact that capitalism flourished once non-capitalist markets had been capitalised and did not require yet more untapped non-capitalist societies for survival suggested that Luxemburg's logic regarding the fallacy of capital was in error. What renders this piece a classic, however, is not its technical detail but its ability to persuasively demonstrate the importance of certain commodity markets within capitalism. For example *The Accumulation of Capitalism* remains one the key texts that provides us with an explanation to why resources and commodities such as oil attract and indeed perpetuate militarism. In addition, over a 100 years since it was written, it provides an equally useful explanation of how and why new capital continues to look for new markets to exploit, no matter how destructive they might be.

The Accumulation of Capital also showcased Luxemburg's views regarding the causes of war and its inevitability under capitalism. According to Luxemburg's reasoning, the imperialist and colonial wars that occurred in the context of the scramble of Africa were a consequence of this pursuit of new markets for capitalist states. This in turn led to a heightened form of militarism that would ultimately snowball into the Great War. As well as being her *Magnus Opus* on political economy, the work did much to shape her approach to pacifism and her famous war-time slogan 'Socialism or Barbarism'—a re-working of Engels's comments on working conditions—to account for the consequences of the Great War and the fate of humanity taken by those who supported it. As such, it remains an important source to any approach to war.

Controversies

Throughout her life and in every assessment of her work since her death, Rosa Luxemburg has attracted controversy. As already discussed, not only has every tradition on the left looked to claim her as one of them, but her work and ideas have become so contested that her controversies have followed her in every facet of her life. Indeed, it could be stated that Luxemburg's career was indeed marked by such controversies but that these same controversies also provide us with an opportunity for having alternative visions of socialist development from the ones characterised by the twentieth century. As both the forms of parliamentary socialism and of state Soviet-inspired socialism failed in their bid to transform and replace capitalism, many of these criticisms as they were thought of at the time do indeed fall down (Cox, 1991). Luxemburg's assault on parliamentary socialism, typified by her attack on Bernstein in *Reform or Revolution*, was obviously one that created significant ripples across the Social Democratic movement. Certainly, at the time of writing, the impact that it had was that Bernstein's revisionism was not accepted; yet, after the establishment of the Weimar Republic, the route towards parliamentary revisionism was taken for granted (see Townsend, 2007 for an overview). This would be formalised after World War II in West Germany with the adoption of the Godesberg Program that adapted this towards post-war mixed economic conditions. In terms of Party socialism, Luxemburg's approach had become a footnote to the history of the SPD and had long been discarded.

The arguments with Lenin when Luxemburg was at the SPD over national determinism were also a source of controversy. Again, by the end of World War II, the era of decolonisation saw self-determination appear as the central expression of national emancipation. As such, many were quick to refer to Luxemburg's interventions in that post-war period as one not just misplaced but ultimately proven to be wrong (Davis, 1976). At the same time, her later criticisms of the form that the Russian Revolution took (Luxemburg, 2004 (1918)) could also be said to have been unfounded, especially considering her failed alternative based upon spontaneity. However, the collapse of twentieth-century socialism in all its national forms has also brought her criticisms back into public focus. Indeed, while many still argue about the context that her work on the national question and party organisation should be taken and understood, the fact that the form of left nationalism that was adopted by twentieth-century socialism ultimately failed by the end of the century shows that Rosa might have had the last laugh (Ryan and Worth, 2010).

There are two other problems that I feel should be mentioned which do not improve with the passing of time. One of the enduring criticisms that remained with Luxemburg's worldview was around her lack of strategic reality in favour for an abstract form of determinism. Again, this depends upon how one interprets her understanding of spontaneity, particularly in relation to the so-called 'vanguard party.' However, the emphasis she put on the mass strike as the main catalyst for revolutionary change does raise several questions. How for example would such tactics lead to a transformation of society? Can a belief that a post-capitalist society will emerge out of the revolutionary process be enough for a political imaginary? Luxemburg was writing in an era when there was a firm belief within socialist circles that capitalism was in its final stages. As such, the conviction in capitalism fallacy was such that a general belief that socialism would emerge from the ashes of capitalism appeared logical and acceptable within such circles. At the time, this claim was criticised for neglecting the importance of strategic organisation—a position that would emerge both from the revisionist position of the SPD and from Lenin. All this fades into insignificance when examined through the lens of western Marxism which was to follow this era and the complexities of Gramscian-inspired understandings of civil society. Indeed, in the contemporary era of global capitalism, autonomous Marxists and more strategic forms of Anarchism have perhaps met this idea of spontaneity better, arguing that the formal transformation of capitalism might not be required through contesting it (Holloway, 2002; Hardt and Negri, 2004).

This brings us back to Luxemburg's own position on war and violence, which is the central concern of this volume. As argued throughout this chapter, Luxemburg's opposition to war was one of her most central principles which she remained committed to. Likewise, her commitment towards humanitarian values, which included an opposition to any sort of violence, was another key element in what we could call a Luxemburgist philosophy. Yet, her approach to unrest and uprisings was far more ambiguous. On one level, Luxemburg was consistent in her opposition to violence. Her attacks on militarism were such that she would criticise Karl Liebknecht over the use of confrontational tactics in the pursuit of the overthrowal of the state. Yet, many historians also stress her support of such violence during the Spartacus uprisings and indeed in the execution of a general strike (Jones, 2016). She certainly did not appear to endorse a Gandhi-style of pacificism. Here, one could question whether such uprisings could qualify for her endorsement of a type of a 'just' war. An assault using certain elements of violence to stimulate revolutionary forces against the bourgeois protectors of capitalism might be understand as justified forms of violence. Taking this further, Gramsci noted that Luxemburg's illustration of the mass strike provided one of the best examples of the 'war of movement' or manoeuvre (Gramsci, 1971, p. 233). That is it represented an apt example of how a frontal assault on the capitalist state could be levelled (he argued that an ideological 'war of position' was required to support it for any form of transformation to succeed). As a general strike does not involve a military-style take-over of a state but is involved in the struggle against the means of production, it represents an assault or 'war' on the bourgeoisie rather than the specific rulers of a state. Thus, any potential violence resulting from this assault would be considered 'just' in nature.

However, we cannot adequately substantiate these claims due to the fact that the confrontations that Luxemburg was alleged to have endorsed did not result in violent exchanges. Her activities in Poland during Tsarist occupation were conducted in secrecy, while the Spartacist Uprising itself saw Liebknecht and Luxemburg disagree over its purpose with Liebknecht pushing for the overthrowal of the *Council of The People's Deputies* and Luxemburg being far more reluctant (Wehler, 1997). Thus, any endorsement for a call to arms from Luxemburg would only have been a largely defensive one in a short-lived altercation that resulted in a one-sided bloodshed—which would eventuate in her own murder. In this instance, at least historic events have shown that any potential controversy should be taken with a pitch of salt. Greater criticisms however can be placed on the observation that Luxemburg's own approach to such confrontation was unclear. Alongside Luxemburg's undoubted commitment to pacifism and significant attack on militarism, ambiguity clouded her views on how much (if any) legitimate 'just' force was required for a revolutionary period to birth a new social order.

Finally, and in contrast from the position of 'pure' pacifists, it is the left's own turn away from pacifism in the 1930s with the rise of fascism. Here, questions must be put to how viable Luxemburg's pacifism was in response to fascism. Certainly, by the time the Spanish Civil War was in full swing, any alternative to a military combative position of the sort that Luxemburg insisted upon was seen untenable in light of fascism militarisation across Europe. As such, later generations of socialists were to dismiss pacifism as a means for engaging reactionary forces.

Legacy

As we have shown, there is no doubt that Luxemburg's legacy is a highly contested one, with a whole array of contested narratives given to the 'real' Rosa and her 'true' contribution to history. In sum, there is no doubt she was a revolutionary, a politician, a political economist, a feminist, and a humanitarian. All would also surely agree that Luxemburg was an independent thinker whose life was defined by a refusal to conform to dominant trends and ideas. In an era where market capitalism has made pre-existing forms of organised socialism go back to the drawing board, Luxembourg's original and alternative approach to socialism is indicative of a different way of thinking about and doing politics. Her willingness to think critically about what was 'possible' as opposed to what is contained within existing structures remains instructive (Worth, 2012). As suggested earlier, the post-Cold War order has been one where movements have emerged to challenge neoliberal capitalism without confronting the nature of the state. As such, many studies of resistance and contestation have focused upon the process of protest and the disruption of production at the civil level rather than at the centre (Bailey et al., 2017). Luxemburg provides us with the very basis for showing why these matter, and why they are relevant.

Significantly, the contemporary era also breathes new life into Luxemburg's position on internationalism. The failure of the twentieth century state to adequately provide a socialist challenge to capitalism in addition with the wider move towards greater integration through globalisation has left partitioned entities less able to viably construct this within the confines of the nation-state. While some cling to the notion that sovereignty still provides the more convenient and indeed most radical process for this challenge (e.g. Mitchell and Fazi, 2017), Luxemburg's stance on rejected self-determination has reminded us of the nation-state's fallacies. Indeed, it is one that remains as important as ever, seen with her persistence in pointing out the fact that no matter how convenient and well-meaning it might appear, the nation-state ultimately divides the working class and forces of nationalism dividing it further. It is also imperative to note that with the left failing to have construct a coherent alternative to neoliberal capitalism, the complex spatial realities of twenty-first-century global politics make Luxemburg's work on internationalism essential for contemporary debate.

Luxemburg also provides us with a humanism that is largely lacking in other socialist theorists of the twentieth century and quite different from the humanism found in, say, Montaigne, who is explored in a previous chapter. Her understanding of revolution and of revolutionary change was one that stressed the totality of emancipation, both societally and individually (Mills, 2020). Her efforts to stress the importance of an ontological revolutionary imaginary for the dialectical development of humanity have proven seminal for the subsequent development of critical theory of all stripes (Brie and Schutrumpf, 2021). For instance her writings on feminism and on democratic empowerment has been pivotal to much later work in this area (see Hutchings, 2020; Muldoon and Booth for respective overviews). In each of these areas, Luxemburg was well ahead of her time.

Yet, and in line with the theme for this volume, it has been Rosa Luxemburg's observations on capitalism and war that attract the most interest. The wars of the late twentieth century and early twenty-first century furnish us with multiple instances wherein capitalism in its many guises has sought to exploit new markets to pursue further growth. It is no surprise, then, that Luxemburg's influence is evident in the literature on the New Imperialism and new Marxist arguments on war (Harvey, 2004; Bieler and Morton, 2018). The trajectory of global capitalism in the last 30 years since the end of the cold war has indeed been one where turbo-charged markets have looked to commodify all areas of natural life. As indeed, the consequences of these developments have led to several factors that have driven us further into crisis. On the one hand, resource wars have led to a further acceleration of the ecological catastrophe that has engulfed the world, and, on the other, technological developments have seen forms of commodification appear at all sorts of spatial levels (whether appearing in material form or imagined) that have provided more arenas for capital accumulation. Luxemburg's conviction that capitalism will always seek new markets no matter how far it puts itself deeper and deeper into crises has thus never been more relevant.

It seems inevitable, then, that Luxemburg's proclamations take on a more significant meaning during times of deep crisis. At the same time, her approach to war, when thinking about questions of just war, is also one which we look for greater clarity. Here, we are left thinking back to her 'socialism or barbarism' maxim. For her, this was a belief—in light of the imperialist wars of the preceding 30 years and the descent into the bloodiest war known to humanity—that capitalism would continue to plunge society into the depth of barbarity in its pursuit of growth. Over 100 years later, considering the collection of crises that have engulfed the contemporary global order, the slogan has regained interest. If not the deterministic conviction that socialism would replace a capitalism that had descended into barbarity, then a realisation that a new form of politics is required to prevent a slide into catastrophe is required.

Conclusion

This chapter has aimed to show how the life and writings of Rosa Luxemburg are a useful addition when looking at the wider philosophy of war. This necessitated a study of her wider political and radical thought. Luxemburg's approach to politics was characterised by her understanding of dialectical materialism. It was this working formula that would shape her understanding of capitalism, contestation, and revolutionary transformation—the conviction that a new world would develop out of the crisis of capitalism and this world would extend far beyond the parameters imaginable to the present. Her approach to war must be understood in this light. She viewed war, alongside nationalism, self-determinism, parliamentary reformism, and insurgency, as part of the wider apparatus of capitalism. War was ultimately a product of the capitalist era that had been normalised and cultured by the bourgeoisie. It also became a necessity in order for capitalism to develop and survive. Thus, pacifism became an essential tool against the acceptance of war and militarism and was a required prerequisite of a socialist opposition. Yet, in practice, despite Luxemburg's opposition to militarism and violence, there is debate over how far she opposed violence when engaging in revolutionary activity. The tantalising ambiguity of her proclamations on these matters and the short-lived experiences of the Spartacus Uprisings do not provide us with any sure answers to these questions. By way of conclusion, then, we might wonder whether, in the final analysis, Luxemburg's pacifism was trumped by a very particular vision of just war. That is a vision of just war as a wider struggle against the existing order and the equal struggle to refuse to accept orthodoxy and conservatism, which she believed were the antithesis of human emancipation.

Works Cited

Bailey, David, Monica Clua-Losada, Nikolai Huke, and Olatz Ribera Almandoz. 2017. *Beyond Defeat and Austerity: Disrupting (the Critical Political Economy of) Neoliberal Europe*. London: Routledge.

Bieler, Andreas and Adam David Morton. 2018. *Global Capitalism, Global War, Global Crisis*. Cambridge: Cambridge University Press.

Brie, Micheal and Jorn Schutrumpf. 2021. *Rosa Luxemburg*. London: Palgrave.

Chen, M. 2015. From Class to Freedom: Rosa Luxemburg on Revolutionary Spontaneity and Socialist Democracy *Archiv für Rechts-und Sozilaphilosophie/Archives
for Philosophy of Law and Social Philosophy* 101(1), pp. 75–86.
Cox, Robert. 1991. Real Socialism in Historical Perspective. In R. Miliband and L.
Panitch. eds. *Socialist Register: Communist Regimes the Aftermath*. London: Mer-
lin Press.
Davis, Horace. 1976. Introduction: The Right of National Self-Determination in Marx-
ist Theory–Luxemburg Vs Lenin. In David Horace. ed. *The National Question:
Selected Writings by Rosa Luxemburg*. New York: Monthly Review Press, pp. 9–45.
Dunayevskaya, Raya. 1991. *Rosa Luxemburg, Women's Liberation and Marx's Phi-
losophy of Revolution*. Chicago: University of Illinois Press.
Evans, Kate. 2015. *Red Rosa*. London: Verso.
Gietinger, Klaus. 2019. *The Murder of Rosa Luxemburg*. London: Verso.
Gramsci, Antonio. 1971. *Selections from the Prison Notebook*. London: Lawrence
and Wishart.
Hardt, Michael and Antonio Negri. 2004. *Multitude*. London: Penguin.
Harvey, David. 2004. *The New Imperialism*. Oxford: Oxford University Press.
Holloway, John. 2002. *Change the World without Taking Power*. London: Pluto Press.
Hutchings, Kimberly. 2020. Revolutionary Thinking: Luxemburg's Socialist Interna-
tional Theory. In Patricia Owens and Katherina Rietzler. eds. *Women's International
Thought: A New History*. Cambridge: Cambridge University Press, pp. 52–71.
Jones, Mark. 2016. *Founding Weimar: Violence and the German Revolution of
1918–1919*. Cambridge: Cambridge University Press.
Lowy, Micheal. 1981. Marxism and Revolutionary Romanticism *Telos* 49, pp. 83–95.
Luxemburg, Rosa. 1900 [1986]. *Reform of Revolution*. London: Militant.
Luxemburg, Rosa. 1906. *The Mass Strike, the Political Party and the Trade Union*.
Available at: www.marxists.org/archive/luxemburg/download/mass-str.pdf.
Luxemburg, Rosa. 1915a. *The Junius Pamphlet: The Crisis of German Social Democ-
racy*. Available at: www.marxists.org/archive/luxemburg/1915/junius/.
Luxemburg, Rosa. 1915b. *An Anti-Critique*. Available at: www.marxists.org/archive/
luxemburg/1915/anti-critique.
Luxemburg, Rosa. 1918. Against Capital Punishment. *Die Rote Fahne* 38(Novem-
ber). Available at: www.marxists.org/archive/luxemburg/1918/11/18c-alt.htm.
Luxemburg, Rosa. 1971. *Selected Writings*. New York: Monthly Review Press.
Luxemburg, Rosa. 1976 [1908]. *The National Question*. New York: Monthly Press.
Luxemburg, Rosa. 2003 [1913]. *Accumulation of Capitalism*. London: Routledge.
Luxemburg, Rosa. 2004 [1918]. The Russian Revolution. In Peter Hudis and Kevin
B. Anderson. eds. *The Rosa Luxemburg Reader*. New York: Monthly Review
Press, pp. 281–310.
Mills, Dana. 2020. *Rosa Luxemburg*. London: Reakiton Books.
Mitchell, William and Thomas Fazi. 2017. *Reclaiming the State*. London: Pluto Press.
Nettl, John Peter. 2019. *Rosa Luxemburg: The Biography*. London: Verso.
Ryan, Barry and Owen Worth. 2010. On the Contemporary Relevance of 'Left
Nationalism'. *Capital & Class* 34(1), pp. 54–61.
Townsend, Jules. 2007. Right-Wing Marxism. In Daryl Glaser and David Walter. eds.
Twentieth Century Marxism. London: Routledge, pp. 46–58.
Wehler, Hans-Ulrich. 1997. *The German Empire*. London: Berg.
Weitz, Eric. 1994. Rosa Luxemburg Belongs to Us: German Communism and the
Luxemburg Legacy. *Central European History* 27(1), pp. 27–64.
Worth, Owen. 2012. Accumulating the Critical Spirit: Rosa Luxemburg and Critical
IPE. *International Politics* 49(2), pp. 136–153.
Ypi, Lea. 2022. Rosa Luxemburg. Stanford Encyclopaedia of Philosophy. Available
at: https://plato.stanford.edu/entries/luxemburg/.

11

LUIGI STURZO (1871–1959)

Gregory M. Reichberg

Introduction

Commentators have not failed to notice that contemporary papal teaching on war omits mention of "just war." The "disappearance of just war" is how French historian René Coste characterizes the absence of explicit appeals to just war within the official pronouncements of the papacy (Coste, 1962, pp. 447–493). As far as I can tell, the last time a pope spoke approvingly of just war was in a 1953 address by Pius XII, when he remarked how "even in a just and necessary war," limits must be observed (Moines de Solesmes, 1956, p. 540). Since then, "just war" has been scrubbed from the papal lexicon, apart from very recently when Pope Francis (see Chapter 19 of this volume) revived the term only to criticize it, namely, to deny its suitability for contemporary moral reflection on armed conflict (Pope Francis, 2020; Reichberg, 2024). "Just war" does figure in the 1992 *Catechism of the Catholic Church* (Catholic Church, 1992, § 2309) where it is placed in scare quotes to describe how the older tradition had framed the conditions of legitimate defense ("the traditional elements enumerated in what is called the 'just war' doctrine"), with the supposition that this nomenclature is no longer embraced by the Roman Magisterium as its own. What accounts for this sidelining of a term that for many centuries had been standard currency in Church discourse? Sturzo provides a missing piece to this puzzle.

Contexts

Preponderant in this move away from reliance on just war in contemporary Catholic thought is the influence of Luigi Stuzo (1871–1959), a Sicilian

DOI: 10.4324/9781003428688-12

Catholic priest, who wrote extensively on war during the years 1926–1928 and intermittently throughout his career. While steeped in the thought of the medieval scholastics (he had studied Thomism in Rome at the Academy of St. Thomas Aquinas and the Gregorian University, where he earned his doctorate), Sturzo adopted a sociological and historical approach to the study of social issues (he held the chair in Political Economy, Philosophy, and Sociology at the Grand Seminary of Caltagirone) in line with the Jesuit Luigi Taparelli d'Azeglio (1793–1862), whose *Theoretical Essay of Natural Right Based on Fact* (Taparelli, 1855) is among the founding documents of modern sociology. The work also contained an extensive discussion of Catholic just war doctrine that Taparelli addressed by reference to Grotius and other internationalist thinkers. Often considered the father of Catholic social doctrine, Taparelli taught for many years in Palermo, which afterwards remained an important center for reflection on natural law.

Sturzo's engagement with social issues had a strongly practical bent. In Sicily, he "organized the first cooperative societies and trade unions for peasants, workers and artisans" (Foreword to Sturzo, 1929, p. 5), and from 1905 to 1920, he was mayor of Caltagirone and served as Provincial Councilor for Catania. In 1919, he founded the Partito Populare Italiano (Italian Popular Party), which quickly became a major force in Italian politics. The Popular Party was a leading inspiration for what later became known as the "Christian Democracy movement" in Italian, European, and Latin American politics, based on the idea that while politics should be informed by Christian principles, it should nonetheless not be organized along confessional lines. At the time, this orientation was held in some suspicion by the Vatican (given its claims of independence from the Church and the fact that Sturzo, the Popular Party's leader, was also a priest), but it was the rise of fascism that led to the demise of the party. A vocal opponent of Mussolini's political program, the Sicilian priest (Sturzo, 1936)[1] coined the term "totalitarianism" to characterize the new tendency. The Popular Party was disbanded in 1923 after the Vatican ordered Sturzo to resign as party leader (Pollard, 1996, pp. 77–82). Sturzo went into exile, settling first in London (1924–1940), and then in the United States (1940–1946). He returned to Italy in 1946 and was subsequently rehabilitated, receiving a permanent seat in the Italian senate.

Texts

It was in London that Sturzo became intellectually engaged on issues relating to war and peace. His first foray into this topic was an article "Il diritto di Guerra" (on the right of war) that appeared in 1926 in Italian and French (both texts are reproduced in Sturzo, 1954). This was followed three years later by *The International Community and the Right of War* (henceforth ICRW). The book was written in 1926–1928 and first appeared in

English translation (1929). Two years later, it appeared in French (Sturzo, 1931), but not until 1954 did Sturzo secure its publication in the original Italian (Sturzo, 1954). Sturzo subsequently concentrated on other aspects of political theory; he published articles inter alia in *Foreign Affairs, The Review of Politics*, and *Blackfriars* as well as several monographs on Italian politics, Vatican diplomatic initiatives, and relations between Church and State. He did, however, publish a collection of essays on inter alia political authority, democracy, and freedom, including one on "Modern Wars and Catholic Thought" (Sturzo, 1942, pp. 29–104). He also wrote an extensive introduction to the Italian edition of ICRW (Sturzo, 1954, pp. xi–xxxvii), where he discussed the historical context of the work and the major events that had taken place in the intervening years—the rise of Hitler, the Second World War, the founding of the United Nations, and the Korean War. He also examined some criticisms that had been leveled against his interpretation of just war theory.

Sturzo drew from Taparelli the idea that social realities should not be understood as timeless essences; rather, these phenomena will be properly conceptualized only when they are set in their appropriate historical contexts (Behr, 2019). By the same token, in so far as they spring from human nature, these social realities, like nature itself, must be viewed in their dynamic and teleological ordering toward higher degrees of perfection. In other words, the teleological law of directedness, of movement from incompleteness to fulfilment, holds for social realities as well as for individual organisms. Error arises when a social form at a particular stage of its historical development is equated with the form itself. This, in Sturzo's eyes, was the error committed by just war theorists of his day. They misidentified a social reality—interstate use of force for the ends of justice—that appropriately had meaning within a determinate historical context—and from this context, they unduly extrapolated wider claims about the inherent connection of war to human nature and its permanency within the human condition.

Sturzo's reason for writing *The International Community and the Right of War* was to promote a new understanding of war in line with beneficial social development. As soon as we understand that no individual state is a "perfect society," complete and sufficient unto itself, then we will see that organized social communities—states—will reach their perfection only as members of a higher international community.[2] And once this international community emerges, it is natural that the task of maintaining order should fall to it. The regime of state self-help will consequently end. Stopping aggression will no longer fall within the mandate of individual states but will become a core task for the international community itself. With the advent of an organized international community, recognition of war as a *right* of states (namely a permission or subjective claim-right that states could exercise at their discretion) will be superseded, and this right will henceforth vest in the international community. The institution of war will thereby cease to exist and will be replaced by another institution, one in which force is exercised solely by the international community, much as national police hitherto had maintained order within single states.[3] When "war" is conceived in this way, it is unsurprising that the very notion of "just war" would appear wholly outdated—and even contradictory—in the contemporary world. Pope Francis' statement in *Fratelli Tutti* (Francis, 2020, §258) that St. Augustine "forged a concept of 'just war' that we no longer uphold in our own day" appears to be a nod in favor of Sturzo's analysis that "just war" is an idea no longer fit for the modern age (Reichberg, 2024).

Police activity as exercised globally by the international community will thus substitute for war. Violence will still exist and will need to be repressed, but this repression, when rightly carried out by the international community, will no longer merit the name "war." "War"—understood as a legally sanctioned institution, a right, that is, an entitlement codified by determinate rules—will no longer pertain to individual states. The institution of war will cede place to a higher form of social existence. But because this shift to a higher state of social organization will not happen instantaneously but will be the outcome of a slowly developing process, it is imperative that right-minded individuals actively promote progress toward it. The elimination of war as a social construct—the natural telos of society—and taken precisely as an end of *rational* agents will require a shift in public *consciousness*. Moving away from war requires a change of mentality, a spiritual renewal of social expectations that permeates the different levels of society.

Controversies

This is the gist of Sturzo's analysis in ICRW. Let us now unpack its main elements. We can start out by framing these in terms of two negations.

Negation # 1. *War is not rooted in human nature.*

To illustrate the contingent linkage of war to human nature, Sturzo draws a comparison to the institution of slavery. For millennia, slavery was thought to be an inescapable and even necessary feature of human existence, exemplified by Aristotle's idea that some people are "by nature" fit to be slaves, due to their inferior mental endowments (*Politics*, bk. 1, chapters 6–8). But with the advancement of human moral awareness, opposition to slavery has spread until becoming quasi universal, and earlier legal support for it has collapsed. With the demise of slavery as an "institution"—a social practice upheld by a system of laws and customs—its "legitimacy" (i.e., the social recognition conferred on it) has likewise declined, to the point that public support for slavery can no longer be aroused. Pockets of slavery might still exist, but these must be hidden from public view and will be subject to sanctions should they come to light.

A recognized practice in ancient Rome, the *jus belli* was supported by a legal, religious, and ceremonial framework—the *ius fetiale* (Reichberg Syse, and Begby, 2006, pp. 47–49)—that stipulated procedures to follow in declarations of war. A related institutional pattern extended into the legal regime of the Middle Ages, when the jurists gave a formal recognition to war against infidels, while recognizing the reciprocal right of princes to declare war against each other, with the added supposition that rights of possession would vest in the victor (Reichberg, Syse, and Begby, 2006, pp. 227–229). This progression reached its apogee in classical international law, wherein an elaborate set of rules were devised by diplomats and jurists on what might count as valid declarations of war and the prerogatives that sovereign states could enjoy in using such force, in contrast to say brigands, pirates, and non-state armed factions, who had no such entitlements (Reichberg Syse, and Begby, 2006, pp. 469–474, 504–518). But with the Covenant of the League of Nations (1920) and the Kellogg–Briand pact (1928), the institutional arrangement whereby states had been accorded freedom, as countenanced in law, to wage war at their discretion, was dismantled. The latter agreement had as its core idea that war would henceforth be prohibited. This meant that the discretionary right to wage war, "as an instrument of national policy," was abrogated by the mutual consent of nations. The contracting parties agreed to settle disputes among themselves "by pacific means," and in so doing renounced resort to war as a method to resolve intractable disagreements (Article 2). For Sturzo, the Kellogg–Briand pact held out the promise that war could effectively be eliminated, not because all armed confrontations would necessarily cease, but rather because these confrontations would lose all legitimacy. States resorting to war would not be able to secure international recognition for their ill-gotten gains (seizure of another's territory, for instance). Resort to war would thereby be disincentivized, hence progressively fewer armed confrontations would take place. The world would be on a stable trajectory to peace as more states renounced resorting to war. That war can be fully de-legitimized, just as slavery has been, demonstrates for Sturzo that, despite appearances to the contrary, the right of war does not pertain to natural law, as Grotius and subsequent Jusnaturalists had supposed (Sturzo, 1929, p. 185).

Negation # 2, *War is never a necessity.*

Leaders of states have an ingrained tendency to believe that when certain situations arise, they have no option but to use force in response. War becomes the "only way out" of a situation that they deem to be intolerable. The appeal to "necessity" provides a justification for war, under the supposition that no viable alternative exists (Sturzo, 1929, pp. 96–97, 215–216). The removal of alternatives entails the suppression of free choice. Sturzo argues, however, that this alleged necessity is in fact an illusion. It arises from an undue conflation of the causes of war and the idea of necessity. He explains this by reference to remote and proximate causes of war.

A "remote cause" refers to the primary motive of a war—the reason that makes it an attractive policy option. This is ordinarily framed by reference to the original wrong that an aggrieved state takes as its primary purpose to overturn. To this end, the state embarks on building up its capacity for response. And in the measure that the ambient culture confers on war the appearance of a legitimate enterprise, states will grow their preparations for war accordingly. To illustrate such remote causes of war, Sturzo cites the case of Alsace-Lorraine for the French (in French eyes wrongfully taken from them by Prussia in 1870) and the Versailles treaty of 1919, which returned this disputed territory to France (a profoundly unjust arrangement in the eyes of many Germans); in each case, the seeds of future conflict were thereby sown by decisions taken years earlier. Sturzo maintains, however, that these remote causes, taken in themselves, can never necessitate (resort to) war because an extension in time opens the possibility of alternative courses of action. Statesmen always have the option of abandoning objectives they previously deemed necessary, once they understand how the underlying interests can be secured by other means.[4]

By contrast, the "proximate causes," of war are punctual incidents that appear to catalyze the start of a war. But here again, Sturzo notes how "there is no necessary connection between responsibility for acts that may become proximate war-causes and the responsibility for war" (Sturzo, 1929, p. 111). On this basis, "to the question whether a State might [ever] find itself in a state of necessity compelling it to make war" (Sturzo, 1929, p. 115), Sturzo answers in the negative:

> In the present organization of the State there can be no necessary war, a nation cannot find itself in a state of necessity obliging it to make war, for there is a juridical system and a permanent state of interstatal relations by which every dispute could be peaceably settled.
>
> *(Sturzo, 1929, p. 115)*

If war is never necessary, at least within the interactions of modern states,[5] at the root of every war, there stands a human will.

> The fact of war, therefore, is necessarily linked with the will of man alone. All the other links exist, but they will not entail necessity.
>
> *(Sturzo, 1929, p. 115)*

In conceptualizing it as an emanation of the human will, Sturzo takes "war" to signify something more than large-scale-organized violence. He admits that war can be understood in this way (as when contemporary political scientists count as "war" any armed confrontation that crosses a certain threshold of battlefield deaths),[6] in which case, it will simply designate *"the use and predominance of material force in conflicts between the various human nuclei"* (Sturzo, 1929, p. 90). However, at a deeper level of historical analysis, "war" has an inherently normative connotation, and here it designates *"the right to settle a dispute between State and State by armed force"* (Sturzo, 1929, p. 89, italicized in the text). Conceptualized in this way as a juridical entitlement (i.e., a so-called "subjective claim-right") by which states are permitted to employ force against each other in settlement of their disputes—the right in question will greatly reinforce the will to war whenever perception of its "necessity" arises.

In modernity, war consequently results from two main factors: (i) the conviction that states possess a right to resort to force (with the implication that they are permitted to do so) and (ii) active preparations for engagement in it (acquisition of arms, establishment of standing armies, and procedures for military training). However, despite appearances to the contrary, the process toward war is neither fatal nor necessary (Sturzo, 1929, p. 113).

Finally, Sturzo admits how for the participants in war, these multiple steps toward war easily take on the coloration of necessity. This illusion is fostered by the linkage they so readily establish between justice and force. Prior to the inception of hostilities, each party will believe its reasons are the stronger, and because justice is on its side, each will consequently believe "the probabilities of victory are in [its] favor" (Sturzo, 1929, p. 97). This psychological mechanism, the surreptitious joining of justice and force, explains why "peoples or their leaders are more ready to seek the terms of settlement of a dispute after a war than before" (Sturzo, 1929, p. 96) with the tragic consequences that so readily follow as wars grind futilely onward toward the exhaustion of the respective peoples. In sum then, for Sturzo, what is now termed the "criterion of last resort" is in fact built on a faulty premise: as alternatives to war can always be found; none should ever be deemed "last" or "final."

The perception of justice in war takes Sturzo into a discussion of the scholastic just war doctrine, which had served as the point of departure for his 1926 article on the right of war.[7] He shows some sympathy for this theory, which he interprets in the line of Alfred Vanderpol, who had conceptualized just war in terms of limiting conditions that states must meet in resorting to armed force. For Vanderpol, war is akin to a criminal legal proceeding; before seeking redress for one's violated rights, an aggrieved state has a moral responsibility to ascertain—according to the standards appropriate for a court of law—that its adversary is guilty (*mens rea*) of intentionally committing an objective wrong. Only then will the aggrieved state be justified in pursuing its claim against the perpetrator state by dint of war. Because the normative bar for resort to force is set so very high on Vanderpol's conception of just war—and Sturzo applauds him for this—should statesmen adopt this framework they will seldom find occasions meriting exercise of their *jus belli*. In today's parlance, we would say that Vanderpol's just war theory was premised on a strong "presumption against war."

Alongside just war, Sturzo recognized several other normative approaches to the problem of war. He focused his attention chiefly on two of them—"war for the reason of State" and "bio-sociological war." The first relates to what today we would term "political realism," and for Sturzo, this reduces to Machiavelli's claim that "war is necessary when it is useful to the state" (Sturzo, 1929, p. 184)—a view, he adds, that was in line with the "age of absolute sovereigns and dominion of reigning houses," when "raison d'État" . . . "substituted for the scholastic *bonum commune*" (Sturzo, 1929, p. 184). The second, by contrast, is the view that might equals right. These three theories represent contrasting attempts at grounding the *jus belli* as an "extreme method of settling disputes between peoples" (Sturzo, 1929, p. 193).[8] All three, likewise, give support to "the tragic conclusion that war . . . is fatal and inevitable" (Sturzo, 1929, p. 193).

Pacifism, by contrast, Sturzo placed in an altogether different category, insofar as it denies any *jus belli* whatsoever, but this viewpoint he rejected as utopian and inconsistent with the due maintenance of order in this fallen world, where force is always needed to hold wrongdoers in check. He accordingly deemed pacifism a patent error,[9] incompatible with social realities of human life and in opposition to natural right itself.

Among the three "systematic theories," Sturzo expressed the most sympathy for the just war approach. He notes how this theory, due to its anchoring in justice and appeals to the "Christian conscience," naturally finds support "among the masses," while raison d'État resonates with "ruling groups and politicians," and bio-sociological war with the "wealthy, and powerful classes" (Sturzo, 1929, p. 192). Indeed, "in the course of a historical process of which the chain of wars has stained the whole world with blood, we always find concrete instances of all three." This results from the fact that:

> [T]he effort to justify war is instinctive among all belligerents and in all wars . . . because of the inborn need of men to justify their own acts; all the more if these acts are public and, as in the case of war, entail terrible consequences.
>
> *(Sturzo, 1929, p. 192)*

Despite this enduring presence of these theories across time, each nonetheless represents only an aspect of our nature, and each, accordingly, calls for subsumption into a higher perspective that more completely fits the telos of our social nature. Sturzo accordingly submits them to critical review. As noted, of the three, he deems just war the superior approach, because it is most consonant with the claims of morality. Nonetheless it too is flawed, and appraisal of its limits enables Sturzo to propose a more integrative approach—"our own theory" as he calls it (the title of ICRW, chapter 3 [Sturzo, 1929, p. 208]) that shows the way toward a world without war.

The theory of just war, he argues, will invariably fail in its application to concrete cases. There is a disjunct between its principles and their application to actual conflicts. Therein lies its principal flaw. Just war theorists of the past always placed the practical efficacity of their teaching on an appeal to the conscience of rulers. But despite the moral significance of attesting to the truth in grave matters of justice, Sturzo acknowledges with discouragement that "the cases when a moral inhibition to make war has served its purpose cannot have been many" (Sturzo, 1929, p. 194). Moreover, he observes how despite the scholastic admonition that war cannot be simultaneously just on the opposing sides (if one is in the right, the other must be in the wrong), it invariably happens that the sovereigns in question will each be incentivized to claim justice for his own cause. And even should they seek the opinion of moralists and jurists, matters will be so arranged that each sovereign will not "fail to obtain replies . . . confirming him in his belief that his cause was justice" (Sturzo, 1929, p. 194). Nor can the citizenry efficaciously object to the sovereign's determination; to whatever criteria of justice they might appeal, he can always respond that they lack direct information about the particulars of the case and ultimately must defer to his superior judgment. After all this has been subtracted, what remains is "only the posthumous judgments of historians," but, by definition, these lack practical value "for the simple reason that they were passed on wars fought and not on wars to fight" (Sturzo, 1929, p. 195).

Alongside this meta-framing of just war in terms of its practical inefficacity, Sturzo also pointed to what he took to be the inherent inconsistencies of the doctrine (in the "just-war theory," he wrote, "we are met by a substratum of illogical elements" [Sturzo, 1929, p. 195]). In this connection, he takes just war to be in its essence a teaching about the exercise of punitive justice. On this understanding, the right to wage war proceeds from the perception of a culpable offense committed by the adversary. The aggrieved party in the person of its sovereign issues a legal judgment of guilt as would a judge in a criminal proceeding against all the people who are morally solidaric with the fault of their leaders, and from this judgment war can be waged by the judiciary state over the guilty state as the punishment due for its wrongdoing. Battlefield victory is the completed form that this punishment takes, and by it the judiciary state receives satisfaction of the wrong done to it (Sturzo, 1929, pp. 195–196).[10]

The inconsistencies are to Sturzo's eyes manifest. "Indeed," he writes:

> Among organized peoples the concept of the extension of individual guilt to the whole population of a state, or of presumed solidarity, has no foundation, whether legal, social, or ethical.
>
> *(Sturzo, 1929, p. 196)*

Moreover,

> [T]he idea that the sovereign or government in declaring war is executing a penal sentence cannot withstand the simplest critical examination. Suarez tries to develop this thesis, but without success. Instead of one authoritative judgment we find two judges, who are the two parties to the dispute, and neither of whom recognizes that the other has any judicial authority, and two contemporary findings, each with the same force. And if war is declared, each of the two ends by formulating a contradictory verdict, one of which cancels out the other.
>
> *(Sturzo, 1929, p. 196)*

At the end of the day, once these inconsistencies are acknowledged, all that remains of the just war theory are some high moral principles that can serve to guide the conscience of rulers. These principles receive added strength from the Christian faith, and that is why the doctrine of just war was embraced most fully during the Middle Ages when the Church had a judicial hold on the conscience of the faithful and stood supreme above states with a capacity to judge of them all, including the wars they waged against each other. But today, Sturzo concludes, it is "impossible for even the Church to raise a doubt as to the justice of a war" (Sturzo, 1929, p. 198), and for this reason, the doctrine of just war cannot emerge from morality into binding law. A new foundation for enduring peace between nations must be found.[11]

As was intimated before, Sturzo locates this peace in an "international organization of civilized peoples in which the right of war is no longer recognized" (Sturzo, 1929, p. 228). The solution does not however consist in establishing a supra-state, "an intolerable hegemonic domination" that "in the long run [will] fall to pieces through the reaction it has aroused" (Sturzo, 1929, p. 229). Nor should we expect this peace to forgo any use of force whatsoever, as "in all organized society there must be coercion and punishment as well as reason and education—that is, a rational use of force, and this both in the state and in international community" (Sturzo, 1929, p. 226). The solution is rather to be found in establishing "zones of war immunity," where progressively larger numbers of states (think, for instance, of the Scandinavian countries today) join together in pacts by which "war is proscribed always, and in every case, as illegitimate" (Sturzo, 1929, p. 228). This will have not only legal but also political bearing, as resort to war becomes definitively unthinkable to states as a method for resolving whatever disputes may arise between them (see Reichberg, 2022).

Legacies

Apart from a few authors (Coste, 1962, pp. 58–59; Minois, 1994, pp. 399–401) who expressly recognize his importance for later developments,

Sturzo's influence on subsequent Catholic thought might best be described as "subterranean."[12] Although his ICRW is rarely cited by leading actors, his advancement of an alternative to just war has had a very real—albeit hidden—impact on how the problem of war came to be framed in the years since ICRW was written. Three legacies merit special mention.

First, there was the "Conventus de bello," a consensus document on "the problem of the morality of war" that a group of French, German, and Swiss theologians published in the form of a declaration (Charrière et al., 1932, pp. 33–47). The fruit of regular meetings that were held in Fribourg, Switzerland, under the auspices of the local bishop, it was understood that they enjoyed the tacit support of the Church hierarchy. The very idea of creating a high-level study group to examine the moral status of war in Catholic teaching came from Eugenio Pacelli (later Pope Pius XII), who was then the Holy See's nuncio in Berlin.[13]

Although Sturzo was not directly cited,[14] the Conventus declaration enunciated a set of points that were in direct continuity with the main theses of ICRW: the very legitimacy of war must be critically reassessed; war should not be considered an unalterable given of the human condition; rather, a proper understanding of the natural sociability of states demonstrates how the purported legitimacy of war as a method for resolving intractable disputes is in fact an accidental feature of relations between states. "[T]o set up the social process of war in such a way that the conflict between the aggressor and his victim is solved only by the arbitrament of the sword," (Charrière et al., 1932, p. 43) and without previous recourse to international institutions is "condemnable by public law and before conscience" they wrote (Charrière et al., 1932, p. 41). The signers acknowledged, however, that a fully organized international community with proper organs of authority was still not yet in existence. The problem, accordingly, was to describe what limits should be placed on the use of force in this interim period. In making decisions about war, individual states should "subordinate national ends to the more general end of the international community"—an obligation of "general or legal justice" (Charrière et al., 1932, p. 41). Moreover, the declaration made clear that engagement in war for punitive ends must henceforth be excluded (Charrière et al., 1932, p. 43). The range of what could count as "just cause" must be narrowed to legitimate self-defense ("to repel force by force"—Charrière et al., 1932, p. 43), and they sharply ruled out appeals to "necessity." In line with Sturzo, the Conventus spoke disparagingly of standing armies ("this disorder that is an armed peace" [Charrière et al., 1932, p. 45]), and to replace this multitude of separate armed forces, it encouraged instead the formation of "a regime of mutual assistance" (Charrière et al., 1932, p. 45) with a system of arbitration at its core. "True security" being assured by these means, circumstances permitting legitimate defense would accordingly decrease, and an expanding "pacific entente" (Charrière et al., 1932, p. 45) would be the result. Finally, the Conventus added the further and related claim that modern warfare, by "a necessity inherent to its nature," and the calamitous destruction it causes, "ceases to be a means proportionate to the very end that alone can justify resort to armed force, namely the establishment of a more human order and peace" (Charrière et al., 1932, pp. 41–43).

The Conventus included no mention of "just war." Instead, its drafters spoke of the "traditional doctrine" (Charrière et al., 1932, p. 45), which they assimilated to the idea of "legitimate defense" against armed attack. This they contrasted nonetheless to "defensive war," which in their eyes must be condemned as bound up with claims to unlimited state sovereignty and "exaggerated nationalism" (Charrière et al., 1932, p. 43). Because "just war" was nowhere mentioned, reading the Conventus in the light of Sturzo's earlier critique of the idea, associated as it was with the notion of "punition," one could easier deduce that it should be deemed a relic of the medieval past, and a fortiori should henceforth be set aside in our contemporary world.

Second, in the years that followed, the language of the *Conventus* declaration would slowly seep into official Papal documents, especially after the Second World War. In this connection, it can be noted that Joseph Delos, a Dominican who had a leading role in drafting the Conventus (Reichberg, 2018, p. 572), resided in Rome from 1944 to 1968, where he served as legal advisor at the French embassy to the Holy See. During these two decades, Delos argued for a very restrictive conception of armed force as used by individual states. Only-on-the-spot repelling of violence may be allowed; and no wider proactive measures can be admitted. Moreover, Delos adhered closely to Sturzo's critique of appeals to necessity (last resort), and with him denied that this notion could justify engagement in "defensive war" (Delos, 1959, p. 318).

A *third* legacy may be identified in the Church writings themselves. When the Second Vatican Council proclaimed how we must "undertake an evaluation of war with an entirely new attitude" (Second Vatican Council, 1965, § 80), it echoed Sturzo's contention that old patterns of thought, including perhaps the scholastic teaching on just war (a term not employed by the Council fathers), should be set aside so as to foster efficacious progress toward a lasting peace. The same Council similarly stated how "It is our clear duty . . . to strain every muscle in working for the time when all war can be completely outlawed by international consent" (Second Vatican Council, 1965, §81). And when many years before Pope Pius II declared how "the idea of war as an apt and proportionate means of solving international conflicts is now out of date" (Pope Pius XII, 1944), these words could have been taken straight from a page of ICRW, and so closely do they track Sturzo's core claim that in the modern age, resort to armed force should no longer serve states as a method to adjudicate their disputes. This in turn was echoed by Pope Paul VI's famous statement at the United Nations: "Never again war, war never again!" (Pope Paul VI, 1965). Like Sturzo, this pope did not mean to imply that a day would come when violent strife on earth would entirely cease, for as he made clear later in the same discourse: into the foreseeable future, some measure of armed force would always be needed to hold wrongdoing in check. The meaning, rather, was that states should never again initiate war to settle their differences or to advance their agendas. Finally, when Pope John Paul II said to the diplomatic corps "*NO TO WAR!* It is never inevitable" (Pope John Paul II, 2003), he was echoing Sturzo's denial that war should ever be necessary, especially in the modern age, when dialogue and diplomacy are always able to provide solutions that preclude appeal to armed force as a "last resort." And finally, when Pope Francis said to the UN's Security Council that "in order to make peace a reality, we must move away from the logic of the legitimacy of war" (Pope Francis, 2023), his words were in direct continuity with Sturzo's contention that war, much like slavery, is an institution that no longer can be validated in our contemporary world.

Notes

1 This article draws on Sturzo's earlier book *Italy and Fascism* (London: Faber and Gwyer, 1926).
2 In a footnote (in Italian) added to the French translation of "Il diritto di Guerra" (Sturzo, 1954, p. 276), Sturzo wrote that as applied to individual states, "perfect society" could have two different meanings. One, fully legitimate, acknowledges that a state will be "complete" when it possesses the full range of legislative, judicial, and executive powers. The other meaning, which he deemed illegitimate, holds that the state is a self-contained entity that precludes subordination to anything higher, whether to the Church or the international community.
3 According to Coste (1962, pp. 58–59), Sturzo borrowed this conception of international order from Taparelli, who had described the organized international community as an "ethnarchy" that requires a distinctive authority whose key task is to regulate relations between sovereign states. This authority will have the responsibility to use force on behalf of the community in addressing the violations of rogue states (Taparelli, 1855, p. 198, §§ 1377–38; Jacquin, 1939).
4 In this vein, against the Athenian generals' supposition that the destruction of Melos was indispensable for the preservation of empire, Michael Walzer comments that this "evades the moral question of whether the preservation of empire was itself necessary" (Walzer, 1992, p. 8). A more recent example of this dynamic could be taken from Finland's abandonment of its national aim, previously deemed necessary, to recover the territories it had lost to Soviet aggression. After trying twice to win them back by force, Finland opted for another path, namely integration with Europe and eventually NATO membership. See Reichberg, Tønnesson, and Syse (2022).
5 Considered as an "abstract hypothesis," Sturzo (1929, p. 115) concedes that such necessity might be possible, but he maintains, regardless, that any concrete instance of it will be hard if not impossible to find, especially under conditions of modernity where diplomacy always represents a viable option.
6 As in the Uppsala Conflict Data Program (https://ucdp.uu.se/).
7 In the opening of "Il diritto di Guerra," Sturzo adverts to a work on scholastic just war theory by one Abbé Pierre Charmetant, whose conceptualization of just war was heavily dependent on Vanderpol (Charmetant, 1925). Sturzo cites Charmetant as a foil to develop a new conception of the right to war.

8 He discusses the three in Chapter 10, "The Three Systematic Theories" (pp. 170–191), followed by Chapter 11, "Criticism of the Three Systematic theories" (pp. 192–207).

9 This is Sturzo's assessment in the opening pages of "Il diritto di Guerra" (Sturzo, 1954, pp. 257–259). Pacifism received little mention in the ICRW, as he had apparently ruled out its relevance in his earlier treatment of the *jus belli*.

10 This conception of just war derives from Aquinas' sixteenth-century commentator Cardinal Thomas de Vio (Cajetan), who took war to be a form of punishment akin to incarceration or confiscation of goods, whereby the subject of said punishment was no longer an individual as in a standard legal proceeding but an entire state that was thereby deemed to be guilty (see Reichberg; Syse, and Begby, 2006, pp. 240–250). Sturzo appears to have borrowed this conceptualization of just war from Vanderpol, and like Vanderpol, Sturzo overlooked how other strands of just war, as represented for instance by Vitoria and perhaps even Aquinas himself, side-stepped the guilt theory and developed an alternative grounding for just war that would be less vulnerable to the criticism Sturzo directed against it (Reichberg, 2017, pp. 162–172).

11 This foundation is presented in Chapter 13, "The Terms of the Problem" (Sturzo, 1929, pp. 223–245).

12 Sturzo's role has been neglected in anglophone historical studies on the ethics of war in the Catholic tradition; he receives no mention in Eppstein (1935) or in the works of James Turner Johnson that I have been able to consult.

13 See Reichberg (2018, pp. 572–573) on the historical background of the "Conventus de bello."

14 A passage by Sturzo on the "state of necessity" was however reproduced as an appendix to the Conventus de bello (Delos and Valensin, 1932, p. 86). In his introduction to the 1954 edition of ICRW, Sturzo acknowledges how the Conventus group sent him a "private message" wherein they expressed their friendship for him, words that were a "comfort to him in a time of great polemical difficulties" (Sturzo, 1954, p. xxxvi).

Works Cited

Behr, Thomas C. 2019. *Social Justice & Solidarity: Luigi Taparelli and the Origins of Modern Catholic Social Thought*. Washington, DC: Catholic University of America Press.

Catholic Church. 1992. *Catechism of the Catholic Church*. Available at: www.vatican.va/archive/ENG0015/_INDEX.HTM.

Charmetant, Abbé Pierre. 1925. *Le droit de guerre*. Paris: Bulletin Catholique International.

Charrière, François, Joseph T. Delos, Franz Keller, Joseph Mayer, Constantine Noppel, Bruno de Solages, Franziskus Stratmann, and Albert Valensin. 1932. Le problème de la moralité de la guerre, [= Conventus de bello]. In *Paix et guerre: La guerre devant la conscience*. Juvisy: Cerf, pp. 33–47.

Coste, René. 1962. *Le problème du droit de guerre dans la pensée de Pie XII*. Paris: Aubier.

Delos, Joseph T. 1959. The Dialectics of War and Peace: Part 1, *The Thomist* 13(3), pp. 305–324; Part 2, 13(4), pp. 528–566.

Delos, Joseph T. and Albert Valensin (eds.). 1932. Documents Annexes. In *Paix et guerre: La guerre devant la conscience*. Juvisy: Cerf, pp. 48–86.

Eppstein, John. 1935. *The Catholic Tradition and the Law of Nations*. Washington, DC: Carnegie Endowment for International Peace.

Jacquin, Robert. 1939. L'ordre international d'après Taparelli d'Azeglio. In Yves de La Brière. ed. *Vitoria et Suarez: Contribution des théologiens au droit international modern*. Paris: A Pédone, pp. 209–223.

Minois, Georges. 1994. *L'Eglise et la guerre*. Paris: Fayard.

Moines de Solesmes. 1956. *Les Enseignements Pontificaux, La paix Internationale, tome 1*. In *La guerre modern*. Paris: Desclée & Cie.

Pollard, John. 1996. Italy. In Tom Buchanan and Martin Conway. eds. *Political Catholicism in Europe 1918–1965*. Oxford: Clarendon Press, pp. 69–96.

Pope Francis. 2020. Fratelli Tutti, On Social Friendship. *Official English Text*, 3 October. Available at: www.vatican.va/content/francesco/en/encyclicals/documents/papa-francesco_20201003_enciclica-fratelli-tutti.html.

Pope Francis. 2023. *Address to the Security Council of the United Nations*. Available at: https://www.vatican.va/content/francesco/en/speeches/2023/june/documents/20230614-consigliosicurezza-onu.html.

Pope John Paul, II, Address to the Diplomatic Corps. 2003. Available at: www.vatican.va/content/john-paul-ii/fr/speeches/2003/january/documents/hf_jp-ii_spe_20030113_diplomatic-corps.html.

Pope Paul, VI. 1965. *Address to the United Nations Organization*. Available at: www.vatican.va/content/paul-vi/en/speeches/1965/documents/hf_p-vi_spe_19651004_united-nations.html.

Pope Pius, XII. 1944. *Christmas Message*. Available at: www.vatican.va/content/pius-xii/en/speeches/1944/documents/hf_p-xii_spe_19441224_natale.html.

Reichberg, Gregory M. 2017. *Thomas Aquinas on War and Peace*. Cambridge: Cambridge University Press.

Reichberg, Gregory M. 2018. Reframing the Catholic Understanding of Just War: Two Contrasting Approaches in the Interwar Period. *Journal of Religious Ethics* 46(3), pp. 570–596.

Reichberg, Gregory M. 2022. From the Nuclear Family to the Family of Nations: Exploring the Analogy. In Pierpaolo Donati. ed. *The Family as Relational Good: The Challenge of Love*. Vatican City: Libreria Editrice Vaticana. Available at: www.pass.va/en/publications/acta/acta_23_pass/reichberg.html.

Reichberg, Gregory M. 2024. The Doctrinal Status of Just War in the Contemporary Teaching of the Catholic Magisterium. *Studies in Christian Ethics* 37(3).

Reichberg, Gregory M., Henrik Syse, and Endre Begby. 2006. *The Ethics of War: Classic and Contemporary Readings*. Oxford: Blackwell Publishing.

Reichberg, Gregory M., Stein Tønnesson, and Henrik Syse. 2022. *Right on its Side, but Not Might? The Lessons and Ancient Greek War Can Teach Ukraine Today*, 22 December. Available at: www.commonwealmagazine.org/ukraine-russia-Thucydides-melian-athens-war.

Second Vatican Council. 1965. Pastoral Constitution of the Church in the Modern World. *Gaudium et Spes*. Available at: www.vatican.va/archive/hist_councils/ii_vatican_council/documents/vat-ii_const_19651207_gaudium-et-spes_en.html.

Sturzo, Luigi. 1929. *The International Community and the Right of War*. Translated by Barbara Barclay Carter. London: George Allen & Unwin.

Sturzo, Luigi. 1931. *La communauté internationale et le droit de guerre*. Translated by Marcel Prélot. Paris. Editions Bloud & Gay.

Sturzo, Luigi. 1936. The Totalitarian State. *Social Research* 3(2), pp. 222–235.

Sturzo, Luigi. 1942. *Les Guerres modernes et la pensée catholique*. Montreal: Editions de l'Arbre.

Sturzo, Luigi. 1954. *La communità internatazionale e il diritto di guerra*. Bologne: Nicola Zanichelli Editore.

Taparelli, Luigi d'Azeglio. 1855 [1840–1843]. *Saggio teoretico di diritto naturale appoggiato sul fatto*. Rome: Civiltà Cattolica.

Walzer, Michael. 1992. *Just and Unjust Wars: A Moral Argument with Historical Illustrations*. 2nd Edition. New York: Basic Books.

12

REINHOLD NIEBUHR (1892–1971)

Eric D. Patterson

Introduction

Reinhold Niebuhr was the most famous American theologian of the twentieth century. However, Niebuhr's fame derived from his application of key Christian concepts to political and social controversies in ways that not only illuminated the underlying causes of the phenomena but also provided policy-relevant guidance at home and abroad. This was particularly true for about four decades from the mid-1930s until ill health consigned him to a much slower pace of activity in the mid-1960s.

Niebuhr is the godfather of a distinctly Augustinian approach to thinking about domestic and, especially, international politics known as Christian realism. Christian realism is a "community of discourse" rather than a formal ideology or disciplined school of thought. According to scholar James A. Herrick, a "community of discourse" is a group "that enable[s] people to think and act with unity" (Herrick, 2021, p. 22) to address a wide range of serious social problems. Niebuhr's friend and colleague, Roger Shinn, describes Christian realism:

> [I]t was Christian in its appropriation of biblical motifs and classical doctrines, such as sin; it was realistic in its criticism of naïve idealism or utopianism, and it was in confrontation with the brute facts and power struggles of the contemporary world. It was alert to both the word of God and the latest news from the European and Asiatic battlefronts.
>
> *(Shinn in Patterson, 2003, p. 4)*

Christian realism was, at its heart, a recovery of older pragmatic and moral ways of thinking about social life rooted in Christian doctrines of

DOI: 10.4324/9781003428688-13

responsibility, justice, and original sin. This framework critiqued the utopian liberalism of Niebuhr's post-World War I milieu and, at the same time, provided a moral argument for action against fascism and communism.

Niebuhr's Christian realism has been an important influence on a number of contemporary just war scholars who are mentioned in this chapter, although Niebuhr himself had reservations about formal just war categories. Niebuhr directly influenced Christian just war thinkers (e.g., Paul Ramsey, Jean Bethke Elshtain, Marc LiVecche) as well as a number of other foreign policy thinkers (e.g., Arthur J. Schlesinger, Hans Morgenthau). Others claim to admire him, including President Jimmy Carter and President Barack Obama. While Obama, who famously referenced the just war tradition in his Nobel Prize acceptance speech in 2009, had his world view also influenced by Niebuhr. In a 2007 interview, when he was a presidential candidate, he remarked on the lessons he learned from reading Niebuhr:

> I take away the compelling idea that there's serious evil in the world, and hardship and pain. And we should be humble and modest in our belief we can eliminate those things. But we shouldn't use that as an excuse for cynicism and inaction. I take away . . . the sense we have to make these efforts knowing they are hard, and not swinging from naïve idealism to bitter realism.
>
> *(Brooks, 2007)*

Niebuhr's Christian realism while theologically Augustinian emphasized the importance of political order in a fallen world and asked that governments take seriously their responsibility—as articulated in Romans 13 and elsewhere—to preserve order, punish wrongdoers, and advance justice. Augustine tells us that what we need to aim for is the "tranquility of order" in this world. It will never be perfect, like heavenly peace in the City of God, but that does not obfuscate the need for political order here. Above all, Niebuhr's teaching communicated to his readers that the political leaders have moral responsibility to promote that order, using force if need be.

Contexts

Karl Paul Reinhold Niebuhr (1892–1971) was born to German immigrant parents in Missouri. Like many immigrant families, German was spoken at home, and the family was part of the typical culture of German–American immigrants of the time, with life associated around the family, community, and local church. The family later moved to Illinois where Niebuhr's father pastored St. John's German Evangelical Synod church. Niebuhr was educated at Elmhurst College in Illinois and Edens Theological Seminary in Missouri and later earned his bachelor of divinity and master of arts degrees from Yale

Divinity School. In the decade before moving to Union Theological Seminary in New York City, Niebuhr served as a pastor in the Evangelical Synod of North America (later called the Evangelical and Reformed Church), first taking the pulpit of his father's church upon his father's death and later serving for a number of years at Bethel Evangelical Church in Detroit.

During World War I, Niebuhr spoke out in favor of German–American patriotism for the United States and criticized the aggression of the Central Powers, attempting to balance a commitment to pacific approaches to security with the need to stop evil (Holder and Josephson, 2013). This balancing act was particularly important in his immediate context, when German immigrants speaking the German language at home and church were seen as potential collaborators with Berlin.

Across his Detroit tenure (1915 to 1928), Niebuhr's congregation grew from being fewer than 100 to over 700 members. It was during his Detroit ministry that Niebuhr saw first hand the practical and ideological problems of power politics. Pastor Niebuhr joined labor leaders in calling attention to the low wages associated with having such a large and willing, low-skilled labor pool. Niebuhr was also critical of the mind- and spirit-numbing nature of the physical work he witnessed when visiting factories. Thus, he came to realize that it was necessary for those without power to organize to counterbalance the power of those at the top, and he became an outspoken friend of labor unions and the civil rights movement and was an influence on Martin Luther King Jr. This came to a head with the growth of the nativist, anti-Catholic, racist Ku Klux Klan (KKK), which openly supported candidates for Detroit's 1925 elections. Niebuhr loudly campaigned against the Klan's candidates.

At the same time, Niebuhr was increasingly influenced by Marxist ideas of power, class competition, and how often political platforms and moral slogans hid deeper economic structures of power. He saw this in the class and racial inequality faced by black Americans, immigrants, and the working poor. Moreover, his pacifist-leaning theological presuppositions and the horrors of World War I convinced him, like many others of his generation, that war was inherently immoral and destructive. Thus, international law and institutions were the best path toward global peace, and he was clearly, at this time, what we would today call a "democratic socialist."

In 1931, Niebuhr married Ursula Keppel-Compton, and the couple had two children. Ursula went on to have a successful career as a professor at Barnard College, while Reinhold taught at nearby Union Theological Seminary from 1928 to 1960. During Niebuhr's long tenure at Union, he continued to be politically active, although transitioning toward a more "realist" posture as what today we would call a "social democrat," including helping establish what became the International Rescue Committee and as well as serving as the first president of the organization that became known as Americans for Democratic Action.

However, by the early 1930s, Niebuhr's theological and political presuppositions that previously leaned toward pacifism, democratic institutions, and the social gospel were mugged by the realities of racial chauvinism and the brutal revanchism of Nazi and Japanese imperialism. In 1939, he famously wrote,

> About midway in my ministry . . . I underwent a fairly complete conversion of thought which involved rejection of almost all the liberal theological ideals and ideas with which I ventured forth in 1915. I wrote a book *Does Civilization Need Religion?* my first, in 1927 which when now consulted is proved to contain almost all the theological windmills against which today I tilt my sword. These windmills must have tumbled shortly thereafter for every succeeding volume expresses a more and more explicit revolt against what is usually known as liberal culture.
>
> *(Niebuhr, 1939, p. 542)*[1]

In hundreds of speeches, essays, and editorials, most notably in *The Christian Century* and then in the magazine he cofounded, *Christianity and Crisis*, Niebuhr helped shape public debates on the issues of war, peace, and security during World War II and the Cold War with his emphasis on classic Christian doctrines of sin, pride, and neighbor love. Throughout the next three decades, Niebuhr took staunch positions against all forms of political chauvinism, whether based on Marxist, Leninist, fascist, racial, ethno-nationalist, or other ideologies. Thus, he typically supported the overall policies of containment and resistance to aggression practiced by Washington and London, while remaining critical of colonialism, racism, and Western hubris. He could be very critical of specific foreign policies and national leaders, but his judicious, pragmatic counsel was sought not only in the academy but also at the U.S. Department of State, the White House, and elsewhere. He had the flexibility of thought to change course when new evidence became available, as he did when his support for U.S. involvement in Vietnam flagged over time.

Niebuhr's frenetic pace of teaching, public speaking, writing, and travel dramatically slowed after he suffered a series of strokes. When he passed away in 1971, it was not immediately clear if there would be a future for his form of Christian realism as his academic partners in early Christian realism, most notably his long-time Union colleague John C. Bennett, had left the Christian realism scene, and there was not a clear intellectual tradition or set of successors (Patterson, 2003). That said, as Patterson and Joustra argued in *Power Politics and Moral Order: Three Generations of Christian Realism—A Reader*, Niebuhr's thought was influential for a second and third generation of Augustinian realists and just war scholars, including Paul Ramsey, Jean Bethke Elshtain, Ernest Lefever, Marc LiVecche, Mark Tooley, Joseph Loconte, and others (Patterson and Joustra, 2022).

Texts and Tenets

It is beyond the scope of this chapter to deal with the massive Niebuhr library, from his speeches and sermons to dozens of books to thousands of published selections. When it comes to Niebuhr's contributions to thinking on political order, international affairs, and the ethics of warfare, two important works stand out. Niebuhr started out as a pacifist or quasi-pacifist in the 1920s and early '30s because he was horrified by the carnage of World War I. Over time, though, Niebuhr realized that responsible, moral action was required to turn back the viciousness of the Nazis, which Niebuhr called for in one of his essays, stating "the Hitlerian imperial will must be broken" (Niebuhr, 1967, p. 52). Quietism of any sort, although supported by some religious groups (notably in the Anabaptist tradition), was seen as an irresponsible behavior because it aided in maintaining the unjust status quo. But his two ground-breaking books that helped to define the field of Christian realism underscore that morally responsible individuals have to stand up to evil. The first was written in the context of World War II, and the second was written just a few years later. Both focus more attention on democracy and the American politico-cultural system, but the presuppositions therein are largely theological and directly relate to how Niebuhr thinks about justice, war, and peace. Those two books are *The Children of Light and the Children of Darkness* (1944) and *The Irony of American History* (1952).

During World War II, Niebuhr gave a series of lectures that ultimately became his volume: *The Children of Light and the Children of Darkness*. The book claims to be a "compelling justification" and "a more realistic vindication" of democracy than the idealistic democratic theories of his day. But the place to begin is his distinction between the types of "children."

For Niebuhr, the children of light are utopians—moral idealists who optimistically believe that reason, progress, and planning will allow society to overcome the challenges that it faces. The children of light assure themselves that individual self-interest can be overcome as men and women willingly submit their unique self-interest to a higher morality (Niebuhr, 1944, p. 9). Niebuhr points to two strands of this idealism in his day. The first is the innocent American bourgeoisie children of light who naively believe that the lesson of American history is that the ideals of the Enlightenment combined with the evolution of society and the invisible hand of the market make the realization of a harmony of interests across society possible (Niebuhr, 1944, p. 7). Niebuhr identifies a more virulent idealism in the utopianism of pre-war Communism, which also envisioned progress toward a Halcyonic society based on progress, science, and planning.

Niebuhr calls the children of light "virtuous, but foolish." In contrast, he calls the children of darkness "wise, but evil" (Niebuhr, 1944, p. 10). The children of darkness are characterized by a moral cynicism that elevates the individual or the tribe by dint of power and prejudice. The children of darkness know no law beyond their own will and interest and therefore are willing to use any means at their disposal to advance their interests, regardless of others. Moreover, the children of darkness are wise in not only advancing their own cause but also exploiting the contradictions in other communities for their own ends.

Hence, the Nazis provided Niebuhr with his contemporary exemplar of the children of darkness. The Nazis were motivated by an exclusionist, absolutist ethic that elevated their tribe to prominence at the expense of others. The Nazis were wise in their use of power for the better part of a decade, ruthlessly punishing some foes on the battlefield (or concentration camps) while cleverly exploiting the weaknesses of the children of light. For instance, Hitler's annexation of the Sudetenland demonstrated the curious mixture of temerity and good intentions on the part of children of light like Neville Chamberlain. Similarly, the Nazis realized that the fractures in Western society between the ruling classes and their subordinates based on economic interests (e.g., "Red" labor unions) and race (blacks, Jews) diminished the effective resistance of the Western powers to German revanchism (Niebuhr, 1944, p. 11).

So, in 1944, Niebuhr avoided classical just war language, but his theological framing of realism versus idealism, focused on moral categories of sin and chauvinism, is clearly informed by the Augustinian (Christian) tradition. Indeed, in 1953, Niebuhr wrote, "[Augustine]. . . proves himself a more reliable guide than any known thinker. A generation which finds its communities imperiled and in decay . . . might well take counsel of Augustine in solving its perplexities" (Niebuhr, 1953, p. 53). Niebuhr's call to action against the children of darkness (Nazis) is reminiscent of Augustine's letter to Boniface (Letter 189), routinely cited by just war thinkers, in which Augustine affirms Count Boniface's military vocation to fight against temporal evil (St. Augustine in Schaff, 1887).

Interestingly, Niebuhr suggests that democracy itself provides a set of solutions to the perennial struggle between the children of light and the children of darkness. Niebuhr begins with the assertion that the Christian doctrine of original sin is the starting point for any theory of politics (Niebuhr, 1953, p. 16). He writes,

> [The book is] informed by the belief that a Christian view of human nature is more adequate for the development of a democratic society than either the optimism with which democracy has become historically associated or the moral cynicism . . .[associated with] tyrannical political strategies.
>
> *(Niebuhr, 1953)*

Niebuhr agrees with classical Christian doctrine that human beings are made in the image of God and therefore have a tremendous creative potential.

However, human beings are also marred by the Fall and therefore are a complex blend of light and darkness. Democracy is responsive to both elements of human nature: "Man's capacity for justice makes democracy possible; but man's inclination to injustice makes democracy necessary" (Niebuhr, 1953, p. xi).

In 1952, Reinhold Niebuhr published his famous book, *The Irony of American History*.

Niebuhr defines irony as "apparently fortuitous incongruities, which, upon closer examination, are not merely fortuitous" (Niebuhr, 1952, p. xxiv). What makes something ironic is the element of the comedic and the unexpected—that which appears to be strength is often the very mechanism of weakness or downfall. For Niebuhr, the ironic is marked by a certain pretension (e.g., strength, virtue) that obscures unconscious weakness. For example, Niebuhr observed the irony that the United States trumpets its prosperity, believing it to be evidence of its virtues, because critics abroad see American wealth and boasting as evidence of imperialism. He observes, "every effort we make to prove the virtue of our 'way of life' is used by our enemies and detractors as proof of our guilt" (Niebuhr, 1952, p. 110).

Niebuhr's central thesis in *The Irony of American History* is that the juxtaposition of American naïveté with inordinate American power in the immediate post-war world was ironic. On the one hand, the United States had existed in splendid isolation and eschewed the immoral politics of European colonialism and continental conflict. On the other hand, the U.S. capacity, especially its economic power, resulted in it achieving superpower status and responsibilities in a very short period time. Hence, Niebuhr observed an adolescent United States motivated by the rhetoric and reality of its own unique history infused with a youthful idealism. Niebuhr did not dispute that there were elements of the American experience that truly made it a "city on a hill," but he critiqued the "Messianic dream" that underscored American exceptionalism. Indeed, it was the pretension that America acted exclusively in terms of a higher morality without regard to self-interestedness that Niebuhr indicted. Niebuhr observed, "our sense of responsibility to a world community beyond our borders is a virtue, even though it is partly derived from a prudent understanding of our own interests" (Niebuhr, 1952, p. 7). The irony of American naïveté was its lack of appreciation for how its own moral discourse obfuscated politics based on national interests (e.g., Manifest Destiny, the Spanish-American War) (Niebuhr, 1952, pp. 15, 18–19, 22–23).

What made the situation in 1952 considerably more complicated was the dawn of American military, political, and economic superiority. Niebuhr feared that it was possible that such power, especially military power, might be combined with the vanity of American moral exceptionalism and result in aggressive policies for the "higher good." This was especially possible, Niebuhr thought, if restless America became frustrated that its goodwill efforts at diplomacy and sharing democracy did little to result in quick fixes for international dilemmas. Indeed, Niebuhr pointed out, American diplomatic engagement combined with awesome Yankee power ironically generated insecurity—the classic security dilemma—rather than trust in some parts of the globe:

> [T]he paradise of our domestic security is suspended in a hell of global insecurity . . . we are the poorer for the global responsibilities we bear. And the fulfillments of our desires are mixed with frustrations and vexations.
>
> *(Niebuhr, 1952, p. 7)*

Today, Niebuhr's *Irony* is usually quoted simply in the context of the dramatic nexus of idealism and power in the American experience. Nonetheless, Niebuhr had more to say on other issues as well. Perhaps the most important was his evaluation of America's competition: Soviet Communism. Niebuhr suggested that Communism was likewise an idealist's delusion—but a far more dangerous and extreme form. What particularly concerned Niebuhr was the lack of institutionalized checks of any kind in the Soviet state, which he called "a vast religious-political movement which generates more extravagant forms of political injustice and cruelty" (Niebuhr, 1952, p. 22).

For Niebuhr, there was a certain irony that the two great idealisms of his day, Communism and capitalist democracy, had defeated nihilistic fascism and were squaring off against one another. Nevertheless, he clearly defined the one as the greater evil, calling Communism "satanic" due to its absolutist ideology and the violent means associated with its ends (Niebuhr, 1952, p. 15).

Niebuhr's *Irony* continues to be read and remains influential. Military historian Andrew Bacevich has called it "the most important book ever written on U.S. foreign policy" (Bacevich, 2008 [1952], p. ix). It is largely due to the arguments in *Irony* that Jimmy Carter, Barack Obama, and John McCain all cite the importance of Niebuhr. And, from a just war perspective, the themes of *Irony* have been picked up again and again, most notably the issue of political authorities responsibly using power while focusing on the motivations ("right intention") for doing so.

Controversies

Classic just war reasoning argues that *political authorities* may decide to utilize force when they are acting on a *just cause* with *right intention*. This helps us distinguish between *force* (lawful, restrained) and *violence* (vengeful, unstrained). When considering the use of force, political authorities should then take into account secondary, prudential criteria, considering whether an armed response is *proportionate* to the threat or grievance, the *likelihood of success,* and taking all reasonable diplomatic measures before employing force (*last resort*).

Once the decision to employ force has been made using the above criteria, battlefield and operational principles further guide and limit the use of force: tactics and weapons should be used robustly with an eye toward victory and troop protection (*military necessity*) in ways that are *proportionate* to the battlefield situation, distinguishing as much as reasonably possibly (*distinction*) between lawful targets, such as foreign military personnel, and unlawful and immoral targets (e.g., houses of worship, hospitals, innocent bystanders). When considering late and post-conflict, a just and enduring peace at war's end is based first on a secure political *order* that seeks *justice* and promotes *conciliation*. Scholars, theologians, philosophers, and practitioners have long called these three dimensions of just war the ethics of going to war (*jus ad bellum*), the ethics of how war is fought (*jus in bello*), and the ethics of post-conflict (*jus post bellum*).

Just war scholarship has had a renaissance in the post-World War II era. Despite the fact that the categories have long been a part of philosophy and statecraft and began to increasingly find their way from customary to positive international law beginning in the last nineteenth century, nonetheless, the destructive nature of the two world wars and the beginning of the atomic age rejuvenated just war scholarship.

Interestingly, however, although Niebuhr used some of the language of just war, he was not a systematic just war thinker, and he was openly skeptical about any categorical framework for foreign policy ethics.

Consider the three deontological principles of *legitimate authority*, *just cause*, and *right intention*. These tend to assume that only political authorities should have force at their disposal, but Niebuhr consistently pointed out the moral flaws of those holding power: power, and thus resources, influence, and security, was typically out of balance. The only way that the oppressed and the vulnerable would have justice would be to rebalance power, and that often requires some form of force. This was Niebuhr's understanding of Gandhi's people power movements in India. Niebuhr called Gandhi's movement "violent" action, because shutting down British textile factories through strikes in India meant that the children of factory laborers went hungry at night. For Niebuhr, this was the application of force to rebalance societal power, and it did cause economic and physical harm. At the same time, Niebuhr problematized the legitimacy of British colonial rule. On the one hand, that rule was imposed by a mixture of force and self-interest. One should not pretend that all of the ethical claims of "the white man's burden" were morally praiseworthy. Moreover, the evolving political situation on the Indian subcontinent called for a better form of justice: the liberation of India's peoples. However, the British Empire was far different from the Third Reich and, according to Niebuhr, thoughtful people can tell the difference in the moral failings of the two systems.

In short, Niebuhr was a dialectician, and he was hostile to applying a sophomoric war checklist of authority, just cause, and right intentions from the safety of the academy. He consistently wanted to pull back the layers of the legitimacy of authority and expose the pretensions that those with power and resources had. He then wanted power to be used forcefully on behalf of justice.

At the same time, however, Niebuhr was deeply skeptical that there could be a virtuous, or just, use of force. The classic just war position, from Martin Luther to C.S. Lewis to J. Daryl Charles, is that statesman and warriors—those with a public service vocation to protect and defend—may be acting morally when employing force. In other words, it is a public good when, as Lewis suggested, a soldier or policeman stops a "homicidal maniac" from harming a child, even if that means killing. Niebuhr, in contrast, routinely spoke of "lesser evils" and suggested what many call the "dirty hands" argument. "Lesser evil" is the idea that killing and the use of force (Niebuhr would have said "violence") are always wrong, always an evil. However, political authorities may have to employ a lesser evil approach to defend their citizens from the greater evil of the rapacious oppression of criminals and adversaries such the Nazis. When soldiers kill in order to protect their fellow citizens, they, and the political authorities who send them to fight, have killed and, therefore, have a level of guilt for killing. In other words, according to Niebuhr, they have "dirty hands" for violating the supreme moral law, what Niebuhr refers to over and over again as "the law of love," which is rooted in the Augustinian notion of *caritas* (i.e., charity, neighbor love).

When it comes to the secondary, prudential just war criteria, Niebuhr was practical. Of course, it is important to utilize all the levers of national power, including diplomacy, sanctions, and the like. Again, he simply had little use for checklist of just war thinking that was really a quasi-pacifism keeping government leaders from doing their duty to justice. Niebuhr was a great contextualist, and so he saw each context in its local and historical context, realizing that although there are some lessons to be learned from history, nonetheless, leaders have a public duty to act in the here and now. Niebuhr realized that those in power had an obligation to action that the academic or editorialist does not, and that leaders had to do so with a host of factors influencing their decisions, from media scrutiny to imperfect intelligence on the motives of adversaries.

When it comes to *military necessity*, *proportionality*, and *discrimination*, Niebuhr has been charged by scholars such as Keith Pavlischek as being an advocate for unrestrained warfare (Pavlischek, 2008). This is in part due to Niebuhr's support for some of the most destructive Allied tactics of World War II, such as the use of the atomic bomb against Japan and his later support for nuclear deterrence against the Warsaw Pact. Niebuhr was not an advocate of entirely unrestrained warfare. His view was that justice compelled the Allied Powers to defeat a diabolical set of foes in the Axis Powers and that it would take extreme measures to bring the war to an end. For

Niebuhr, what was most important in all of this was the motivating factors for the decision to bomb factories and cities. If aerial raids in Germany and Japan were designed to exact vengeful retribution from German and Japanese citizens, then it was a vindictive, sinful policy. But if these activities were designed to break what Niebuhr once called the "Hitlerian will" and end the war, then they were appropriate forms of attack.

When it comes to the post-war principles of *order*, *justice*, and *conciliation*, Niebuhr wrote a great deal about order and justice. The basic imposition of a just and secure political order was a frequent topic of his writing, and it naturally flowed from his opposition to the "demonic" systems of Communism and National Socialism. When considering domestic and international political arrangements, his balance-of-power approach naturally made him a proponent of democratic institutions with their checks-and-balances. This, too, was his justification for a form of United Nations organization as a form of international collective security. In short, he was no idealist claiming that democracy was a good in and of itself, a post-war ideal to be achieved. He saw democratic mechanisms in the U.S. context (federalism, separation of powers, checks, and balances) and on the international scene (e.g., collective security, international law, the UN) as procedural checks on unequal power.

In conclusion, the mature Niebuhr was always skeptical of three things when it came to the just war tradition. The first was the moral grandstanding by governments, particularly when it came to issues of war and national security policy. Thus, he would be one to talk about the "justified" use of force in a given situation rather than calling any conflict a "just war." The second was that he was forever uncomfortable with frameworks, such as the formal just war criteria, as they being too limiting both for statesman and for intellectual inquiry. Third, Niebuhr always emphasized the concepts of "lesser evil" and "dirty hands" in his analyses of force, and thus he differed from the traditional just war view that leaders and warriors can use force virtuously to restrain evil and promote justice.

Legacy

Reinhold Niebuhr was not the only Christian realist of his time, but he had an outsized influence on those around him and future intellectuals who may be called the second and third generations of Christian realism (Patterson and Joustra, 2022). That first generation included Niebuhr's close friend and colleague at Union Seminary, John C. Bennett, author of books such as *Christian Realism* and an author at the magazine Niebuhr co-founded, *Christianity and Crisis*. U.S. Secretary of State John Foster Dulles, an important figure in interfaith work and author of the Six Pillars of Peace that outlined a moral framework for the post-World War II world, was also a Christian realist. British figures such as historian Herbert Butterfield, diplomat Adam Watson, and influential international relations scholar Martin Wight, who is often associated with the English School, fit into this camp not just due to their thinking but also due to their associations and interactions, such as with the British version of the Rockefeller Foundation-funded Committee on Theory of International Politics (Patterson and Joustra, 2022)

This first generation of Christian realism roughly runs from about 1932 to 1965 and corresponds to the struggles against militant fascism and then Communism. By the late 1960s, the first generation of Christian realists were leaving the scene due to age and health, but the focus also shifted. Although Niebuhrian analysis continued, a new generation applied that analysis to the evolving character of international affairs (Patterson and Joustra, 2022).

The second generation of Christian realists typically interacted with Niebuhr at some point. Among the most prominent in the new generation were Princeton University ethicist Paul Ramsey (1913–88), foreign policy expert Ernest W. Lefever (1919–2009), and political scientist Kenneth W. Thompson (1921–2013). Ramsey's emphasis on Christian just war thinking, in the context of wars of national liberation, Communist insurgencies, WMDs, and Vietnam inspired a generation of new just war thinking rooted in Christian tradition, most notably that of his student James Turner Johnson (b. 1938). Thompson supported the work of Niebuhr and others from his position leading the Rockefeller Foundation's International Relations Program and then went on, while at the University of Virginia, to publish dozens of books and monographs of his own on the ethics of domestic and foreign policy. Some Catholics, most notably the young George Weigel (b. 1951), rose to prominence making Augustinian and just war arguments about the Cold War. Weigel went to work at Lefever's Ethics and Public Policy Center (Patterson and Joustra, 2022). Jean Bethke Elshtain (1941–2013) increasingly cited the influence of Augustine and Niebuhr, whom she called the "greatest theologian of his time" (Elshtain, 2003, p. 106).

Niebuhr's influence continued into a third generation of Christian realism, that of the post-Cold War era. This generation of Christian realists was heavily influenced by a renaissance in a strand of Augustinian-inspired thinking: Christian just war statecraft, influenced particularly by Paul Ramsey and James Turner Johnson. Over time, self-identifying Christian realists grew more theologically conservative, committed to the orthodoxy of Scripture and their theological traditions than to Niebuhr and his Union Seminary friends. Does this matter? Many of today's Christian realists think so. One example is a well-known essay by Keith Pavlischek, a Ph.D. scholar in religious ethics who also served in the U.S. Marine Corps and intelligence community. Pavlischek provides an insightful critique of Niebuhr's thinking. Pavlischek argues that the Idealist-turned-Realist Niebuhr never fully lost his youthful idealism, which corrupted his approach to war and security, allowing him to justify terrible things—such as the fire-bombing of Dresden—as "lesser evils" because Niebuhr lacked the discipline imposed by the Augustinian just war (Pavlischek, 2008).

Today, the theological conservative trend has meant a far more disciplined approach to matters of war and peace framed by Augustinian just war thinking, and thus a more critical stance toward pacifism and the shoddy thinking that cannot tell the difference between nonviolent direct action in a democracy (such as Martin Luther King Jr.'s marches and sit-ins) and terrorism and insurgency or justified collective self-defense (war). Contemporary Christian realism is also characterized by a friendly conversation between like-minded Catholics (George Weigel, Joseph Capizzi), Christian public intellectuals (Oliver O'Donovan, Jean Bethke Elshtain), and conservative Protestants (Eric Patterson, J. Daryl Charles, Nigel Biggar, Mark Tooley, Marc LiVecche, Keith Pavlischek). Contemporary Christian realism, with its Augustinian roots and moral-historical methodology, already had some overlaps with the English School of international relations theory, from Adam Watson to Scott Thomas. As Robert Joustra and Simon Polinder observe, Christian realism has found new international expressions in Reformed circles, with a growing appreciation of Dutch theologian-statesmen like Abraham Kuyper and Hermann Bavinck as early influencers of a nascent "Amsterdam School."

Nevertheless, what one finds at the end of the day is that almost every single one of the third-generation thinkers cite Niebuhr as crucial to their intellectual development. In my own case, it was the discovery of his writing as a Christian thinker on topics from original sin to ethno-religious nationalism that captured my attention, resulting in three edited volumes on Christian realism's past: *The Christian Realists: Reassessing the Contributions of Niebuhr and His Contemporaries* (Patterson, 2003), *Christianity and Power Politics Today: Christian Realism and Contemporary Political Dilemmas* (Patterson, 2008), and *Power Politics and Moral Order: Three Generations of Christian Realism—A Reader* (Patterson and Joustra, 2022). The establishment of *Providence: A Journal of Christianity and American Foreign Policy* seeks to fill the gap that Niebuhr's *Christianity and Crisis* did during World War II. A new release in 2023 brings more theologians to the discussion: *The Future of Christian Realism: International Conflict, Political Decay, and the Crisis of Democracy* (Gingles et al., 2023). Ironically, Niebuhr's work continues to inspire scholars and students in international relations and related social sciences although his impact on theological studies is negligible. Thus, he remains an important figure as we consider the ethics of statecraft and force.

Note

1 Note, for the contemporary reader, Niebuhr's use of the term "liberal" does not mean classical liberal but rather what he would have called "idealist" and today would be labeled "progressive."

Works Cited

Augustine, St. 1887. Letter to Boniface (Letter 189). In Philip Schaff. ed. *Nicene and Post-Nicene Fathers, First Series, Vol. 1.* Translated by J. G. Cunningham. Buffalo, NY: Christian Literature Publishing Co. (Revised and edited for New Advent by Kevin Knight. Available at: www.newadvent.org/fathers/1102189.htm.)

Bacevich, Andrew J. 2008 [1952]. Introduction. In Reinhold Niebuhr. ed. *The Irony of American History.* Chicago and London: University of Chicago Press.

Brooks, David. 2007. Obama, Gospel and Verse. *The New York Times,* 26 April. Available at: http://select.nytimes.com/2007/04/26/opinion/26brooks.html.

Elshtain, Jean Bethke. 2003. *Just War against Terror: Ethics and the Burden of American Power in a Violent World.* New York: Basic Books.

Gingles, Dallas, Joshua Mauldin, and Rebekah L. Miles. 2023. *The Future of Christian Realism: International Conflict, Political Decay, and the Crisis of Democracy.* New York & London: Lexington Books.

Herrick, James A. 2021. *The History and Theory of Rhetoric: An Introduction.* 7th Edition. New York and London: Routledge.

Holder, R. Ward and Peter B. Josephson. 2013. Obama's Niebuhr Problem. *Church History* 82(3), pp. 678–687. https://doi.org/10.1017/S000964071300070X.

Niebuhr, Reinhold. 1939. Ten Years that Shook My World. *The Christian Century* 56(17), pp. 542–546.

Niebuhr, Reinhold. 1944. *The Children of Light and the Children of Darkness.* New York: Charles Scribner's Sons.

Niebuhr, Reinhold. 1952. *The Irony of American History.* Chicago and London: University of Chicago Press.

Niebuhr, Reinhold. 1953. *Christian Realism and Political Problems.* New York: Charles Scribner's Sons.

Niebuhr, Reinhold. 1967. Christian Faith and Natural Law. In D. B. Robertson. ed. *Love and Justice: Selections from the Shorter Writings of Reinhold Niebuhr.* Cleveland and New York: Meridian Books, pp. 46–54.

Patterson, Eric D. (ed.). 2003. *The Christian Realists: Reassessing the Contribution of Niebuhr and His Contemporaries.* Lanham, MD: University Press of America.

Patterson, Eric D. (ed.). 2008. *Christianity and Power Politics Today: Christian Realism and Contemporary Political Dilemmas.* New York: Palgrave Macmillan.

Patterson, Eric D. and Robert J. Joustra (eds.). 2022. *Power Politics and Moral Order: Three Generations of Christian Realism—A Reader.* Eugene, OR: Cascade Books.

Pavlischek, Keith. 2008. Reinhold Niebuhr, Christian Realism, and Just War Theory: A Critique. In Eric Patterson. ed. *Christianity and Power Politics Today: Christian Realism and Contemporary Political Dilemmas.* New York: Palgrave Macmillan, pp. 53–72.

13

G.E.M. ANSCOMBE (1919–2001)

Chris Brown

Introduction

Elizabeth Anscombe was one of the most influential British philosophers of the second half of the twentieth century.[1] Her career was stellar; after a First in Greats at Oxford in 1942, she was awarded a postgraduate fellowship at Newnham College in Cambridge, which she took up in order to become a student of Wittgenstein. She returned to Somerville College, Oxford, in 1946 while still travelling frequently to Cambridge to visit Wittgenstein whose friend she had become. After his death in 1951, she was one of his two literary executors and was entrusted by him with the translation into English of what became the *Philosophical Investigations* (1953), one of the most influential works of philosophy of the last hundred years.[2] As well as propagating Wittgensteinian ideas, her study of *Intention* (1957) garnered extraordinary praise ('the most important treatment of action since Aristotle'—Donald Davidson) and was the basis of her reputation among academic philosophers, on the strength of which she was elected Professor of Philosophy at the University of Cambridge in 1970, which post she held until retirement in 1986. Of perhaps greater importance for the general reader and to this essay was her role in revitalising the study of ethics and, in particular, in the development of what came to be known as 'virtue ethics'.

Context

To grasp what was involved here, a bit of context may be helpful. In the years around the Second World War, Oxford philosophy and, in particular, the study of ethics were largely founded on the proposition that terms like 'good'

DOI: 10.4324/9781003428688-14

or 'bad' had no substantive meaning—the older generation of English Hegelians, Bradley, Green, and Bosanquet, whose work could have been employed to resist this position were decidedly out of fashion, and to follow their lead was to invite derision. Instead, key current figures were A.J. Ayer and R.M. Hare. Ayer's (1936) classic introduced to the Anglo-Saxon world the logical positivism developed by the Carnap Circle in Vienna in the 1920s. The central move of *Language, Truth and Logic*, owing much to the work of David Hume, was to draw an absolutely rigid distinction between statements of fact and statements of value and to deny that the latter could be deduced from, or verified by, the former. To say that something was 'good', on this account, was simply to say 'I like it' and 'bad' meant 'I don't like it'—any attempt to expand the meaning of these terms was described not as 'wrong'—because that would imply there was a standard against which claims could be measured—but as 'meaningless'. Ayer's many disciples killed any discussion that broadened the meaning of good or bad by claiming not to understand what was being said—the frequently heard statement 'I don't understand' was not a request for enlightenment but an accusation of sloppy thinking directed against anyone whose position hinted at metaphysics. Hare's *The Language of Morals* (1951) offered a slight but significant modification to Ayer's logical positivism by regarding the use of terms like 'good' and 'bad' as essentially prescriptive. To say of something that it was good was to say 'I like it, and so should you'. This is a genuine modification of the so-called 'hooray–boo' theory of good and bad, but it preserves the basic idea that there is no substantive reference point here and no fact about the world, or about human beings, that could be relevant.

The implications of this position were hard to take for those who had lived through the era of concentration camps and genocide. Was it really the case that there was no objective sense in which one could say that the Nazis who had perpetrated such atrocities were wrong to have done so? Perhaps there was not, and we have to have the courage to face the unfortunate fact that there are no external standards of right and wrong which would allow us to make such a judgement. In this respect if in no other, Oxford philosophy was similar to Jean-Paul Sartre's existentialist philosophy in contemporary France. Sartre (2007) and his disciples understood the defeat of fascism as occurring because people were prepared to take a stand against it, with no guarantee of victory or of the rightness of their action.[3]

In any event, although exposed both to Oxford orthodoxy and, via Iris Murdoch, to French existentialism, Anscombe was one of a number of Oxford philosophers, mostly women, who strongly resisted this limited account of ethics.[4] Of the four women identified as important to this resistance in recent books by Benjamin Lipscomb (2021) and Clare Mac Cumhaill and Rachael Wiseman (2022), the two key figures for this context are, I think, Anscombe and Philippa Foot. Iris Murdoch, whose profile as a novelist and public intellectual was higher than any of the other three, made important philosophical contributions but was less involved with the academic discipline than Anscombe or Foot, while Mary Midgley's important work on the human animal came to fruition much later in the period. Simply as a moral philosopher, Philippa Foot's work was perhaps the most important of the four, her studies re-establishing an Aristotelian account of the virtues as being central to the task of combating the absence of substance in the work of Ayer, Hare, and their disciples, and her account of natural goodness are the key antidotes to logical positivism and prescriptivism.[5] But if Foot was, in the end, the key figure in the resistance, Anscombe made one vital contribution early in the game, in the form of a short but very influential essay entitled 'Modern Moral Philosophy'.[6]

In this paper, Anscombe advanced three theses: first, that moral philosophy was not possible in the absence of a viable philosophy of psychology; second, that terms like 'moral duty' and 'moral obligation' are hangovers from an earlier conception of ethics, which no longer survives; and, third, that the differences between the various British moral philosophies developed over the previous century or more are of little, if any, significance. The importance of psychology follows from the fact that the dominant approach to ethics asks of any situation, 'what ought *we* to do?' Anscombe's point is that trying to answer this question without an account of who or what '*we*' are is futile. In the past, such an account existed, inherited from the Greeks, especially the Stoics, and the Old Testament and developed by Christianity, and words such as 'duty' and 'ought' stem from that inheritance, but modern secular society, while continuing to use this vocabulary, has jettisoned the framework—the 'form of life,' as Wittgenstein might say—that made sense of it.[7] The modern world tries to operate a legal form of ethics but without a legislator. A law conception of ethics requires a God as lawgiver—Jews, Stoics, and Christians have such a belief, but contemporary moral philosophy, while continuing to use the language such a conception generated, no longer accepts the foundation upon which that language is built. Instead, modern moral theory is 'consequentialist' (a term coined by Anscombe in this essay) and as such is incompatible with the:

> Hebrew-Christian ethic which teaches that '. . . there are certain things forbidden whatever consequences threaten, such as; choosing to kill the innocent for whatever purpose however good; treachery . . . idolatry, sodomy, adultery, making a false profession of faith.'

Modern philosophers believe that such prohibitions do not operate in the face of some consequences. 'But of course the strictness of the prohibition has as its point *that you are not to be tempted by fear or hope of consequences*' (Anscombe, 1981, p. 34). This ethic is, of course, incompatible with classical utilitarianism as well as emotivism and prescriptivism, but it also runs contrary to Kantianism—Kant's 'categorical imperative' is ultimately as ungrounded as Bentham's pleasure principle, as was illustrated by the fact that some Nazi war criminals felt able to call in aid Kantian principles to explain their actions.

An Aristotelian approach to the virtues was important to Anscombe, but not because he offered a law conception of ethics—on the contrary, in the classical world, it was the Stoics who envisaged a divine legislator; Aristotle's approach is different and does not rely on words like 'duty' or 'ought'.[8] Rather, Aristotle's *Ethics* is based on the notion that a good human being is one who acts in accordance with the virtues, and it follows that human beings are capable of recognising that certain types of actions are contrary to those virtues—to reiterate, he does not use 'ought' language to talk about morals but rather, for example, characterises action incompatible with the virtue of justice as 'unjust'. Anscombe holds that this language is still available to us—we can say that, for example, to convict an innocent person of a crime is 'unjust' without invoking the kind of thinking that would be involved in calling it 'wrong'. This is an important point, as we will see, because it means that ethical statements can make sense in the absence of a lawgiver, but in fact, as the above list of prohibitions may have indicated to the reader, Anscombe actually *does* believe in the existence of a divine legislator and so has no difficulty herself in using a law conception of ethics. She converted to Roman Catholicism while still at school in the late 1930s and for the rest of her life was a practising Catholic—indeed, like many another convert, her Catholicism was particularly rigid and demanding, strikingly more so than the Catholicism of the 'cradle-Catholic'. In later life, much of her energies went into opposing changes to the law prohibiting abortion and into defending traditional doctrine on contraception. In the context of this chapter, though, of the greater interest is the small number of papers she wrote defending a Thomist version of the just war.

Texts and Controversies

The three papers are: 'The Justice of the Present War Examined' (hereafter 'JPWE'), the first part of a self-published pamphlet written with Norman Daniel in 1939; 'Mr Truman's Degree', a pamphlet published in Oxford by Anscombe in 1957 containing a version of the speech made to Convocation opposing the award of an honorary Doctorate to President Truman; and, 'War And Murder,' a contribution to a collection *Nuclear Weapons: A Catholic Response* edited by Walter Stein in 1961.[9] It may be worth noting that JPWE was originally subtitled 'A Catholic Response' until the authors were required by the Bishop of Birmingham to remove this description as the essay did not in fact have the necessary *imprimatur*. Two of the three papers were self-published squibs, but Anscombe obviously felt sufficiently committed to them to include them in her *Collected Philosophical Papers*, and rightly so, because in these pages are to be found both an excellent short summary of the Thomist account of justice in war and, contrary to the intent of the author, equally excellent illustrations of the ways in which such an account can be problematic.

JPWE offers the most comprehensive account of her thinking about the just war. She begins by making the point that the Thomist arguments she advances are based on natural law and do not require the support of revealed religion to be obligatory, clearly wishing for her argument to be compelling for non-Catholics—yet, the reader will note that throughout the essay, she refers to the importance of avoiding the sin of violating these principles, which suggests she doesn't take this separation as seriously as one might have expected. It is worth noting also that at the time of writing this very early essay, she appears not to have engaged seriously with Aristotelian thinking, so perhaps it is unsurprising that she doesn't make as much as she might have of a virtues-based approach to the just war.

In any event, she begins the essay conventionally by listing the seven conditions of the just war[10]—in summary, (1) there should be a just occasion for war, (2) it should be declared by a lawful authority, (3) with right intention and (4) employing right means, (5) there being no alternative to war and (6) a reasonable hope of victory, and the (7) probable good must outweigh the probable evil effects of the war (p. 73). She immediately allows that conditions (1) and (2) are met—there is a just cause and the British state is a lawful authority—and she is willing to give the benefit of the doubt on (5) and (6)—this leaves (3), (4), and (7) as problematic. And, she stresses:

> If these [latter conditions] are not fulfilled, *this* war is rendered wrong, however just the occasion, however desirable that we should fight *a* war. Nor, if we know that a war is wrong, may we take part in it without sin, however grievous it may seem to stand apart from our fellow countrymen.
> *(p. 73, emphasis in original)*

Before moving to look at 'right intention' and 'right means' it is worth noting that she closes one potential escape route for someone who opposes the government but believes in the justice of their own interpretation of what the war is about, which was a common position among those who were both anti-Fascist and against the Chamberlain government. A just war, she says—accurately following the tradition—can only be declared by a lawful authority, and if the war that is so declared is unjust, then participation in that war is also unjust, irrespective of the private intentions of the individual.

'If a war is to be just, the warring state must intend only what is just, and the aim of the war must be to set right certain specific injustices' (p. 74).

Anscombe suggests that we have reason to question the sincerity of our government, arguing that the pact with Poland was made for dubious reasons and the undoubted injustice done to Poland seems more like a pretext than a reason for the war. The Treaty of Versailles was, she argues, patently unjust, and the British government has made no attempt at a just and reasonable settlement of the disputes it has caused.[11] The government's war aims now may seem vague and if so should be condemned. Vagueness is unacceptable 'For it is a condition of a just war that it *should* be fought with a *just* intention; not that it should *not* be fought with an *unjust* intention' (p. 75). But in fact, she continues, the government's professed intentions are not merely vague but unlimited. 'They have not said "when justice is done on points A, B and C, then we will stop fighting." They have talked about "sweeping away everything that Hitlerism stands for" and about "building a new order in Europe"' (p. 75). This, she suggests, means our intentions are unlimited, and as a result, 'we are fighting against an unjust cause, indeed; but not for a just one' (p. 75). There is much, of course, to be said about this, but for the moment it is worth noting that this critique is not directed against the later policy of unconditional surrender but against the earlier aim of removing Hitlerism.

So much for 'just intention'—on 'just means', Anscombe argues that an attack on civilians could never be just, and yet the allies have said that, although they will adhere to international law in this matter, they reserve the right to adopt appropriate measures if the Germans should break it, which indicates a willingness to attack civilians in reprisal, which, she states, would be unjust and sinful. Moreover, the already existing blockade on Germany in effect attacks civilians. Anscombe does not accept the argument that in modern war, whole populations are involved and could become legitimate targets. 'No one may be deliberately attacked in war unless his actions constitute an attack on the rights which are being defended or restored' (p. 77). Civilians are not committing such wrongs even if their actions add to the strength of the nation which is unjustly fighting.[12] She argues—accurately as it happens—that as the war progresses, the willingness to attack civilians will increase, effectively predicting the shift to area bombardment by the Allies later in the war.

Finally, Anscombe addresses the issue of whether the probable good effects of the war would outweigh its probable evil ones and decides that they would not. Putting an end to the injustices of Germany would be the probable good, but the probable evil is, she argues, already apparent in the desire that is being expressed to turn Germany into a pariah nation; moreover, the curtailment of liberty, destruction of property, and general corruption that war brings would outweigh the limited good that might happen.

Leaving aside the issue of 'just means' until after considering 'Mr Truman's Degree', what is to be made of Anscombe's account of right intention? It seems to me that her account of the standard meaning of this element of the just war is essentially correct, and the logic of her application of that element to the situation in 1939—spelled out in the essay in more detail than can be conveyed here—is difficult to fault. And yet, it also seems to me that the end result is totally devoid of both common sense in general and of any genuine engagement with the nature of German National Socialism in particular. Perhaps there is some excuse for the latter, given that the worst atrocities of Nazism took place later in the war, but there was enough known at the time to make it clear to most intelligent observers that 'sweeping away everything that Hitlerism stands for' was a reasonable objective. What we are seeing here is the logical working through of the consequences of ignoring consequences—although, interestingly, in her account of the balance between probable good and probable evil, Anscombe does seem to be considering consequences, concluding that on balance, it would be better not to fight a war, even though the cause be just. It is, I think, at this point, that one would want to challenge Anscombe's belief that the demands of natural law do not require religious support—on the contrary, it seems to me that one could only believe in an absolute refusal to consider consequences if one also believed that a benevolent God would ensure that all will be well whatever we do. Without such an assurance, Michael Walzer's notion of 'supreme emergency' makes sense; sometimes, there are going to be situations where we will want to say that if that is what the law says, the law is an ass. As Justice Jackson famously declared in the U.S. Supreme Court, 'the Constitution is not a suicide pact', and on the same line of reasoning, the notion that the criteria for a just war preclude steps that would lead to the defeat of Nazism suggests that flexibility of the sort refused by Anscombe might, in fact, be necessary.

Similar thoughts arise when considering just means, and, once again, Anscombe's essay on 'Mr Truman's Degree' inadvertently illustrates the point. Doubtless, the Oxford authorities initially regarded their decision to nominate Harry. S Truman for an honorary degree as uncontroversial—Truman had been President of the United States in the immediate post-war years, overseeing the Berlin Airlift, the formation of NATO and the creation of the Marshall Aid program, and directing the Western response to North Korean aggression in 1950 and as such seemed obviously qualified for the award. Anscombe disagreed on the basis that his decision to use nuclear weapons on civilian targets made him a war criminal, and his other positive qualities (if such they were) could not wipe out this stain. Contrary to normal practice, she pushed the issue of Mr Truman's degree to a debate and vote in Convocation—the historian Alan Bullock led for the defence, and the degree was voted through by a substantial majority, but her point was made.[13]

Anscombe's argument against Truman rested on an absolute commitment to non-combatant immunity under all circumstances. She acknowledges that the Allied policy of unconditional surrender (which in any event she opposed for the reasons considered earlier) meant that without the use of such weapons, a land invasion of Japan would be needed and would certainly lead to a greater loss of life than the bombings but, predictably, argued that this was not relevant. Doing something intrinsically wrong cannot be justified by the good it produces. There is an interesting argument here, and many writers who otherwise would disagree with Anscombe's approach actually agree with her conclusion—Michael Walzer, for a significant example, regards the use of nuclear weapons at Hiroshima and Nagasaki, and indeed area bombing as a practice, as war crimes. What is distinctive about Anscombe's approach is its insistence that the decision to use nuclear weapons was a sin and a sin by a particular individual, Harry S. Truman. Truman certainly took the final decision to use the bomb, but there seems something disproportionate about laying the act on his shoulders in this way. The Manhattan Project which produced the bomb had been authorised by President Roosevelt and Prime Minister Churchill, and the practice of area bombing of cities had flourished in Europe and Japan before Truman became President. Once the weapon was available, it was overwhelmingly likely that it would be used, and it would have taken an extraordinary act on the part of the President to stop things taking their course, an act that would, as Truman was well aware, probably doom half a million U.S. soldiers who would die in the invasion of Japan, not to mention the civilians who would inevitably die in the cross fire. Truman himself never had any doubt about that he had taken the right decision—one suspects part of the animosity that Anscombe expresses stemmed from that unwillingness to admit to what she saw as his guilt.

The absolute prohibition on the killing of innocents is central to the essay 'War and Murder', the focus of which is on why people in general, and Catholics in particular, don't seem to understand why this absolute prohibition rules out reliance on nuclear weapons and the obliteration of cities. A large part of the answer, she argues, lies in a false belief in pacifism which has led people to condemn the use of force under any circumstances and thereby fail to see the moral difference between the legitimate and the illegitimate use of force. She describes a pacifist ethic as essentially non-Christian, based on a misreading of the New Testament and the view that Christianity is 'an ideal and beautiful religion impracticable except for a few rare characters. It preaches a God of love whom there is no reason to fear' (p. 55) whereas 'the truth about Christianity is that it is a severe and practicable religion not a beautifully ideal and non-practicable one' (p. 56). The Christian God is the God of the Old Testament who makes rules for living which He expects to be followed. Legitimate force exercised by state authorities is a necessary and desirable feature of social life, and force may sometime be necessary internationally—and not simply in self-defence as some interpreters of the just war now argue. She instances the suppression of the slave trade under Lord

Palmerston as an exercise of force that was wholly excellent.[14] Anscombe is characteristically bullish on this;

> The present-day conception of 'aggression' like so many strongly influential conceptions, is a bad one. Why *must* it be wrong to strike the first blow in a struggle? The only question is, who is in the right, if anyone is.
>
> *(p. 52 emphasis in the original)*[15]

The point is that by failing to understand that force was sometimes necessary, pacifists don't make the crucial distinction between those who may, in some circumstances, rightfully be the subject of violence and those against whom violence may not be inflicted. A second source of confusion is 'double-think about double effect' (p. 58). Whereas a distinction between the intended and the merely foreseen effects of a voluntary action is essential, the principle here has been, she argues, repeatedly abused to suggest that somehow the killing of innocents is taking place by accident; 'it is nonsense to pretend that you do not intend to do what is the means you take to your chosen end' (p. 61). A complicated account of the meaning of intention is being employed in order to obfuscate something that is clearly immoral, the large-scale massacre of the innocents, which would be the result of the use of nuclear weapons, indeed already has been the result of such use in 1945.

What of the notion of deterrence? Some defend the threat of nuclear annihilation as necessary to prevent the conquest of Europe by Soviet Russia—but predictably she has no sympathy for this consequentialist argument. Those Catholics who are prepared to use this argument reveal their lack of faith in God's promise to the church. Powerfully she proposes:

> Those, therefore, who think they must be prepared to wage a war with Russia involving the deliberate massacre of cities, must be prepared to say to God: 'We had to break your law, lest the Church fail. We could not obey your commandments, for we did not believe your promises.'
>
> *(p. 61)*

The same argument was made at greater length in one of the major contributions to the debate about nuclear deterrence in the 1980s—John Finnis, Joseph Boyle, and Germain Grisez's (1987) *The Morality of Nuclear Deterrence*—and involved a more realistic approach to the consequences of abandoning deterrence than was commonplace at the time. Still, as with Anscombe's other absolutist statements, the idea that consequences can be so disregarded requires the support of a strong religious faith.

Legacy

Iris Murdoch referred once to Anscombe's 'ruthless authenticity', her absolute sincerity, which was as visible in her work as perhaps in her rather unorthodox lifestyle (quoted in Conradi, 2001, p. 273).[16] Speaking of one of their early mentors, Philippa Foot described Donald MacKinnon as 'holy' while defining holiness as 'an absolute lack of sense of proportion' and the same could well be said of Anscombe (Conradi, 2001, p. 127). She was an extremist in the literal sense that she took arguments to the extreme—she followed what she saw as the logic of a position to its natural conclusion and whereas others might shy away if that conclusion were to be unpalatable, she regarded such backsliding as unacceptable. In a comparison that would horrify both parties, she was in this one (and only this one) respect similar to today's leading utilitarian thinker, Peter Singer; thus, the logic of Singer's support for abortion rights led him to support the infanticide of new-born children with severe disabilities—at this point, most people would retrace their steps and look for what had gone wrong to lead to such an unacceptable conclusion, but Singer stood by his logic. This is, I think, comparable with Anscombe's belief that Hitlerism could only be opposed if every last sub-clause of the traditional account of the just war were met—someone less ruthlessly authentic might have had second thoughts about rushing into print with such a view, but Anscombe had convinced herself that her reasoning was correct and so pressed on regardless.

The result is that she presents a very clear account of the Catholic just war tradition, but where her account falls down is when she attempts to argue that natural reason without revelation is capable of arriving at the same conclusions. It makes sense to disregard the consequences of action if one is sure that in all circumstances, a loving God will shield those who follow his commands. Without such a belief, the reliance on a code that disregards consequences is, I think, impossible to defend. This is of some significance when it comes to Anscombe's modern legacy.

Tradition Catholics who broadly share her theology, such as the aforementioned trio of Finnis, Boyle and Grisez, will have no major problems with her conclusions, but it is interesting that contemporary Vatican thinking on just cause seems to be focused entirely on self-defence, contrary to Anscombe's robust defence of humanitarian action.[17] Anscombe would probably have been amused by the idea that she was more Catholic than the Pope but would have been less amused by the fact that the modern writers on the just war who are closest in their analysis to her approach are the lineal descendants of the moral philosophers that she, Foot, Murdoch, and Midgley criticised so heavily in the 1950s and thereafter. The so-called 'revisionist' just war theorists—as it happens are mostly based in Oxford, for example, Jeff McMahan, Cecile Fabre, David Rodin—are analytical philosophers who employ the kind of thinking criticised in Anscombe's 'Modern Moral Philosophy' in order to reach the kind of conclusions she reached in her writings on the just war. Like Anscombe, they follow the logic of the traditional categories and are deeply suspicious of attempts to depart from that logic to take into account the realities of modern warfare. They adopt her refusal to acknowledge that sometimes the consequences of following the rules might be contrary to the public good—the big difference, though, is that they hold this position without the belief in God which underpinned Anscombe's commitment. They are doing precisely what Anscombe warned against—operating a legal form of ethics without a legislator. They are, in effect, testing to destruct the idea that one can build a tightly argued framework for the just war, that will provide a set of answers rather than a set of questions, without relying on the word of God to provide the underlying legitimation, without using the concept of sin or other essentially religious formulations.

Anscombe believed that such a framework could indeed be constructed on the basis of natural reason, but her work is littered with examples illustrating that this simply doesn't work. Interestingly, although in 'Modern Moral Philosophy,' she identifies Aristotle and the virtues as a way of being ethical without a law model of ethics, she doesn't take this up in her writings on the just war. As noted earlier, at the time of writing JPWE, she had not studied Aristotle in depth, but later she did, though she chose not to revise her thinking. Rather than seeing the Thomist categories as starting points for the exercise of judgement—*phronesis*—she turns them into a set of incredibly restrictive rules of precisely the sort that could only be justified if they were the commands of a law-maker.

The swashbuckling vigour with which she argues and her willingness to take up unpopular positions are commendable and deserve to be emulated by modern writers, but, as the revisionists inadvertently illustrate, the full package of her thinking on the just war makes sense only for those who can follow her in some version of the religious commitment that was central to her thinking. The moral seriousness with which she worked through the logic of her commitment to Thomist reasoning may attract our admiration and her unwillingness to compromise her beliefs sharply, and I think it favourably contrasts with the way in which so many of our contemporary Christian leaders have adapted their faith to the mores of the age. But, in the end, those of us who cannot share her beliefs cannot commit to her heroic struggle with the world.

Notes

1 Gertrude Elizabeth Margaret Anscombe (1919–2001), who published as G.E.M. Anscombe, was Elizabeth to friends and family and, at her insistence, 'Miss Anscombe' to the rest of the world. This was sometimes a source of confusion since she was married (to the philosopher Peter Geach) with seven children.
2 The current version, Oxford: Wiley-Blackwell, 2009, still is at root based on her translation, somewhat revised. Wittgenstein subsidised her for a period studying

German in Vienna to get her language skills to the point where such a translation was possible, a sign of how much he valued her contribution.

3 Based on a lecture he gave in 1945, *Existentialism is a Humanism* (2007) is the easiest way into Sartre's philosophy.

4 Why this resistance was led by women is an interesting question; perhaps because the absence of conscription-age men between 1939 and 1945 opened up opportunities for women which wouldn't have been there otherwise, or perhaps, conversely, it was because they were marginalised already at Oxford in this period and so had less reason to conform to the prevailing philosophical consensus.

5 Philippa Foot *Virtues and Vices and Other Essays in Moral Philosophy* (1978) and *Natural Goodness* (2001).

6 'Modern Moral Philosophy', *Philosophy: The Journal of the Royal Institute of Philosophy*, vol. XXXIII No. 124, January 1958. Cited here from Anscombe (1981, pp. 26–42).

7 The argument here is similar to that of Alasdair MacIntyre's *After Virtue* (1981). MacIntyre (also a Catholic convert, although in his case in his 50s) is more pessimistic than Anscombe about the prospects for what he calls an Augustinian–Thomist approach to moral philosophy—she puts her faith in the Church in the world, he looks to a new monasticism as the way to preserve the virtues. For more on MacIntyre, see Chapter 15.

8 Aristotle is treated in Chapter 1 of this book.

9 Reproduced, respectively, in Anscombe (1981, pp. 72–81, 62–71, 51–61).

10 In fact, strictly speaking, only the first three criteria relate specifically to the just war, the other four are applications of general principles of right conduct to the circumstances of war.

11 Both parts of this sentence are, of course, contestable.

12 Anscombe notes the doctrine of 'double effect' but sees it as irrelevant—the doctrine concerns, she argues, the accidental killing of civilians which is not the issue here. More on this is discussed later in the chapter.

13 It was during the lobbying to get people to attend Convocation and vote down Anscombe that the phrase 'the women are up to something' was used, somewhat inaccurately, since the other 'women' though supportive of Anscombe were not directly engaged with the issue.

14 However, she characterises Palmerston's prosecution of the Opium War against China as diabolical, so would no doubt have opposed the award of an Honorary Degree to him by Oxford University in 1862. In fact, the suppression of the slave trade began well before Palmerston was in office, so he doesn't really merit the credit she gives him.

15 The 'if anyone is' here is crucial—as she acknowledges, the probability is that warfare is injustice.

16 As to Anscombe's lifestyle, anecdotes abound. Her unusual dress sense would attract attention even in the more easy-going 2020s, and her approach to child-rearing, which was based on the principle that dirt didn't matter and that children should be allowed to run riot, would probably attract the attention of modern social workers. Was her eccentricity an affectation? Bernard Williams, a colleague and astute observer, thought so, but the question is unanswerable at this distance.

17 See Christian Nikolaus Braun's essay on Pope Francis in this volume, Chapter 19.

Works Cited

Anscombe, G. E. M. 1957. *Intention*. Oxford: Blackwell.

Anscombe, G. E. M. 1981. *Ethics, Religion and Politics: Collected Philosophical Papers, Vol. III*. Oxford: Blackwell.

Ayer, A. J. 1936. *Language, Truth and Logic*. London: Victor Gollancz.

Conradi, Peter J. 2001. *Iris Murdoch: A Life*. London: HarperCollins.

Finnis, John, Joseph Boyle, and Germain Grisez (eds.). 1987. *The Morality of Nuclear Deterrence*. Oxford: Clarendon Press.

Foot, Philippa. 1978. *Virtues and Vices and other Essays in Moral Philosophy*. London: Blackwell.

Foot, Philippa. 2001. *Natural Goodness*. Oxford: Oxford University Press.

Hare, R. M. 1951. *The Language of Morals*. London: Oxford University Press.

Lipscomb, Benjamin J. B. 2021. *The Women Are Up to Something: How Elizabeth Anscombe, Philippa Foot, Mary Midgley and Iris Murdoch Revolutionised Ethics*. Oxford: Oxford University Press.

Mac Cumhaill, Claire and Rachael Wiseman. 2022. *Metaphysical Animals: How Four Women brought Philosophy Back to Life*. London: Vintage.

MacIntyre, Alasdair. 1981. *After Virtue*. Notre Dame, IN: University of Notre Dame Press.

Sartre, Jean-Paul. 2007. *Existentialism is a Humanism*. New Haven: Yale University Press.

Wittgenstein, Ludwig. 1953. *Philosophical Investigations*. Oxford: Blackwell.

14

FRANTZ FANON (1925–1961)

Gabriel Mares

Introduction

If political theory was late to the study of empire (Pitts, 2010), just war theory was until recently simply absent. On the one hand, this may be because critics of just war theory's relationship with empire situate themselves outside the tradition (e.g. Maldonado-Torres, 2008; Asad 2009); on the other hand, even after the emergence of such critiques, many in the field resisted—or simply dismissed out of hand (e.g. Bellamy, 2017)—a "turn to empire."

Jessica Whyte uncovers an irony in this lack of curiosity about empire in just war thinking: while the "just war revival" is generally dated to the publication of Michael Walzer's *Just and Unjust Wars* in 1977, the language of "just war" was used by representatives of postcolonial states at the preparatory meetings[1] in advance of the Additional Protocols to the Geneva Conventions between 1974 and 1977. And it was the states of the Global North who resisted this language, with the U.S. representative warning of "the dangerous concept of the just war" (Whyte, 2019).

We can treat the Additional Protocols as a rupture, a Pocockian "lost moment" from which to think differently about our present moment. What might an alternative to the "just war revival" that originates with Walzer look like? There is an emerging wave of scholarship which seeks to deepen just war's engagement with colonialism, rather than simply reject the tradition because of its complicity with empire (e.g., Dussel, 2007; Finlay, 2015; Hutchings, 2019; Whyte, 2019; Mares, 2021).

It is in this spirit that I turn to Frantz Fanon, the oft-maligned theorist of the Algerian revolution. He argued that decolonization was necessarily a violent event precisely because it resisted the quotidian violence of the colonial

DOI: 10.4324/9781003428688-15

world—a violence that was obfuscated by many Western scholars who chastised anti-colonial movements for their "descent" into violence. When Fanon has been treated by just war theorists and specialists in international law, he is generally treated as a theorist of (terrorist) violence (Elshtain, 2007; Saul, 2008). Rather than using just war theory to condemn Fanon, or Fanon to reject the just war tradition, in this chapter I seek to use Fanon to complicate just war thinking, pulling in the direction of an anti-colonial "lost moment" of just war.

Contexts

Fanon was born in Fort-de-France, Martinique. Living through World War II as well as multiple anti-colonial struggles, Fanon died at the peak of Algeria's revolutionary struggle against colonial France—in Bethesda, Maryland, with his transportation there facilitated by the CIA.

Fanon enlisted in the Free French forces in 1943, fighting in Africa in World War II—experiencing anti-black racism from French soldiers.[2] After the war, he studied psychiatry in Lyon, France. Though a medical student, he regularly attended lectures by the philosopher Maurice Merleau-Ponty. Fanon's thesis was interdisciplinary: "drawing on ideas from the anthropologist, philosopher, and sociologist Lucien Levy-Bruhl, Fanon argued that one should, as a psychiatrist, reach to the patient's humanity instead of the material nexus of effects or symptoms" (Gordon, 2015, p. 15).[3] Fanon returned to Martinique, then left for Algeria, joining Blida-Joinville Hospital as a psychiatrist in 1953. Here, Fanon encountered the search for a "nexus of symptoms" as undertaken in a colonial setting. Describing the interactions between French doctors and Algerian patients he witnessed, Fanon wrote, "Fairly soon the doctor . . . worked out a rule of action: with these people you couldn't practice medicine, you had to be a veterinarian" (Fanon, 1965, p. 127). Colonial doctors treated patients as unreliable narrators of their symptoms, and in his own work, Fanon posited an ontological grounding for the practice of engaging colonial patients as zoological specimens.

Fanon resigned from Blida-Joinville in 1956, writing,

> Although the objective conditions under which psychiatry is practiced in Algeria constituted a challenge to common sense, it appeared to me that an effort should be made to attenuate the viciousness of a system of which the doctrinal foundations are a daily defiance of an authentically human outlook.

But finding his role as a psychiatrist as simply an agent in a colonial institution, he concluded, "What is happening is the result neither of an accident nor of a breakdown in the mechanism. The events in Algeria are the logical consequence of an abortive attempt to decerebralize a people" (Fanon, 1988, pp. 52–53). After his resignation, Fanon traveled widely as a doctor, writer, and representative of the *Front de liberation nationale* (FLN). After refusing treatment in France for cancer, he left for the United States for treatment, dying in 1961.

Fanon's thinking cannot be easily "contextualized" in a single cultural milieu; he studied as a psychiatrist, but engaged Hegel, Marx, and Sartre in his writings. He was influenced by Aimé Césaire but critiqued the Négritude movement, ultimately leaving the Caribbean to work in Algeria. Efforts to pin him down and define him in a singular way prove notoriously difficult. His work exists at the intersection of multiple discourses, meaning that defining a "problem space" around which we can organize analysis and critique is necessarily an interdisciplinary effort.

Texts and Tenets

In his brief life, Fanon wrote three complete books, *Peau noire, masques blanc* (*Black Skin, White Masks*, 1952); *L'an Cinq de la révolution Algérienne (A Dying Colonialism,* 1959); and *Les Damnés de la terre (The Wretched of the Earth,* 1961), as well as numerous essays, clinical reports, and plays. A collection of Fanon's essays, *Pour la revolution Africaine (Towards the African Revolution,* published posthumously in 1964), explores and expands on themes in the three monographs. A new collection of his out-of-print work, *Écrits sur l'aliénation et la liberté (Alienation and Freedom,* 2015), greatly expands our access to his medical, dramatic, and political writings.

Broadly speaking, Fanon confronts the problem of colonialism. Contrary to how much scholarship at the time defined colonialism, Fanon understood colonialism to be an ontological phenomenon as much as a phenomenon of sovereignty and cartography. "Anti-Blackness" was, in Fanon's account, an idea that structured any society dominated by European powers. To engage Fanon's thinking, just war theorists must start from his critiques of language and (false) universality—as language and universality are also central to the just war revival.

Michael Walzer begins *Just and Unjust Wars* by positing a shared moral vocabulary as the basis—and significance—of reviving just war theory (2000, p. 20). He admits that he does not believe he inhabits the moral world of Genghis Kahn, but holds that this vocabulary held in common allows discussions of justice and war to be meaningful both transhistorically and transculturally. While different standards of justice may prevail in different times and places, the shared moral vocabulary means that discussions across contexts are not incommensurable. The possibility of justice, then, begins in language.

Walzer does not go to great lengths to defend this proposition. Recent work in just war theory (e.g., Hutchings, 2019; Mares, 2021) draws attention to the ways in which supposedly universal categories ("police" or "industrial laborers") utilized in just war thinking can mask ideas that reproduce civilizational hierarchies.

For Fanon, language and universality centered on European man are sites of anti-blackness. In *Black Skin, White Masks*, Fanon probes the problem of anti-blackness in medicine, Eurocentric humanist thought, and society more broadly—namely, how ontologizing whiteness creates a falsely universal "human" against which blackness is always found wanting. Black becomes the anti-human, defined through its contrariness to the universal model. This found expression not only in literature and philosophy but also in the medical sciences and all forms of measure and judgment. Lewis Gordon frames Fanon's alternative through the critique of universal systems based on an unassailable centrality of European Man: it is a "demand often imposed upon people of color . . . [to] accept the tenets of Western civilization and thought without being critical of them. Critical Consciousness," which Gordon identifies as part of Fanon's project, "asks not only whether systems are consistently applied but also whether the systems themselves are compatible with other projects" (Gordon, 2015, p. 20). Fanon writes derisively that "today's Blacks want desperately to prove to the white world the existence of a black civilization" (Fanon, 2008, p. 17). Thus, Fanon rejects vindicationist political narratives in which an oppressed group gains "recognition" by demonstrating that their achievements are "as good as" those held to be universally great.

In this complex context, just war theorists should be cautious about how they engage thinkers from anti-colonial traditions. To simply claim that various anti-colonial thinkers and revolutionaries "actually" embodied or complied with aspects of just war thinking would be to risk constructing a vindicationist narrative. What is needed is not to demonstrate that Third Worldist or anti-colonial thinkers can be used to construct a cosmopolitan canon of just war thinkers—that they, too, shared the "universal moral vocabulary"—but rather to allow just war thinking to be profoundly *affected* by critical engagement with anti-colonial thought.

Fanon begins *Black Skin, White Masks* with a proclamation that "a Black is not a man." Blackness is "a zone of nonbeing," and in this work he is "aiming at nothing less than to liberate the black man from himself" (Fanon, 2008, p. xii). The zone of nonbeing is a "hell" simultaneously constructed and obfuscated by the anti-black world. It creates an inferiority complex through "a double process: first, economic. Then, internalization or rather epidermalization of this inferiority" (Fanon, 2008, p. xv). Locating blackness in the epidermis, the skin, makes blackness an experience which can be examined phenomenologically, contrary to colonial scientific attempts to define blackness as akin to species differentiation. The wearing of and shedding of skin are important images in Fanon's construction of the "lived experience" of blackness.

For Fanon, blackness means one is to be defined through an "epidermal racial schema." Blackness makes impossible the romantic striving for equality or an end to alienation through effort, self-improvement, and association; Fanon refers to such striving as "a psychological phenomenon that consists in believing the world will open up as borders are broken down" (Fanon, 2008, p. 5). In a reflection on language, he writes that the educated black man believes that "he proves himself through his language" (Fanon, 2008, p. 8). Yet, the black man speaking perfect French becomes an oddity precisely because he remains a black man: "The fact is the European has a set idea of the black man, and there is nothing more exasperating than to hear: 'How long have you lived in France? You speak such good French'" (Fanon, 2008, p. 18). To be black in an anti-black society means that one's skin is paramount; both the fear and the praise of blackness in such a society reinforce that one is, first and foremost, black.

The epidermal racial schema is key to understanding the most famous moment in *Black Skin, White Masks*; in Chapter 5, "The Lived Experience of the Black Man," Fanon writes of being "seen" by a young French boy on a train:

"Look! A Negro!"[4] It was a passing sting. I attempted a smile.

"Look! A Negro!" Absolutely. I was beginning to enjoy myself.

"Look! A Negro!" The circle was gradually getting smaller. I was really enjoying myself.

"*Maman*, look, a Negro; I'm scared!" Scared! Scared! Now they were beginning to be scared of me. I wanted to kill myself laughing, but laughter had become out of the question.

I couldn't take it any longer, for I already knew there were legends, stories, history, and especially the *historicity* that Jaspers had taught me. As a result, the body schema, attacked in several places, collapsed, giving way to an epidermal racial schema.

(Fanon, 2008, pp. 91–92)

The epidermal racial schema forces Fanon to live outside of his own body and in the "skin" created through countless images, drawn from "legends, stories, history," that the young boy and his mother have already received and internalized. To the child, Fanon is an object of fascination and horror *because* the child already "understands" what black skin in an anti-black society means.

Reflecting on the legitimation of state violence against black bodies, Robert Gooding-Williams draws on Fanon to argue that such legitimations draw attention to—rather than away from—images of violence, "to affix these images to that body, as if to say repeatedly, "Look, a Negro!". . .[the] black body [becomes] that of a wild "Hulk-like" and "wounded" animal whose every gesture threatened the existence of civilized society" (Gooding-Williams, 2006, p. 10). *Embracing* images of state violence "proves" the black body's potential aggression; defenders of such violence point to every fist or bent knee as a potential blow to be legitimately feared. Juries and publics are reminded that it is right to fear the black body, even as it is subjected to violence by agents of the state. Fanon's (and Gooding-Williams') focus on sight, image, and how these project prejudice onto the body helps us to cast a critical eye on ways in which the search for violent threats may validate the projection of threats onto Othered bodies—even as claims of universality and respect for "humanity" abound.

Fanon's rejection of a falsely universal "humanity" is particularly relevant for thinking about noncombatants. Alternately referred to as "discrimination" and "distinction," *jus in bello* is premised on a claim that certain people may be legitimately killed in war while others may not. Terrorism (whether by sub-national groups or state terror) is the rejection of the principle of distinction. People are killed "indiscriminately," not in the sense of "randomly" but in the sense that there is no discrimination between "legitimate" and "illegitimate" targets—groups may be defined as "objective enemies" and that alone justifies their killing (Elshtain, 2004, p. 18). Walzer (2000) cites dramatized scenes of bombing in the film *The Battle of Algiers* to describe a tactic that targets "people for who they are," while Elshtain (2007) names Fanon as the "theorist of terrorism" who seeks liberation for some (the "colonized") by the indiscriminate killing of others ("colonizers"). This is an impoverished view of Fanon's theories of complicity, which both relies too heavily on Jean-Paul Sartre's reading of Fanon, as well as an uncharitable reading of a single chapter, "On violence."

More recent just war theorists (McMahan, 2009; Mares, 2021) probe the question of "morally liable civilians" by investigating whether civilians who would not traditionally be classified as "direct participants in hostilities" (the standard set forth in the Additional Protocols to the Geneva Conventions) might be legitimate targets in certain contexts, such as settler colonialism. Fanon offers different understandings of responsibility and collaboration, thinking more broadly about their significance for the foundation of new political communities.

Beginning instead with Chapter 5 of *A Dying Colonialism*, "Algeria's European Minority," readers gain an important insight into Fanon's views on collaboration, political foundations, and the meaning of "colonizer" in his texts. In this, Fanon distinguishes himself from Albert Memmi's formulation in *The Colonizer and the Colonized* (1955) and demonstrates the shortcomings of his ungenerous interpreters.

Memmi argued that it was impossible for the European living in the colonies to *not* be a colonizer. Colonial privilege could not be relinquished short of physically abandoning the colony; affective commitments to colonialism were not required for one to be a colonizer.[5] Sartre's preface to *The Wretched of the Earth* appears to make such a categorical distinction: "For in the first phase of the revolt killing is a necessity: killing a European is killing two birds with one stone, eliminating in one go oppressor and oppressed: leaving one man dead and the other man free" (Sartre, 2004, p. lv). Sartre certainly grasps the generative act of violence—but to reduce Fanon's theorization to the act of an African killing a European simply misrepresents Fanon. Fanon's essay "Algeria's European Minority" demonstrates the shortcomings of understanding "colonizer" and "European" as easily interchangeable.

In 1959, newly elected president Charles de Gaulle called for a "democratic" solution to the ongoing Algerian civil war. Writing in response to this political development, Fanon notes that "In Algeria, democracy is tantamount to treason" (Fanon, 1965, p. 150). Political life is controlled by the settler establishment in a system that continental fascism imitated.[6] While de Gaulle's call might sound reasonable to outsiders, settler colonial fascists would never allow a truly democratic resolution to unfold, thus making "European democrats" a fugitive minority.

"European democrats"—Fanon's term for those who would support popular rule in Algeria, overthrowing the settler state—act in secret. "Drowned in the European mass, they live in a world of values that their principles reject and condemn. . . . This democratic European, accustomed to semi-clandestine contacts with Algerians, unwittingly learns the laws of revolutionary action" (Fanon, 1965, pp. 150–151). Fanon recognizes this democrat as party to the struggle, insisting that "not a single Frenchman has revealed to the colonialist police information vital to the Revolution" (Fanon, 1965, p. 151) and noting a variety of ways they have assisted, from withholding information during interrogation to providing (embargoed) medicine and food to anti-colonial forces.

What does this make the European democrat? Fanon insists,

> *For the FLN, in the new society that is being built, there are only Algerians. From the outset, therefore, every individual living in Algeria is Algerian. In tomorrow's independent Algeria it will be up to every Algerian to assume Algerian citizenship or to reject it in favor of another.*
>
> *(Italics in original; Fanon, 1965, p. 152)*

Thus, neither colonized and black, nor nationality and blackness are interchangeable in Fanon's formulation. The nation must be open to all who would accept it.[7] Later, in *The Wretched of the Earth*, Fanon writes that "the colonist is no longer interested in staying on and coexisting once the colonial context has disappeared" (Fanon, 2004, p. 9). Reading *Wretched* together with this essay clarifies that "the colonist" is simply one who is wedded to the colonial system; this is not a call for ethnic cleansing, for the European democrat will choose to stay. The contrast between European democrat and the colonizer may be read as parallel to Germans who resisted Nazi rule and those who supported or acquiesced to Nazi rule—these are not simply neighbors who hold "different political beliefs."

Resistance to settler colonial fascism is a moment when the European democrat transcends the "mask of whiteness" and embraces the *idea* of a new nation not structured on anti-blackness. The struggle is generative of Fanon's "new man," meaning this essay captures how the European democrat may strive toward liberation. It thus serves as a parallel to the more famous parts of Fanon's *oeuvre*, which focus on the black man's struggle toward liberation.

Fanon's European democrat allows him to distinguish between those who recognize and support justice and those who do not. Analytic/revisionist just war theorists ask, "can a soldier know the justice of their own cause?" Among just war theorists, "the epistemic problem" has been a central feature since the analytic turn (McMahan, 2004, 2009). Against both Walzerians and international lawyers, who insist that the matter of the justice of a particular war is a matter for states while soldiers cannot be expected to opine or act on that basis, revisionist just war theorists hold that soldiers (and civilians) must be expected to make moral evaluations of wars waged by their states. A corollary to this is that there could be "just and unjust warriors." Against the "moral equality of combatants" thesis, which holds that every soldier fighting for a legitimate authority has a right to fight, soldiers fighting for an unjust cause (which they could reasonably know is unjust) are "unjust warriors" who have no moral right to kill "just warriors."

For Fanon, violence introduces its own epistemic problem; in fighting, the colonized comes to recognize himself. In the first chapter of *The Wretched of the Earth*, "On Violence," Fanon makes clear that one cannot assume such knowledge will be held *prior to* any fighting. Violence in this way is generative, epistemically speaking.

Fanon begins from the premise that "decolonization is always a violent event" (Fanon, 2004, p. 1). But that is largely because the colonial world is a violent one:

> The colonized world is divided in two. The dividing line, the border, is represented by the barracks and the police stations. In the colonies, the official, legitimate agent, the spokesperson for the colonizer and the regime of oppression, is the police officer and the soldier.
>
> *(Fanon, 2004, p. 3)*

It is the experience of violence and counter-violence, attacks and reprisals, which awakens the colonized to the true nature of the regime and his exclusion from "the human:"

> The most alienated of the colonized are once and for all demystified by this pendulum motion of terror and counterterror. They see for themselves that

any number of speeches on human equality cannot mask the absurdity whereby seven Frenchmen killed or wounded in an ambush . . . sparks the indignation of civilized consciences, whereas the sacking of the Guergour *douars*, the Djerah *dechra*, and the massacre of the population behind the ambush count for nothing.

(Fanon, 2004, p. 47)

This experience of alienation forecloses the possibility of a search "for justice in the colonial context" (Fanon, 2004, p. 43).

Violence is ever-present in the colonial situation: the colonized are ruled by a quotidian violence called "order." Because the colonized are alienated from "the human," they are defined as an ever-present threat to that "order." Robyn Marasco clarifies how Fanon's reading of colonial violence and counter-violence escapes many academic analyses:

Political historians and analysts often tell the story of violence . . . [depicting] resistance movements that begin in nonviolence and "fall" or "lapse" into violence when initial hopes are disappointed. Fanon presents . . . an alternative temporality of violence. . . . He describes a social structure built on systematic and institutionalized violence, a resistance movement that begins in sporadic and volatile fits of violence, a political organization that emerges to give form and direction to violence that is spontaneous and unpredictable, and *then* the introduction of nonviolence as a reactionary and desperate appeal for compromise.

(Marasco, 2015, p. 159)

By Fanon's reading, violence is always present in the colonial situation; it is through the experience of violence that the colonized comes to recognize both the impossibility of justice in the colonies *and* the possibility of his own liberation. Fanon frames the generative process thusly: "Decolonization is truly the creation of new men. But such a creation cannot be attributed to a supernatural power: The "thing" colonized becomes a man through the very process of liberation" (Fanon, 2004, p. 2). "Consciousness raising" cannot be done *prior to* violence.

This construction of anti-colonial violence challenges how "the epistemic problem" is constructed in revisionist just war theory. Approaching war from the perspective of ethical theory, the epistemic problem assumes an antecedent state of peace in which deliberation occurs. Just war as a *political* theory cannot posit this. Fanon's thinking in this chapter, as well as in Chapter 5, "On Colonial War and Mental Disorders," shows how these ideas and positions emerge within a violent context, not antecedent deliberation.

But it would be incorrect to say that anti-colonial thinkers, and Fanon in particular, embraced a moral equality of combatants. Whyte notes that Vo Nguyen Giap's writings on war (which Whyte identifies as an alternate form of just war thinking) bear a striking resemblance to recent revisionist work on the moral *in*equality of combatants.

> "Justice," the North Vietnamese delegation contended, "demands that there should not be equal treatment between war criminals and their victims.". . . the Vietnamese delegation introduced a draft article that aimed to deny [POW] status to "war criminals" —defined as all those who fought on the side of the aggressor. . . . They argued that combatants on the US side waged an unjust war of aggression and were thus, by definition, war criminals who did not deserve the same rights.
>
> *(Whyte, 2019, p. 878)*

Fanon would likely have agreed; the French paratroopers deployed to crush FLN resistance to colonial French rule were unjust fighters, deserving no particular protection.

The final chapter of *The Wretched of the Earth*, "Colonial War and Mental Disorders," is dominated by a series of (anonymized) clinical case studies. If the argument of Chapter 1, "On Violence," is about the transformative and generative power of violence in anti-colonial struggle and in making "the new man," then "Colonial War and Mental Disorders" can be read dialectically as the radically unpredictable effects of violence on individuals. The case studies include both Algerians and French-Algerians traumatized by the violence of colonialism, revolution, and counterterrorism: "We believe that in the cases presented here the triggering factor is principally the bloody, pitiless atmosphere, the generalization of inhuman practices, of people's lasting impression that they are witnessing a veritable apocalypse" (Fanon, 2004, p. 183). Fanon's first example is of a "militant, who never for a moment had thought of recanting, [who] fully realized the price he had had to pay in his person for national independence" (Fanon, 2004, p. 185). This chapter should temper romantic readings of "On Violence" that laud the raw emancipatory potential of anti-colonial violence and cause the reader to question whether the triumphal tone Fanon used in chapter one was truly his own voice, or whether he was ventriloquizing at times for rhetorical purposes.

Fanon recounts his therapy sessions with a French-Algerian police inspector who had tortured suspected FLN fighters and collaborators. "He has lost his appetite and his sleep is disturbed by nightmares. . . . At home he has a constant desire to give everyone a beating. And he violently assaults his children, even his twenty-month-old baby" and eventually turned on his wife (Fanon, 2004, p. 197). Fanon reflects,

> This man knew perfectly well that all his problems stemmed directly from the type of work conducted in the interrogation rooms. . . . As he had no intention of giving up his job as a torturer . . . he asked me in plain language

to help him torture Algerian patriots without having a guilty conscience, without any behavioral problems, and with a total peace of mind.

(Fanon, 2004, pp. 198–199)

Fanon contrasts this with another French policeman he treated (Case No. 4), who had sworn, "Doctor, I'm sick of this job. If you can cure me, I'll request a transfer to France. If they refuse, I'll resign" (Fanon, 2004, p. 196). Both policemen had similar experiences, and one could recognize the injustice of his actions. So, it is not so simple as to say that Fanon "disproves" (revisionist) just war thinking, or that he confirms it; rather, we should take this as an opportunity to *complicate* just war thinking by asking whether the construction of "the epistemic problem" and the moral *in*equality of combatants is ultimately counter-productive? Does this binary framing—moral equality versus inequality—ultimately result in an intractable stalemate? Is there a way to think *with* Fanon to embrace a both/and approach to moral liability?

Here, a question arises for the vocation of the just war theorist. The police inspector discussed earlier came to Fanon looking for scientific, secular absolution for his work as a torturer; he wanted Fanon to "cure" him, such that his work as a torturer did not affect his "normal" life and so that he would no longer beat his children or tie up his wife. He knew his cause to be just, that extreme measures were warranted, and expected Fanon to aid him in carrying out these duties. Fanon, in understanding that "normalizing" colonial violence was an indelible part of his job as a psychiatrist in a colonial context, could not reconcile this with his calling to psychiatry. Just war theorists need to reflect on their calling as well.

Moral authority is called upon to justify torture. In Argentina's *guerra sucia*, the Catholic church offered absolution and encouragement to soldiers fighting the scourges of "communism" and "terrorism," up to and including condoning torture. The Argentine Catholic Church openly opined that the *junta* was fighting a just war (Osiel, 2001). In America's War on Terror, Jean Elsthain argued that "torture lite" (techniques defined as torture by the Convention Against Torture that Elshtain sought to make ambiguous) was a "tragic necessity" in the fight against Islamist terrorism (Elshtain, 2005). In the Israeli context, Michael L. Gross wrote, "Torture is permitted as a last resort to save innocent lives as long as the innocent are not tortured. Even ticking bombs do not override the life of the innocent" (Gross, 2010, p. 137). The Argentine Catholic Church, Elshtain, and Gross provide a cautionary tale about deeming "unjust" fighters to be without moral standing.

Fanon resigned rather than participate in the charade of scientifically "normalizing" colonial brutality. What is the responsibility of the just war theorist when institutions in which they are engaged, or colleagues and fellow just war theorists, offer such absolution for torture, repression, or atrocities? Is there more required of this vocation than scholarly debates?

Controversies

If Fanon is known for his defense of violence, it is important not to give a sanitized account of his work. His writings on violence certainly *appeal* to those who would violently resist domination. Some of this appeal comes from reading "On Violence" in isolation from his other works, but there is no doubt that for Fanon, violence *in itself* is something to be grappled with, and sometimes embraced in its messiness, rather than simply condemned—and certainly not moralized. Fanon thus rejects both the liberal attempt to engage in violence with "clean hands" through moralizing its own violence, as well as offering a deeper engagement with the messiness of violence than Sartre's (1948), Walzer's (1973), and Rawls' (discussed in this volume) "dirty hands" justifications, which recognize the necessity of exceptional moments but fail to grapple with the trauma of extreme violence.

The embrace of extreme violence against colonialism may be at odds with the just war revival, with modern just war's emphasis on restraining violence, but it bears striking parallels to the justification of extreme violence against "barbarians" by some classical just war theorists like Vitoria and Vattel (Brunstetter, 2021). Fanon argues that the violence of the colonized is repurposing the violence of the colonist—in this way, Fanon's *theorizing* violence may be repurposing the violence of the just war tradition. "The challenge now is to seize this violence as it realigns itself" (Fanon, 2004, p. 21).

One controversial aspect of Fanon is his insistence that decolonization requires the destruction of all traces of the colonial world, rather than creating a new nation on the boulevards built by European conquerors:

> To dislocate the colonial world does not mean that once the borders have been eliminated there will be a right of way between the two sectors. To destroy the colonial world means nothing less than demolishing the colonists' sector, burying it deep within the earth or banishing it from the territory.
>
> *(Fanon, 2004, p. 6)*

If the destruction of all traces of colonialism is the *telos* of true decolonization, can post-conflict reconciliation be possible, or will "true decolonization" simply reject that as another colonial residue?

In not shying away from violence, Fanon also rejects peace as an imperative in itself: "Enlightened by violence, the people's consciousness rebels against any pacification" (Fanon, 2004, p. 52). Nonviolence in the colonial context is often, in Fanon's account, an exhortation by the colonial bourgeoisie and outsiders made against the colonized, which aims to preserve the broader colonial system. While many associate nonviolence with Martin Luther King (discussed in a later chapter in this volume) or Gandhi's *satyagraha*, that is not what Fanon is critiquing: his despised "nonviolence" is the premature end of struggle, appointing the colonial bourgeoisie to be the representatives of "the people," and negotiating how this privileged class will assume responsibility for the colonial state. Ending the struggle will maintain the subordinated position of the colonized, which violence can end: "For the last can be the first only after a murderous and decisive confrontation between the two protagonists" (Fanon, 2004, p. 3).

Reading Fanon as a theorist of violence, rather than engaging his writings on language and ontology, reduces him to a revolutionary pedagogue or tactician—an itinerant foreigner-revolutionary. Paired with his Marxism, this approach confuses him for a Martinican Che Guevara. Evidence of this confusion can be seen in Christopher Finlay's sympathetic pairing of the two in the final chapter of *Terrorism and the Right to Resist*; asking whether terrorism could ever be justified, Finlay examines "On Violence" with Guevara's *Guerrilla Warfare*. Focusing on a "pedagogic function" of revolutionary violence, Finlay argues "Fanon recognized [violence's] dramaturgical importance too but also imagined that it could help to restore lost agency to the oppressed through their participation" (Finlay, 2015, p. 295).

Finlay demonstrates the dangers of reading Fanon primarily for an account of violence, when he claims that for Fanon, "[T]he individual could be cured of the traumatic scars of violent colonization. 'At the level of individuals,' Fanon writes, 'violence is a cleansing force. It frees the native from his inferiority complex and from his despair and inaction; it makes him fearless and restores his self-respect'" (Finlay, 2015, p. 296).

The cited passage, however, does not refer to *individual trauma*, which will leave lasting scars, but rather to the mindset of inferiority cultivated by colonialism. The cleansing, Fanon argues a few sentences later, will prepare the newly emancipated people for an egalitarian society, against those who would centralize authority in themselves:

> [T]he people have come to realize that liberation was the achievement of each and every one and no special merit should go to the leader. Violence hoists the people up to the level of the leader. . . . When they have used violence to achieve national liberation, the masses allow nobody to come forward as "liberator".
>
> *(Fanon, 2004, p. 51)*

Against Finlay, a careful reading of *Wretched* reveals that the violence of colonialism and decolonization *traumatizes* all sides, which importantly means that the anti-colonial rebels must yield to a subsequent generation to lead the newly independent state. The insistence that *neither* the heroes of anti-colonial war nor the colonized bourgeoisie should assume leadership of the new state also means that Fanon, though popular among revolutionaries, is often hated by leaders of postcolonial states.

Legacy

Fanon has a powerful global legacy among both revolutionary movements (Gibson, 2011; Gordon, 2015; Ciccariello-Maher, 2017) as well as in academic circles. Particular examples include Steve Biko and the resistance movement against Apartheid in South Africa and the Black Panthers in the United States. In academia, Fanon has confounded postcolonial theorists, served as a canonical figure in the founding of Caribbean philosophy, and exerts a strong influence on Black Studies programs. I want to depart from Fanon's particular legacy and think about what sort of legacy Fanon, and anticolonialism more broadly, *could* have on just war thinking.

As explored in *Just War Thinkers*, just war theory before Walzer experienced a revival among Catholic theologians through the work of Paul Ramsey, and his student James Turner Johnson, in the 1960s and early 1970s (Ramsey, 1968; Johnson, 1975). A parallel aborted revival, however, exists, which is revealing. Whyte (2019) recovers the attempt by postcolonial states to use just war language to challenge international humanitarian law in the preparatory meetings to the Additional Protocols (AP I) in the early 1970s. These new participants in international order reasoned that major powers had dominated previous conventions and that existing international law reflected the preferences of those states in ways that confirm Charles Mills' understanding of the racial world order discussed in the final chapter of this volume; in particular, colonial powers were not bound to treat captured anti-colonial fighters as POWs because these were not "international conflicts." Instead, captured anti-colonial fighters could be simply labeled "terrorists" and denied POW protections. Newly independent states at the convention instead turned to an older language of justice— just war—to challenge existing international law. The U.S. representative to the Additional Protocols warned of "the dangerous concept of the just war" invoked by postcolonial states to undermine the moral and legal authority of Great Powers. Ultimately, the United States did not become a signatory to AP I.

Walzer's *Just and Unjust Wars* appeared shortly after the conclusion of the Additional Protocols but did not reference them or the 1973 UN resolution 3103 regarding the status of combatants in conflicts against colonial, alien, and racist regimes. Colonialism does not figure prominently in *Just and Unjust Wars*; indeed, the only sustained treatment of colonial contexts in the book occurs in the section on terrorism. Walzer analyzes four examples of terrorism: a 1910's IRA bomber who abandoned his bicycle bomb in the wrong location, killing civilians instead of destroying the intended target; Zionist Stern Gang assassins killing a British imperial official in Egypt in 1944; Viet Cong assassins during the wars against first France and then the United States; and (dramatized) scenes of bombing in the film *The Battle of Algiers* (Walzer, 2000, pp. 197–206). It is the final example that, for Walzer, captures the essence of what would become "modern terrorism" and most clearly demonstrates what is irredeemably unjust about such a tactic.

Walzer locates the problem of terrorism in struggles against imperial powers; and while he is ambivalent about the early twentieth-century cases, he is unreservedly repulsed by the tactics deployed by the African and Southeast Asian anti-colonial resistances. In this way, Walzer's framing casts decolonization as an original sin of the postcolonial state rather than an indictment of the racist state-system. In turning to Fanon, I demonstrate the poverty of a Walzerian analysis that reduces decolonization to the question of terrorism. Fanon offers just war theory a different entry point: we must learn to theorize *with* those whose causes were presumptively unjust, those who were prohibited (by earlier just war theorists, even!) from fighting back against the violence of European civilizing missions.

Notes

1 The Diplomatic Conference on the Reaffirmation and Development of International Humanitarian Law Applicable in Armed Conflicts.
2 See the essay "West Indians and Africans" in *Toward the African Revolution* for Fanon's account.
3 In modern therapy, this is referred to as "patient-in-context."
4 In the original French, an anti-black slur is used. Multiple translators opted to render it as "Negro."
5 Memmi abandoned these positions over the anti-Semitism that emerged in many postcolonial states. *Decolonization and the Decolonized* (2008) documents his frustration with the corruption and anti-Semitism of the postcolonial world.
6 Fanon, echoing Césaire's *Discours sur le colonialism,* held that Nazism brought to Europe a system that was pioneered in Europe's overseas colonies. Recent scholarship confirms that colonial rule was far more brutal than "official" accounts let on (Elkins, 2022).
7 That the independent Algerian government did not implement this tolerant approach is a tragedy of decolonization, perhaps demonstrating Fanon's naivete, but it is not *reflective* of Fanon's theorizing.

Works Cited

Asad, Talal. 2009. *On Suicide Bombing.* New York: Columbia University Press.
Bellamy, Alex. 2017. Francisco de Vitoria (1492–1546). In Daniel Brunstetter and Cian O'Driscoll. eds. *Just War Thinkers: From Cicero to the Twenty-First Century.* New York: Routledge, pp. 77–91.
Brunstetter, Daniel. 2021. *Just and Unjust Uses of Limited Force: A Moral Argument with Contemporary Illustrations.* Oxford: Oxford University Press.
Ciccariello-Maher, George. 2017. *Decolonizing Dialectics.* Durham, NC: Duke University Press.
Dussel, Enrique. 2007. Alterity and Modernity (Las Casas, Vitoria, and Suarez: 1514–1617). In Nalini Persram. ed. *Postcolonialism and Political Theory.* Lanham, MD: Lexington Books.
Elkins, Caroline. 2022. *Legacy of Violence: A History of the British Empire.* New York: Alfred A. Knopf.
Elshtain, Jean Bethke. 2004. *Just War Against Terror: The Burden of American Power in a Violent World.* 2nd Edition. New York: Basic Books.
Elshtain, Jean Bethke. 2005. Reflection on the Problem of 'Dirty Hands'. In Sanford Levinson. ed. *Torture: A Collection.* Oxford: Oxford University Press, pp. 77–89.

Elshtain, Jean Bethke. 2007. Terrorism. In Charles Reed and David Ryall. eds. *The Price of Peace: Just War in the 21st Century*. Cambridge: Cambridge University Press, pp. 118–155.

Fanon, Frantz. 1965. *A Dying Colonialism*. Translated by Haakon Chevalier. New York: Grove Press.

Fanon, Frantz. 1988. *Toward the African Revolution*. Translated by Haakon Chevalier. New York: Grove Press.

Fanon, Frantz. 2004. *The Wretched of the Earth*. Translated by Richard Philcox. New York: Grove Press.

Fanon, Frantz. 2008. *Black Skin, White Masks*. Translated by Richard Philcox. New York: Grove Press.

Finlay, Christopher. 2015. *Terrorism and the Right to Resist: A Theory of Just Revolutionary War*. Cambridge: Cambridge University Press.

Gibson, Nigel C. 2011. *Living Fanon: Global Perspectives*. London: Palgrave-MacMillan.

Gooding-Williams, Robert. 2006. *Look, a Negro! Philosophical Essays on Race, Culture, and Politics*. New York: Routledge.

Gordon, Lewis. 2015. *What Fanon Said: A Philosophical Introduction to his Life and Thought*. New York: Fordham University Press.

Gross, Michael L. 2010. *Moral Dilemmas of Modern War: Torture, Assassination, and Blackmail in an Age of Asymmetric Conflict*. Cambridge: Cambridge University Press.

Hutchings, Kimberley. 2019. Cosmopolitan Just War and Coloniality. In Duncan Bell. ed. *Empire, Race, and Global Justice*. Cambridge: Cambridge University Press, pp. 211–227.

Johnson, James Turner. 1975. *Ideology, Reason, and the Limitation of War: Religious and Secular Concepts, 1200–1740*. Princeton, NJ: Princeton University Press.

Maldonado-Torres, Nelson. 2008. *Against War: Views from the Underside of Modernity*. Durham, NC: Duke University Press.

Marasco, Robyn. 2015. *The Highway of Despair: Critical Theory after Hegel*. New York: Columbia University Press.

Mares, Gabriel. 2021. Just War Theory after Empire and the War on Terror: Re-Examining Non-Combatant Immunity. *International Theory* 13(3), pp. 483–505.

McMahan, Jeff. 2004. The Ethics of Killing in War. *Ethics* 114(4), pp. 693–733.

McMahan, Jeff. 2009. *Killing in War*. Oxford: Oxford University Press.

Osiel, Mark. 2001. *Mass Atrocity, Ordinary Evil, and Hannah Arendt: Criminal Consciousness in Argentina's Dirty War*. New Haven, CT: Yale University Press.

Pitts, Jennifer. 2010. Political Theory of Empire and Imperialism. *Annual Review of Political Science* 13, pp. 211–235.

Ramsey, Paul. 1968. *The Just War: Force and Political Responsibility*. New York: Charles Scribner's Sons.

Sartre, Jean-Paul. 1948. Dirty Hands. In Lionel Abel. trans. *Three Plays by Jean-Paul Sartre*. New York: Alfred A. Knopf.

Sartre, Jean-Paul. 2004. Preface. In Frantz Fanon. ed. and Richard Philcox. trans. *The Wretched of the Earth*. New York: Grove Press.

Saul, Ben. 2008. *Defining Terrorism in International Law*. Oxford: Oxford University Press.

Walzer, Michael. 1973. The Problem of Dirty Hands. *Philosophy & Public Affairs* 2(2), pp. 160–180.

Walzer, Michael. 2000. *Just and Unjust Wars: A Moral Argument with Historical Illustrations*. 3rd Edition. New York: Basic Books.

Whyte, Jessica. 2019. The 'Dangerous Concept of the Just War': Decolonization, Wars of National Liberation, and the Additional Protocols to the Geneva Conventions. *Humanity* 9(3), pp. 313–341.

15

ALASDAIR MACINTYRE (1929–)[1]

Anthony F. Lang, Jr.

Introduction

In 1981, Alasdair MacIntyre, a British moral and political philosopher, published *After Virtue*. The book soon took on the status of a classic, setting out a provocative and thoughtful argument about how moral language had lost its underlying social and political context. As such, MacIntyre argued, we live in an age which is "after virtue" —a time when the sociological foundations of our moral language have disappeared. The book brought to fulfilment a project on which MacIntyre had been working since he published his article "Notes from the Moral Wilderness" in 1958. The project, which progressed through various stages, including the publication of his (1966) book, *A Short History of Ethics*, tried to link up the social background to the moral virtues. For MacIntyre, separating moral and political philosophy from social science had contributed to our inability to understand why we could no longer make considered judgments about the world around us.

MacIntyre found a solution to this problem when he embraced a Thomistic Aristotelianism, a framework that he believed enabled judgments about moral matters. This positioning of his ideas has been the focus of his publications since *After Virtue*. He has critiqued both the Enlightenment project and the postmodern project (as he calls them). His position is not a simple one of Christian faith (though it is very much reliant on the Christian tradition); rather, his position is that Thomism demonstrates how one can bring into conversation very different theoretical traditions, enable them to enrich each other, and create a clear and sustainable moral framework.

It is odd that as a tradition of thought which is indebted to Aquinas in many ways, that is, the just war tradition, is not a body of thought to which

DOI: 10.4324/9781003428688-16

MacIntyre has contributed directly. Even more oddly, as a tradition of thought that seeks to combine a set of practices with moral judgement is not one that has been the focus of MacIntyre's work. His work does, admittedly, tend to be more at the level of metatheory than the applied theory that is often the subject of just war theorizing, and he has critiqued a parallel body of applied ethics—business ethics. And, he has written one piece on military ethics, which is examined later in the chapter. But the overarching point I want to make in this chapter is that in light of his theoretical development, MacIntyre would be an obvious person to be part of the just war tradition.

Perhaps the way he is most helpful for the tradition is in his metatheoretical contributions to how we think about ethics in the contemporary world. As I will note later in the chapter, he does intervene on practical ethics at times, but his more important insights come in how he frames the disconnect between ethics and social and political practices. This is directly relevant to the just war tradition, particularly in the debates between revisionists and traditional just war theorists. MacIntyre has little time for ethical theory that remains in the abstract, always insisting that ethical insights must be linked to social and political life. This is precisely the point made by theorists such as Michael Walzer.[2] MacIntyre, then, helps us to see how military ethics should be more closely linked to what military officials think and do in the conduct of their everyday lives.

Contexts

Alasdair MacIntyre was born in Glasgow in 1929.[3] Soon after, his parents, both doctors, moved to London. His father died soon after this, and his mother moved to Belfast. He was first educated at Epsom College, which was created to board and educate poor members of the medical profession. He studied Classics at Queen Mary College in London, then became a postgraduate student at Manchester University at the age of 21, becoming a lecturer in philosophy only three years later.

While this iterant background would not seem to ground him in any particular cultural context, MacIntyre himself explains that he was influenced by living in what he calls a "Gaelic culture," learning Scots Gaelic from one his aunts (Lutz, n.d.). He was raised:

> [I]n a Gaelic oral culture of farmers and fishermen, poets and storytellers, a culture that was in large part already lost, but to which some of the older people I knew still belonged with part of themselves. . . . On the other hand, I was taught by other older people that learning to speak or to read Gaelic was an idle, antiquarian past-time, a waste of time for someone whose education was designed to enable him to pass those examinations that are the threshold of bourgeois life in the modern world.
>
> *(Knight, 1998, p. 255)*

This quote hints at MacIntyre's later work which finds conceptual conflicts between a broadly defined 'modernity' and traditions of practice and culture that stand opposed to that modern world.

As an undergraduate, MacIntyre focused not only on classical texts but also on modern philosophy, attending lectures by A.J. Ayer and Karl Popper among others (Lutz, n.d.). He also encountered the anthropologist Franz Steiner, whose work introduced him to the link between cultures and their moral frameworks (Knight, 1998, p. 259). While being a student in London, he became more aware of the poverty and deprivation of those living around him, which partly turned him to Marxism. His initial scholarship focused on the intersection of Marxism and philosophy; while he is now more critical of Marxism, his concern with matters of economic justice and injustice has continued throughout his career (Blackledge and Davidson, 2008). The other major influence on his intellectual life was his Christianity. He was baptized in the Presbyterian tradition, and he notes in an interview that he tried to keep his Christian faith fenced off from his intellectual life. For a time, he rejected Christianity altogether, but, as he notes "parts of Thomism survived in my thought from those times" (Knight, 1998, p. 257). Sometime in the 1970s, he converted to Catholicism, the tradition that has grounded his intellectual life ever since.

MacIntyre's academic career was very much an itinerant one, moving from the United Kingdom to the United States and then back and forth. He has taught at the following institutions: University of Exeter, University of Leeds, Oxford University, Brandeis University, Boston University, Wellsley College, Vanderbilt University, University of Notre Dame, and Duke University, along with visiting professorships at numerous other institutions. He retired from teaching and has ended up primarily at the University of Norte Dame. He has deposited his papers with London Metropolitan University. Indeed, these two universities seem to represent the two poles of his intellectual development: London Metropolitan representing his early interest in Marxism and his upbringing in the UK, and Notre Dame representing his turn to a Thomistic Catholicism. While he no longer publishes as actively as he once did, he continues to give lectures both through Notre Dame's Center for Ethics and Culture (https://ethicscenter.nd.edu/) and the annual conference of the International Society for MacIntryean Inquiry (www.macintyreanenquiry.org/).

Texts

MacIntyre has published numerous works over the course of his career, spanning a wide range of topics in moral and political philosophy. In this section, I will first provide an overview of his works, then turn to the trilogy of books for which he is perhaps best known, starting with his 1981 book *After Virtue*.

MacIntyre's work falls into different stages, the first of which focused largely on debates in and around Marxist thought and practice. In these works, he both engages in specific debate about Marxist theory but at the same time critiques Marxism on a number of grounds. In one of his earliest works, he argues that Marxism and Christianity share a number of features, disputing the claim of many Marxists that it is a social science. Engaging in this question on the level of metaphysics, he then turns to how this impacts moral claims from both traditions of thought. Interestingly, he sees here a 'fusion of horizons', where Marxism and Christianity might come together to encourage the Church to move away from a focus solely on the afterlife, and Marxism can acknowledge the importance of religious belief and prayer, both of which he sees as beneficial to moral life. So, unlike his later works, in this work, MacIntyre concludes that two traditions of thought can come together and create a productive moral framework (MacIntyre, 1953).

His engagement with Marxism also moved him toward having a more critical perspective on the possibility of moral critique. In an important paper from 1958, MacIntyre starts by asking how to go about critiquing Stalin, a question that generated a great deal of debate among Western Marxists. In trying to stay true to Marxist ideas yet also confronting the horrors of Stalin's rule in the Soviet Union, many Marxists felt unable to critically assess what had happened. MacIntyre uses the different positions in the debate around Stalin (the "Stalinist," the "moral critic," the "revisionist") to think critically about what it means to engage in moral analysis. As he notes, this debate among these different figures may help us to "find a way out of own wilderness" (MacIntyre, 1958–1959 [reprinted in Knight, 1998, p. 31]).

MacIntyre's scholarship moved from Marxism toward a wider critical engagement with the social sciences. He has long argued that to understand moral questions requires an understanding of society and politics. But while he sees the importance of understanding social and political context, he also does not accept many of the major tenets of contemporary social science, either in the mid-twentieth century or today. We can see this partly in his critique of Marxism when he argues it is more of an ideological belief system than a social science. He makes a different but related critique of Freud's idea of the unconscious, arguing that Freud's theories, while brilliant and helpful in understanding why we do what do, also reduce us to something not truly human: "to portray the specifically human as human, and not as nervous system plus muscles, or as chain molecules, or as fundamental participles, is not to explain at all" (MacIntyre, 1958, p. 98). He brings many of these concerns together in his (1971) book, *Against the Self-Images of the Age*, which is a critique of both the social sciences and analytic philosophers, both of whom he sees as having a failed conception of intentions, desires, and actions (MacIntyre, 1971).

MacIntyre developed his critique of analytic philosophy indirectly in his 1966 history of ethics. In this overview of Western philosophy, he starts with the pre-Socratics and moves through mid-twentieth-century philosophy. In so doing, he makes clear throughout, particularly in the early chapters on the Greeks, that philosophy and social norms cannot be kept separate: "Moral concepts are embodied in and are partially constitutive of forms of social life" (MacIntyre, 1966, p. 1). It is not simply that moral concepts have a location in a time in place; more importantly, perhaps, they have a history, and "to understand this is to be liberated from any false absolutist claims" (MacIntyre, 1966, p. 269).

These critical perspectives lead to the trilogy of books for which MacIntyre is most famous: *After Virtue* (1981); *Whose Justice, Which Rationality* (1988); and *Three Rival Versions of Moral Enquiry* (1990). In fact, the books are not a trilogy in the sense that they were planned to be written together; in a 1984 postscript to the second edition of *After Virtue*, he does indicate that he is working on a book on justice, though this is not really spelled out in any detail. And, the lectures that compose *Three Rival Versions of Moral Inquiry* move his argument from a more critical one toward a stronger embrace of a particular form of inquiry.

After Virtue is the book for which MacIntyre is most well-known, and it is the book that made his fame as a scholar. It picks up on his critiques of social science and philosophy and puts them into a larger critique of what he calls the Enlightenment project. He begins with an interesting thought experiment in which he imagines a world where scientific concepts and terms are still used, but the background conditions that made sense of them disappear. He suggests that we are facing the same situation with moral discourse and that we use words like right, good, and virtue but we no longer understand the social contexts in which those words emerged. He argues that analytic philosophers and social scientists both assume a universal discourse that reflects a kind of liberal enlightenment project, but that the claims they make do not function as they should because there is no location for them in a social or political context.

With this critique set out in the first half of the book, he then turns to an exploration of the virtues, arguing that Aristotle provides the best theorization of such virtues. As Aristotle argued, so MacIntyre argues that virtues are best understood through the ancient Greek word *arete*, which means excellence. This excellence is defined by the practice in which one is engaged, which can range from professions to games to life itself. Aristotle moved from the question "what does it mean to be a good carpenter" to "what does it mean to be a good person" in order to elaborate on what it means to be morally good (Aristotle, 2000). However, unlike Aristotle, MacIntyre does not confine his ideas of the good of the human person and the related virtues to a single time and place. Rather, MacIntyre argues that practices and the virtues associated with them have a history; that is, they develop through time. But this history, if it is to truly reflect a set of practices, should be understood within a tradition, a key part of MacIntyre's approach to moral philosophy.

Traditions provide the standards and authorities by which judgments about a practice and the good associated with it can be made. So, MacIntyre is not just arguing that moral philosophy simply changes; virtues change in accordance with standards which also evolve, but which evolve through traditions of thought and practice. From this follows MacIntyre's critique of liberalism, which he argues gives us a false sense of choice about our life plans and moral ideas. Instead, he encourages us to see that our individual choices alone are inadequate as a source of judgment about what is right or wrong; instead, we must include insights from traditions of thought which we may or may not understand we are part of (MacIntyre, 1984, p. 190).

His 1988 book, *Whose Justice, Which Rationality*, continues this focus on the contextual nature of moral concepts and argument. As the title suggests, however, it focuses on the concepts of justice and rationality. The book argues not just that moral concepts have a history but that rationality does as well (MacIntyre, 1988, p. 9). For instance, in Chapter 2, MacIntyre explores the nature of Homeric rationality as it relates to justice. He notes that the understanding of rational deliberation in the world of the *Iliad* and the *Odyssey* finds little parallel with our efforts to understand our decision-making. MacIntyre demonstrates his overall argument by highlighting how the classical Greek world (fourth and third centuries BC) had to grapple with ideas of reasoning as they related to ideas of justice and virtue. The challenge they faced was how to translate not only these moral ideas around justice but also the assumptions around reason that underlie them in a very different social context, but a context in which Homer's poems still carried a great deal of cultural weight.

One way to see this point, in a way that overlaps with debates in just war, concerns the relationship between virtue and victory. MacIntyre notes that within the Homeric poems, particularly the *Iliad*, to be a virtuous warrior was generally to be a victorious one. But this is qualified by ensuring that the competition in which the two agents are participating is a just one, one with rules that ensure fair play (MacIntyre, 1988, pp. 26–29). Importantly, however, these rules are not about protecting civilians, being proportional, or any of the standards by which we judge just wars today. Rather, they are rules concerning a level playing field between combatants, something closer to the rules of a sporting contest rather than moral rules. That is to say, a just war for a Homeric warrior is more directly linked to victory than it is for us in the contemporary world.[4]

This book provides further evidence for MacIntyre's argument through a careful examination of Aristotle, Augustine, and Aquinas. He is moving here toward a much closer embrace of the Thomist tradition which frames many of the books which follow this one. He puts these ideas in conversation with David Hume, his representative of the Enlightenment. The latter third of the book carefully develops the concept of a tradition, particularly how traditions of thought embody notions of rationality and, as a result, justice and virtue. MacIntyre is particularly drawn to Aquinas in these discussions because of Aquinas' achievement of being able to combine two traditions—the

Aristotelian and the Augustinian—to create a new tradition of thought. He argues that the liberal/Enlightenment tradition has been unable to achieve this because of its singular assumptions about what constitutes reason and ethics. Aquinas, instead, demonstrates how traditions can be assessed and brought together, creating something that is not only faithful to the traditions combined but also new and insightful.

The final work in this central trilogy is *Three Rival Versions of Moral Enquiry*, which were first delivered as the Gifford Lectures at the University of Edinburgh in 1988 and then published in 1990. Having set out what constitutes a tradition and how traditions can and cannot relate to each other, this book compares three different approaches to moral enquiry: the Encyclopedic, the Genealogical, and Tradition. As one might guess, MacIntyre concludes that it is only what he calls Tradition that really can provide us with true moral understanding. The Encyclopedic is a broadly liberal one, taking the encyclopedia as a model for how the liberal Enlightenment mind approaches moral questions (along with all other questions). He focuses on the Ninth Edition of the *Encyclopedia Britannica*, published in the late nineteenth century, as the ideal type of this approach. It reflects a particularly British Enlightenment idea of how knowledge is to be gathered and understood, reducing everything to a form of practical rationality that leads to a liberal political structure (though only for the civilized and not the colonized). The Genealogical is reflected in a combination of Nietzsche and Foucault, both of whom sought to understand history not as a progressive liberal process but as a series of disruptions and changes which reveal power dynamics rather than an improvement of moral sensibility. MacIntyre finds both of these approaches problematic, and so he spends the remaining lectures trying to develop an Aristotelian/Thomistic alternative, which he simply calls Tradition.

As developed in *Whose Justice, Which Rationality*, this Thomistic alternative is privileged as the only one that can rationally encounter and bring together divergent approaches, as Thomas did with Aristotle and Augustine. MacIntyre addresses the problem of having 'too many Thomisms', which is the idea that there are competing uses of Aquinas in the history of political and moral enquiry, even within the Catholic Church. He explores Pope Leo XIII's encyclical *Aeterni Patris*, issued in 1879, which made a case for all Catholic scholars to use the work of Aquinas in their understanding of morality and theology. As MacIntyre notes, however, this did not lead to a single Thomism but created a number of competing Thomisms.[5] Despite these contrasting Thomistic approaches, MacIntyre concludes that all the Thomists share an understanding of narrative and its relation to moral enquiry. In this passage, he compares the three approaches:

> So, the encyclopaedists' narrative reduces the past to a mere prologue to the rational present, while the genealogist struggles in the construction of his or her narrative against the past, including that of the past which

his perceived as hidden within the alleged rationality of the present. The Thomists' narrative, by contrast with both of these, treats the past neither as mere prologue nor as something to be struggled against, but as that from which we have to learn if we are to identify and move towards our telos more adequately and that which we have to put to the question if we are to know which questions we ourselves should next formulate and attempt to answer, both theoretically and practically. This reappropriation of the past in a way which directs the present towards a particular—and yet eternal—future takes place at two interrelated levels, that of theoretical enquiry and that of the practical embodiment of that enquiry.

(1990, p. 79)

Once again, history and narrative are intricately linked for MacIntyre in his understanding of morality. His conclusion that only the Thomistic tradition understands this and can deal with changes in history forms a key element of his later works.

Controversies

The previous section demonstrates how much MacIntyre has invested in large metaethical debates. He has not completely neglected practical ethical questions, including those related to the just war tradition. In this section, let me highlight two interventions that raise importance questions about the way we understand military ethics.

The first is his 1984 lecture, "Is Patriotism a Virtue?" which he delivered as the Lindley Lecture at the University of Kansas. The second is a keynote address he delivered at the University of Notre Dame in 2013 entitled "Military Ethics: A Discipline in Crisis." The first text provides some insight into questions around *jus ad bellum*, for it addresses a core claim of the just war tradition: that national interests are worth defending with military force if necessary. The second focuses far more on *jus in bello* questions, with a focus on how such *in bello* rules are formulated and taught at military institutions around the world.

In his 1984 lecture, MacIntyre continues his critique of what he calls here liberal morality or the Enlightenment morality which he has criticized throughout his career. He contrasts the liberal morality that would argue for a single moral standard to be applied to all persons with a morality that arises from a particular nation-state. While he does not use the term, liberal morality contrasted with patriotism can perhaps best be described as liberal cosmopolitanism. Though he is critical of this cosmopolitanism, he does not conclude that patriotism is to be preferred; indeed, he never really answers the question of whether or not patriotism is a virtue. Instead, he concludes that neither cosmopolitanism nor patriotism is fully defensible on its own.

The point he does make in his conclusion is that claims by countries like the United States and France which combine their patriotism with a call to universal liberalism do not make sense. No country can advance a moral universalism while also advancing its own patriotism. And, for the purposes of this chapter, he highlights the dilemma this causes for anyone asked to use force to advance such universalist claims. Specifically, he notes that any soldier who believes in a liberal position of moral universalism will not be able to use military force to defend a particular country, for to take that liberal position will lead to the "dissolution of social bonds" that make it possible to kill and die for a community (MacIntyre, 2008 [1984], p. 186).

How is this relevant to just war thinking? MacIntyre is obliquely (and perhaps not knowingly) critiquing the revisionist position that individual moral soldiers should be capable of morally evaluating their country's actions (McMahan, 2009). It is not simply that individual combatants will disagree with their country at times; it is that to put them in a position of making a moral evaluation of their country by means of universal standards will undermine their commitment to making the ultimate sacrifice in situations of extreme danger. MacIntyre's paper highlights a fundamental tension within the analytical and revisionist tradition; that the use of military force by individuals in the service of their country relies much more on the social bonds that patriotism engenders than it does on universal ideals of right and wrong. In some sense, this is a sociological point rather than a moral one, but, as with all of MacIntyre's thought, one cannot leave the sociological behind when understanding moral principles.

Michael Walzer's account of just war is close to MacIntyre here, in that Walzer makes a strong case that wars are fought by communities rather than individuals. While Walzer's claim is based on a liberal social contract idea, his description of how individual rights move to state rights sounds somewhat like MacIntyre:

> Over a long period of time, share experiences and cooperative activity of many different kinds shape a common life. . . . The moral standing of any particular state depends upon the reality of the common life it protects and the extent to which sacrifices required by that protection are willingly accept and thought worthwhile.
>
> *(Walzer, 2015, p. 54)*

There is a difference between the two theorists, in that MacIntyre is famously dismissive of rights, saying belief in them is like belief in "unicorns and witches" (MacIntyre, 1984, p. 69). While MacIntyre is also sometimes called a communitarian, his approach differs very distinctly from Walzer's (and most others) because of his resistance to traditional liberalism. Clearly, however, he sits much closer to the traditionalist than to the revisionist approach to just war theorizing.

The second article I wish to highlight is his 2013 lecture on the crisis in military ethics. The lecture responds more directly to the military conflicts in which the United States was engaged in the Middle East following the attacks of 9/11. This is the one instance I could find in his scholarship where he directly engages contemporary military policy and practice. The article focuses on U.S. military operations in Iraq, with a particular focus on the agreement that the United States made with Sunni tribal sheikhs in support of U.S. efforts to defeat al-Qaeda. The lecture elucidates military virtues in a holistic sense. He begins with the idea of courage, asking if a Nazi soldier could be considered courageous. Rather than simply viewing courage on a scale from brashness to cowardice (a typical Aristotelian understanding of it, i.e., finding the mean between extremes), MacIntyre highlights the fact that practical virtues cannot be disconnected from the goods to which they are devoted: "To be courageous is not only for an agent to risk or to endure dangers and harms, but to do so in a way and to an extent that is proportionate to the goods at stake in that agent's situation" (MacIntyre, 2015 [2013], p. 4). MacIntyre reminds us that Aristotle and other virtue ethicists argued that it is through phronesis or prudence that we can connect our wider goals to the specific virtues we enact in military actions.

MacIntyre argues that there is another, perhaps equally important, virtue that military officers need to enact: justice. This might seem obvious to those writing about just war, but MacIntyre makes clear that the term justice is rather anodyne in the just war tradition. He uses an example of a military commander in the field in Iraq making the decision to ally with Sunni militants against Al-Qaeda, an act that went beyond his political mandate. In so doing, however, the commander was able to effectively remove Al-Qaeda from a particular location in Iraq. But once the United States withdrew from Iraq, the largely Shiite government in Baghdad failed to uphold that agreement with Sunnis. This failure has had devastating consequences in Iraq and elsewhere. He then brings in another example, which is the U.S. and French approach to combating Islamic extremism in North Africa, particularly in Mali. In this case, MacIntyre highlights the Sufi community of Mali and their role in saving Islamic treasures located in Timbuktu. He points out that, rather than working with these Sufi communities, the U.S. and French approach to defeating militants in the region has been one of primarily military operations with little regard for allies from cultural and religious groups.

MacIntyre uses both these examples to argue that a concern with justice and what he calls political prudence would suggest that U.S. military actions have missed the larger picture concerning the justice of its various engagements with Islamic communities around the world. He is not making an argument for ignoring the military operations that might be necessary. But he is highlighting the success of the one commander in Iraq and the failures elsewhere to assert that military ethics needs to avoid an overreliance on rules and embrace a virtue-oriented approach to military ethics. Rather than creating checklists of what can and cannot be done, MacIntyre presents us here with a need to educate commanders and perhaps politicians as to how to not only enact specific virtues but also to see those virtues in a larger social and political context.

Legacy

MacIntyre's work has not generated controversy among just war thinkers because he does not really engage directly with them. His work has, of course, generated controversy and left some important legacies about how we understand the use of force.

MacIntyre's transition from Marxism to Christianity, particularly Thomistic Christianity, certainly limits the appeal of his position to those who wish to make a case from a wider and/or secular perspective. Of course, the insights of Aquinas have been central to many in the just war tradition; most recently, both Gregory Reichberg and Christian Nikolaus Braun have demonstrated how Aquinas' insights can provide new ways of seeing aspects of contemporary military activity (Reichberg, 2017; Braun, 2023). None of these authors, however, assume the existence of God to make their arguments. While his earlier work assumed that different moral frameworks existed as a result of different social contexts, his later work has moved much more toward a transcendent notion of moral good. As one commentator notes, "There exists for MacIntyre, as for Thomas, an absolute criterion of truth, and it is tradition-transcendent; it is, in fact, the divine intellect" (Lutz, 2004, p. 126). While some scholars say MacIntyre's reading of Aquinas is inaccurate or problematic (Haldane, 1994), these seem more like quibbles in intellectual history; it is clear that MacIntyre relies on a Christian perspective to orient his thinking. So, one legacy would be that for those who write from a secular just war perspective, MacIntyre's approach would be less helpful.

Admittedly, there are enough just war thinkers who have a Christian perspective, such as Jean Bethke Elshtain, Nigel Biggar, and Oliver O'Donovan. Most of these works, however, tend to rely on an intellectual history approach to Christianity rather than a faith-based one. That is, they draw on the insights of figures such as Augustine and Aquinas, but they do not make their case in terms of the Christian faith. MacIntyre does not link his brief interventions on war to Christian faith, but his overall approach to ethics and traditions does rely on a notion of truth which makes the concept of God necessary. If just war thinking is about trying to reach a set of truth claims about whether or not force can be just, then MacIntyre's reliance on God may well be a stumbling block for many.

A related point is how just war is understood within the Catholic Church, where MacIntyre finds his orientation. He has written about the role of the

Catholic university in today's world, though the focus of that work is more on the holistic approach to knowledge that a Catholic university can provide rather than reflections on the just war tradition (MacIntyre, 2011). But what is most relevant here is how the just war tradition currently sits within the Catholic Church, and how this relates to MacIntyre's understanding of a tradition. In his encyclical, *Fratelli Tutti*, issued in 2020, Pope Francis does not quite dismiss the just war tradition, but he certainly does not embrace it (Paragraph 258). Francis states clearly: "Let us not remain mired in theoretical discussions, but touch the wounded flesh of the victims" (Paragraph 261). Francis is building upon a shift in the Vatican's approach to the just war tradition which is grounded in international law and the United Nations Charter rather than the intellectual tradition that stretches back to Augustine and Aquinas and which the Catholic Church has often claimed as its own. MacIntyre's focus on the authority of a tradition, particularly the Catholic tradition, suggests that a Catholic might need to follow the authority of the pope rather than an individual's insights into whether or not war can be considered just today. Because MacIntyre does not engage with just war thinking directly, this is less an issue for his own work. But, one might see how a Catholic may face difficulties in confronting the just war tradition when following MacIntyre's approach.[6]

Outside of these challenges arising from his faith-based approach to moral philosophy, I think MacIntyre leaves us with some important legacies. Let me highlight three here.

First, MacIntyre makes a strong case that virtue and morality more broadly must arise from the goods internal to practices. That is, to understand what it means to be a virtuous person means understanding what goods persons seek. The fact that just war thinking arose not from philosophical reflection but from the practices of those using force makes a virtue-oriented approach such as MacIntyre's very helpful. James Turner Johnson has clearly explained how the medieval European origins of the just war tradition emerged through a combination of philosophical insights and the practices of knights whose focus on chivalry shaped various *in bello* rules (Johnson, 1981). This combination of philosophy and practice might well benefit from the refinements and extensions of the virtue ethic tradition that has been the focus of MacIntyre's scholarship.

Second, MacIntyre has highlighted the problems of a liberal and cosmopolitan ethos. He has debunked, convincingly I think, the idea that there can be a simple form of rationality which leads to the conclusions necessary for understanding and navigating the complexities of the contemporary world. In *Whose Justice, Which Rationality,* he explains how there is not one single conception of justice and not one conception of rationality. Instead, he argues for what he calls a tradition-oriented approach. This puts him firmly on the side of those who find just war theorizing less helpful than a historical approach to the just war tradition. His fully developed idea of a tradition demonstrates how traditions force us to understand where authoritative voices and texts should play a role in our efforts to make moral judgments. But for MacIntyre, a tradition is not an ossified relic that does not change. Instead, it is a set of ongoing arguments which respond to new practical developments and new theoretical insights. MacIntyre can, in other words, help those who write from a just war tradition perspective to better understand what it means to work within a tradition and continue to make arguments in response to new world developments, as Christian Nikolaus Braun does in his use of Aquinas to understand drone warfare (Braun, 2023).[7]

Finally, MacIntyre has moved toward making narratives central to his way of understanding ethics. For MacIntyre, narratives of lives give us the means to better understand how agents navigate the complexities of political life. In his 2016 study of different figures, including the U.S. Supreme Court Justice Sandra Day O'Connor and the Trinidadian Marxist historian and political activist C.L.R. James, MacIntyre links specific narratives to his Aristotelian and Thomistic tradition of thought (MacIntyre, 2016). Just war thinkers could learn much from this approach. So many of the theories of just war highlight specific moments or single wars. Combined with his insights on justice in his lecture on military ethics, this emphasis on narrative should help us to understand that ethical judgments require a longer-term approach—one that can encompass more than just single moments in time. Perhaps a focus on narratives, including narratives of grievance between combatants, can be a new source of insight for just war thinkers.

Notes

1 Thanks to Daniel Brunstetter, Cian O'Driscoll, and Christian Nikolaus Braun for helpful feedback on an earlier draft of the chapter.
2 MacIntyre is sometimes linked with Walzer as a fellow communitarian. But, as I note later in the chapter, their communitarian frameworks have very different starting points.
3 Some of these details come from (Cornwell, 2010).
4 Though, see Cian O'Driscoll's nuanced work on the relationship between victory and just war. O'Driscoll does address this relationship in the Ancient World but does not present the relationship in the way that MacIntyre does (O'Driscoll, 2020).
5 There is something of a parallel here in Greg Reichberg's analysis of the different Thomisms that have informed just war thinking, from Catetan's emphasis on punishment to de Molina's emphasis on liability (Reichberg, 2017). I am indebted to Christian Nikolaus Braun for bringing this point to my attention.
6 I have made this point in relation to James Turner Johnson's critique of the American Catholic Bishop's letter on nuclear weapons; see Lang (2009).
7 Many others draw on historical traditions to make their points, with James Turner Johnson being the most obvious example.

Works Cited

Aristotle. 2000. *The Nicomachean Ethics*. Cambridge: Cambridge University Press.

Blackledge, Paul and Neil Davidson (eds.). 2008. *Alasdair MacIntyre's Engagement with Marxism: Selected Writings, 1958–1974*. Leiden: Brill Publishers.

Braun, Christian Nikolaus. 2023. *Limited Force and the Fight for the Just War Tradition*. Washington DC: Georgetown University Press.

Cornwell, John. 2010. MacIntyre on Money. *Prospect Magazine* 176(20 October). Available at: web.archive.org/web/20110125084039/www.prospectmagazine.co.uk/2010/10/alasdair-macintyre-on-money/.

Haldane, John. 1994. MacIntyre's Thomist Revival: What Next? In John Horton and Susan Mendus. eds. *After MacIntyre: Critical Perspectives on the Work of Alasdair MacIntyre*. Notre Dame: University of Notre Dame Press, pp. 91–107.

Johnson, James Turner. 1981. *Just War Tradition and the Restraint of War: A Moral and Historical Inquiry*. Princeton, NJ: Princeton University Press.

Knight, Kelvin (ed.). 1998. *The MacIntyre Reader*. Cambridge: Polity Press.

Lang, Anthony F., Jr. 2009. The Just War Tradition and the Question of Authority. *Journal of Military Ethics* 8(3), pp. 202–216.

Lutz, Christopher Stephen. n.d. Alasdair MacIntyre. *Internet Encyclopaedia of Philosophy*. Available at: https://iep.utm.edu/mac-over/.

Lutz, Christopher Stephen. 2004. *Tradition in the Ethics of Alasdair MacIntyre: Relativism, Thomism, and Philosophy*. Lanham, MD: Lexington Books.

MacIntyre, Alasdair. 1953. *Marxism: An Interpretation*. London: Student Christian Movement Press.

MacIntyre, Alasdair. 1958. *The Unconscious: A Conceptual Study*. London: Routledge and Kegan Paul.

MacIntyre, Alasdair. 1958–1959. Notes from the Moral Wilderness. *The New Reasoner* 7, pp. 7–8.

MacIntyre, Alasdair. 1966. *A Short History of Ethics*. New York: Collier Books.

MacIntyre, Alasdair. 1971. *Against the Self-Images of the Age: Essays on Ideology and Philosophy*. London: Duckworth.

MacIntyre, Alasdair. 1984 [1981]. *After Virtue*. 2nd Edition. Notre Dame: University of Notre Dame Press.

MacIntyre, Alasdair. 1988. *Whose Justice? Which Rationality?* Notre Dame: University of Notre Dame Press.

MacIntyre, Alasdair. 1990. *Three Rival Versions of Moral Enquiry: Encyclopaedia, Genealogy, and Tradition*. Notre Dame: University of Notre Dame Press.

MacIntyre, Alasdair. 2008 [1984]. Is Patriotism a Virtue? Reprinted in Igor Primoratz. ed. *Military Ethics*. London: Routledge.

MacIntyre, Alasdair. 2011. *God, Philosophy, Universities: A Selective History of the Catholic Philosophical Tradition*. London: Rowman and Littlefield.

MacIntyre, Alasdair. 2015 [2013]. Military Ethics: A Discipline in Crisis. In George Lucas. ed. *The Routledge Handbook of Military Ethics*. London: Routledge, pp. 3–14.

MacIntyre, Alasdair. 2016. *Ethics in the Conflicts of Modernity: An Essay on Desire, Practical Reasoning, and Narrative*. Cambridge: Cambridge University Press.

McMahan, Jeff. 2009. *Killing in War*. Oxford: Oxford University Press.

O'Driscoll, Cian. 2020. *Victory: The Triumph and Tragedy of Just War*. Oxford: Oxford University Press.

Reichberg, Gregory M. 2017. *Thomas Aquinas on War and Peace*. Cambridge: Cambridge University Press.

Walzer, Michael. 2015. *Just and Unjust Wars: A Moral Argument with Historical Illustrations*. 5th Edition. New York: Basic Books.

16

MARTIN LUTHER KING JR. (1929–1968)

Juan M. Floyd-Thomas

Introduction

On 4 April 1967, Reverend Dr. Martin Luther King Jr. delivered his legendary address "Beyond Vietnam: A Time to Break the Silence" at the historic Riverside Church in New York City. Written in consultation with scholars and close advisors Vincent Harding and John Maguire, the speech marked the culmination of King's thoughtful, intentional, and consistent anti-violence stance to address the great American military and moral quagmire of his lifetime, namely the Vietnam War.

Unquestionably, King's message was a major media event particularly because, as the internationally renowned leader of the nonviolent civil rights movement in the United States, King deliberately chose Riverside Church and its prestigious pulpit as the prime venue to address his stance on U.S. foreign policy in general and his views on the Vietnam conflict in particular. Even though he had briefly spoken out against U.S. involvement in Southeast Asia in 1965, it was the increasing pressure from aides and confidantes such as Reverend James Lawson, Reverend Andrew Young, and Vincent Harding that finally convinced King to end his silence about America's military conflict with Vietnam. More importantly, King's "Beyond Vietnam" sermon was a risk of faith in the most Kierkegaardian sense because he knew full and well that opposing the war would certainly create bad blood between him and President Lyndon Johnson precisely at the time that the civil rights movement needed more presidential support to guarantee social and economic gains at home (Branch, 2007).

Much of King's tenure as a domestic civil rights leader was also spent gauging the successes and failures of historic events on the world stage,

DOI: 10.4324/9781003428688-17

including decolonization efforts and armed liberation movements across the Global South. Suffice it to say that King's considerations on internationalism, anti-racism, anti-imperialism, nonviolence, and ending poverty were inter linked, and while he shares a Christian moral foundation with certain veins of the just war tradition, he utilized a moral compass that was quite distinct. The insights from King's moral stance should be instructive to all of us as we are still dealing with our generational war, that is, the ongoing aftermath of our military engagement in Iraq and Afghanistan over the 20-year period spanning the George W. Bush, Obama, Trump, and Biden presidencies. Toward that end, this chapter will emphasize how King's "Beyond Vietnam" both in its historic context and in its overall philosophical and hermeneutical content illustrates a nonviolent moral framework that challenges the just war doctrine and, in doing so, provides a prophetic witness in the quest for peace while also holding a more meaningful and truthful definition of patriotism to bear.

Contexts

As one of the most heralded and documented figures of the twentieth century, Dr. Martin Luther King Jr. rose to fame with his advocacy of nonviolent civil disobedience to advance social justice (Lewis, 1970; Carson, 1998; Eig, 2023). Born Michael King on 15 January 1929, in Atlanta, Georgia, he was the second child and first son of Baptist minister Michael Luther King, Sr. and his wife, the former Alberta Williams, who herself was the daughter of the Rev. Adam Daniel Williams, pastor of the Ebenezer Baptist Church. When King was two years old, his maternal grandfather died, and his father became pastor. Four years after that—in 1935—his father changed his and his son's name from Michael to Martin in tribute to Martin Luther, the sixteenth-century German leader of the Protestant Reformation. Now known as Martin Luther King Jr., he was enrolled at the all-black Young Street Grade School. Because of their elite position as prominent church leaders, the King family was considerably shielded from the full magnitude of the Great Depression's impact in his racially segregated community. Nevertheless, the family did feel the severity and savagery of racial segregation, a dire situation which the elder King condemned at every possible opportunity.

King's early upbringing was rooted in strong African American civic institutions of family, church, school, and community. After completing his elementary education, King attended the Laboratory High School at Atlanta University until its closure in 1942 and then transferred to Booker T. Washington High School, where he excelled academically. In 1944, King graduated from high school early and enrolled at Morehouse College, a prestigious all-male historically Black college in Atlanta, Georgia, at the age of 15. King greatly benefited from the exposure to a comprehensive liberal arts education at Morehouse from 1944 to 1948. Initially, he had been resistant to pursuing a career in the ministry, mostly because he was embarrassed by the perceived emotionality of congregations within the Black Christian tradition. While at Morehouse, though, he was deeply influenced by the college's president, Benjamin Mays, as well as his philosophy teacher, George D. Kelsey, both of whom were also ordained ministers. While still an undergraduate in 1947, King became an ordained minister in his father's church and preached his first sermon.

From 1948 to 1951, King went on to study at the racially integrated Crozer Theological Seminary in Chester, Pennsylvania, as only one of six African American seminarians. It was during his time at Crozer when King first became captivated with the teachings of Mahatma Gandhi. As King's academic studies began to shape his activism, he steadily embraced Gandhi's perspective that nonviolent resistance could be used to harness anger and frustration, channeling them into a more positive force for societal transformation (Watley, 1985). From 1951 to 1955, King was awarded a fellowship for graduate studies and enrolled in the doctoral program at Boston University where he earned a Ph.D. in systematic theology.

Shortly after receiving his doctorate, King was able to put his ideals to the test. On 1 December 1955, seamstress Rosa Parks refused to give up her seat on a Montgomery bus to a white male passenger. Her subsequent arrest sparked a boycott of the public transportation system by the city's African American populace. Although the boycott was already in place, King assumed a leadership position, thus putting into practice the teachings of civil disobedience he embraced. This thrusted him into national prominence, especially after he was arrested. The boycott lasted for more than a year, during which time his home was twice bombed, and his life was repeatedly threatened. His standpoint was vindicated when the U.S. Supreme Court eventually ruled that the state laws on segregation on buses were unconstitutional in 1956.

From when he emerged as a leader during the Montgomery Bus Boycott in 1955–1956, to the famous protest march in Birmingham, Alabama, in 1963 (during which he was arrested and famously wrote the "Letter from Birmingham Jail"), to the March on Washington for Jobs and Freedom which culminated in King's famous "I Have a Dream" speech delivered at the Lincoln Memorial, King was both admired and reviled in his crusade to achieve racial equality. He was an eloquent and potent figure bridging societal divides, as evidenced by his access to the halls of power in both political (the White House) and religious (the Vatican) spheres. In the fall of 1964, he became the youngest recipient of the Nobel Peace Prize at the age of 35, largely for his key role as leader of the Southern nonviolent civil rights movement. But his efforts effectively put a target on his back.

Like many just war thinkers, King's reflections on war were driven by the specific controversies and conflicts of his day, specifically the Vietnam War.

While Paul Ramsey refused to condemn the war, King, like Michael Walzer, situated himself in the anti-war camp. For Walzer, just war theory provided:

> [A] moral doctrine readily at hand, a connected set of names and concepts that we all knew . . . without this vocabulary, we could not have thought about the Vietnam war as we did, let alone communicated our thoughts to other people.
>
> *(2006, p. xix)*

Yet King did not rely solely on the prescribed moral vocabulary of the just war tradition. When he began to speak out in opposition to American involvement in Southeast Asia in 1967, he had a different moral point of reference best articulated in his "Beyond Vietnam" speech.

King did not have time to see the long-term impact of his nonviolent anti-war moral stance. He was assassinated on 4 April 1968, in Memphis, Tennessee, where he had traveled to support the striking sanitation workers, at the age of 39 years.

Texts and Tenets

While quite visible and vocal on various matters of U.S. domestic policy throughout the 1960s, King was initially reluctant to speak out against military conflict in Vietnam and the larger Southeast Asian region. When asked his opinion about Vietnam in March 1965 by journalists, King cautiously stated that he was "sympathetic" to President Johnson's predicament but did not believe that "violence can solve the problem" (King, 2 March 1965). King's final phone conversation with Johnson took place in late 1966, and the chief topic of their discussion was America's involvement in Vietnam. In the ensuing months, Johnson arranged to meet with King on two separate occasions, but King canceled both scheduled meetings. Bewildered by this situation, Johnson asked his aides to find out why King seemed to be avoiding him. Yet, when King delivered his "Beyond Vietnam" speech on 4 April 1967, the answer to Johnson's query became abundantly clear. It is worth listening to a recording to fully grasp King's oratory power (King, 1967).

In his hour-long speech, King said that he was moved to "break the betrayal of my own silences and to speak from the burnings of my own heart" against the U.S. war in Vietnam (King, 2013, p. 80). When King stood behind the sacred desk that is the pulpit of the Riverside Church in the full glare of the media spotlight, he had to address the great question that swirled above him like so many storm clouds, namely "why break his silence" when supposedly "peace and civil rights don't mix" (King, 2013, p. 80). In response to this speculative concern, Dr. King offered seven key reasons why ending the

Vietnam War, and inevitably all wars, had become a centerpiece of his prophetic ministry: because he fundamentally believed waging war abroad and making domestic civil progress were incompatible (Fairclough, 1984, p. 19). To garner the scope of his elocution, it is important to recognize King's audience: "Tonight, however, I wish not to speak with Hanoi and the National Liberation Front, but rather to my fellow Americans" (King, 2013, p. 81). He was, in a sense, asking his fellow citizens to follow his lead by "bringing Vietnam into the field of my moral vision" (King, 2013, p. 81). The speech is, as I have argued elsewhere, King's attempt to redress the critical shortcomings of Augustinian just war theory (Floyd-Thomas, 2011, p. 151).

King's first reason casts the spotlight on the monetary costs of war and what nobler endeavors get sacrificed to wage war. He argued that President Johnson's Great Society initiatives to transform the United States would be compromised by the nation's military expenditures in Southeast Asia. While he had high hopes for the program, "then came the buildup in Vietnam and I watched this program broken and eviscerated as if it were some idle political plaything of a society gone mad on war" (King, 2013, pp. 81–82). Waging war is expensive, and the Vietnam War was taking away vital resources that could have been spent on the War on Poverty instead. From the civil rights leader's perspective, Congress was spending more wealth and legislative clout on the use of military force overseas, leaving virtually nothing left to spend on domestic programs: "A nation that continues year after year to spend more money on military defense than on programs of social uplift is approaching spiritual death" (King, 2013, p. 93).

As prescient as King's critique was more than half a century ago, his words are even more timely now, in the first decades of the twenty-first century. The United States has spent trillions of dollars in the twin wars in Afghanistan and Iraq that have led to little change on the ground, while taking funding away from domestic programs in dire need. King would have us ask before we undertake any current or future wars: how might that money be better spent?

The war's cost in lives, both on their side and ours, was also a significant factor that entered his moral vision. Regarding the Vietnamese, he bemoaned that our violence leaves no basis for trust:

What do they think as we test out our latest weapons on them, just as the Germans tested out new medicine and new tortures in the concentration camps of Europe? Where are the roots of the independent Vietnam we claim to be building? Is it among these voiceless ones? We have destroyed their two most cherished institutions: the family and the village. We have destroyed their land and their crops. We have cooperated in the crushing—in the crushing of the nation's only non-Communist revolutionary political force, the unified Buddhist Church. We have supported the enemies of the peasants of Saigon. We have corrupted their women and children and killed their men. Now there is little left to build on, save bitterness.

(King, 2013, pp. 86–87)

Regarding U.S. troops, he lamented the government "was sending [the sons, brothers, and husbands of the poor] to fight and die in extraordinarily high proportions relative to the rest of the population" (King, 2013, p. 82). For those who survived, this was causing something akin to moral injury: "We are adding cynicism to the process of death, for they must know after a short period there that none of the things we claim to be fighting for are really involved" (King, 2013, p. 89). Taking that point even further, King put the spotlight on the underlying racial inequalities in "taking the [Black] young men who had been crippled by our society and sending them eight thousand miles away to guarantee liberties in Southeast Asia which they had not found in southwest Georgia and East Harlem" (King, 2013, p. 82). America watched black and white boys on television as they fought, killed, and died together on the battlefield but could not seat them together in the same school classroom. According to King's *Where Do We Go From Here: Chaos or Community?*, twice as many Black troops as whites were in combat at the beginning of 1967, and twice as many Black soldiers were killed in proportion to their numbers in the general population (King, 1968; Darby and Rowley, 1986, pp. 43–44). Given the prospect of being asked to fight unjust wars, King proposed a moral alternative: "As we counsel young men concerning military service, we must clarify for them our nation's role in Vietnam and challenge them with the alternative of conscientious objection" (King, 2013, p. 91).

A third reason reflects the moral consistency of his philosophy of nonviolence. King tailored his remarks to highlight his reflexive engagement with the roots of lethal violence not only within American culture but also within the premise of human nature itself. His domestic nonviolent credentials were well known, but the Vietnam War called on him to extend these to the international realm as well. King reasoned that he "could never again raise [his] voice against the violence of the oppressed in the ghettos without having first spoken clearly to the greatest purveyor of violence in the world today: my own government" (King, 2013, p. 82). This recalls historian Charles Beard's farsighted comment during the early days of the Cold War when he noted that war-makers would pursue "perpetual war for perpetual peace" (Barnes, 1953). The synergy between King's and Beard's views on modern warfare indicates that, from the view of government, there will always be an urgent reason to go to war to secure peace, but the reality on the ground is such that war does not in fact beget peace.

King's rejection of violence makes him an oft-ignored thinker when it comes to exploring the moral assumptions key to the Christian just war tradition. Unlike historical figures such as Augustine and Aquinas, and contemporary figures like Paul Ramsey, James Turner Johnson, and Jean Bethke Elshtain (as well as Christian realists such as Reinhold Niebuhr), King rejects the premise that war is a "necessary evil" aspect of the human condition. Instead, he believes that the society can overcome its proclivity to see war as the solution to the problem of evil—just war for the sake of peace—by focusing on people instead of profit and ideology: "A true revolution of values will lay hand on the world order and say of war, 'This way of settling differences is not just.'" While this may sound naive and tap into what just war scholars and Niebuhrian Christian realists view as the naivety of pacifists who turn a blind eye to true evil in the world, King's stance has an introspective bent to it that should be emphasized:

> Here is the true meaning and value of compassion and nonviolence, when it helps us to see the enemy's point of view, to hear his questions, to know his assessment of ourselves. For from his view we may indeed see the basic weaknesses of our own condition, and if we are mature, we may learn and grow and profit from the wisdom of the brothers who are called the opposition.
>
> *(King, 2013, p. 88)*

Niebuhr, also discussed in this volume, purportedly was an important influence on Barack Obama's view of war. King was also influenced by the German–American theologian and pastor but felt that Niebuhr distorted the prospects of pacifism because he "interpreted pacifism as sort of passive nonresistance to evil expressing naïve trust in the power of love" (King, 1958, p. 98). King was far from naïve. If anything, he believed those who put their trust in war were the naïve ones. His Vietnam speech forced those who saw violence as the solution to confront war's long-term destructive impact. King decisively broke with Niebuhr's perspective, informed by his embrace of Gandhian principles of nonviolent struggle and broader nonviolence philosophy:

> [T]rue pacifism is . . . a courageous confrontation of evil by the power of love, in faith that it is better to be the recipient of love than the inflicter [sic] of it, since the latter only multiplies the existence of violence and bitterness in the universe, while the former may develop a sense of shame in the opponent, and thereby bring about a transformation and change of heart.
>
> *(King, 1958, pp. 98–99)*

If nonviolence was the moral path to combat racial injustice domestically, war was a manifestation of that injustice internationally. With parallels to Charles Mills—who is discussed in the final chapter of this volume—a fourth reason points to the global racial inequalities manifested in America's warmongering. King realized that the domestic sickness of America has an international dimension—its imperial wars rooted in racism, militarism, and materialism. In his own words:

> Now, it should be incandescently clear that no one who has any concern for the integrity and life of America today can ignore the present war. If America's soul becomes totally poisoned, part of the autopsy must read: Vietnam. It can never be saved so long as it destroys the deepest hopes of men the world over.
>
> *(King, 2013, p. 83)*

Placing this statement into a more contemporaneous purview, pursuing an unjust war, whether in Vietnam back then or the twin wars in Afghanistan and Iraq more recently, is toxic to the very soul of America.

Fifth, criticizing war was part of the vocation of being a Nobel Peace prize recipient. King used the validation of the international community to reject American bellicose overtures against those framed, in the context of the Cold War, as mortal enemies, what he refers to as "that strangely anonymous group we call 'VC' or 'communists.'" He drove this point home by saying: "I cannot forget that the Nobel Peace Prize was also a commission, a commission to work harder than I had ever worked before for 'the brotherhood of man.' This is a calling that takes me beyond national allegiances" King, 2013, p. 83). In no uncertain words, he sought to remind the entire nation that he had earned the Nobel Peace Prize and not the Nobel *Negro* Prize—my phrase—as so many of his critics were trying to insinuate. As a Nobel laureate, King indicated he had an undeniable right and unquestionable authority to speak out against militarism and imperialism even if it irked fellow citizens and outraged political leaders, up to and including the U.S. president.

While acknowledging his worldly bona fides, King's sixth reason situates his moral choices in the realm of divine responsibility. Even if he had not received the Nobel Prize, he observes: "I would yet have to live with the meaning of my commitment to the ministry of Jesus Christ" (King, 2013, pp. 83–84). Taking this point even deeper, King rightly says "to me, the relationship of this ministry to the making of peace is so obvious that I sometimes marvel at those who ask me why I am speaking against the war" (King, 2013, p. 84). Simply put, according to King's assertion, true Christianity— much like true democracy—should cease wars rather than cause them. Perhaps anticipating his Christian critics, he rhetorically asked:

> Could it be that they do not know that the good news was meant for all [people]—for Communist and capitalist, for their children and ours, for black and for white, for revolutionary and conservative? Have they

forgotten that my ministry is in obedience to the One who loved his enemies so fully that he died for them? What then can I say to the Vietcong
or to Castro or to Mao as a faithful minister of this One? Can I threaten
them with death or must I not share with them my life?

(King, 2013, p. 84)

Finally, King's bold antiwar stance moved beyond the confines of race,
class, or nation to his divine calling as a child of God, that is, "the privilege and the burden of all of us who deem ourselves bound by allegiances
and loyalties which are broader and deeper than nationalism and which go
beyond our nation's self-defined goals and positions" (King, 2013, p. 84).
King was driven by what he considered the absolute truth of Christian
thinking linked to its pacifist roots. These outshined the just war ideas that
emerged during the historic shift away from Christian pacifism at the time
of the Roman empire, via Augustine, Gratian, and beyond to become what
we know call the just war tradition. King had a different moral foundation:
"We are called to speak for the weak, for the voiceless, for the victims of
our nation, for those it calls 'enemy,' for no document from human hands
can make these humans any less for our brothers [and sisters]" (King, 2013,
p. 84). In his own inimitable fashion, King offered the insight that, regardless of whatever patch of earth one might find him or herself, we are all
inhabitants of God's creation.

By looking beyond Christianity's just war past, King's moral stance
offers a modern inspiration for Christians and non-Christians alike to
endorse its original pacifism. As the just war scholar James Turner Johnson detailed, the pacifist elements that dominated early Christianity were
ultimately discarded in favor of the classic view of just war to secure a just
peace, a shift which gave birth to the just war tradition (Johnson, 1981).
The Christian church has since debated the merits of just war, with seemingly every generation offering renegotiated interpretations. In the post-
9/11 era, this has led to just war thinking that has loosened the restraints
on war by re-embracing the classic just war tradition found in Augustine,
as epitomized by the works of James Turner Johnson and Jean Bethke
Elshtain that address the War on Terror (Brunstetter, 2014). The curious
reader might explore other Christian thinkers who hold a deeper, if not
quite pacifist, presumption against war, such as Erasmus and Bartolomé
de las Casas (Brunstetter, 2020). King's anti-war stance also offers interesting points of parallels to the presumption against war found in Pope
Francis, as well a critical alternative to Fanon's turn to necessary anti-
colonial violence—both discussed in later chapters. King's moral stance
is not a brand of naive humanism; it comes from a profound engagement
with racial and social injustice and harnesses a different vein of Christianity than the principles of the just war tradition.

Living in the long shadow left by Dr. King as a pacifist prophet who spoke uncomfortable yet uncompromising truths to the American empire, let us remember "Beyond Vietnam" as a noteworthy sermon that depicts a vital theological and moral framework for evaluating the turn to war that is equally part patriotic and part prophetic. "America," King professed,

> [T]he richest and most powerful nation in the world, can well lead the way in this revolution of values. There is nothing except a tragic death wish to prevent us from reordering our priorities so that the pursuit of peace will take precedence over the pursuit of war.
>
> *(King, 2013, pp. 93–94)*

Controversies

After revealing his sevenfold reasoning for opposing the American war in Vietnam, Dr. King suffered possibly the worst backlash of his career. Even though he is now lionized as the American patron saint of peace and nonviolent social change, it is important to remember that he was initially demonized and skewered by the federal government, the national media, and the public. In many regards, the most alarming backlash to King's comments came from members of the Black middle class who had previously been the chief beneficiaries, if not the strongest supporters, of King's prophetic ministry. King was denounced in many circles as "traitorous" for his remarks. While long hated by many white southern segregationists, this speech turned the more mainstream media against him. For example, even a typically moderate periodical such as *Life* magazine called the sermon nothing less than "demagogic slander that sounded like a script for Radio Hanoi." Likewise, *The Washington Post*, one of the nation's more venerable and thoughtful newspapers, flatly declared that King had "diminished his usefulness to his cause, his country, his people." Even the president turned against King. Whereas Johnson was clearly viewed as a political ally of the civil rights movement and was observed on several occasions referring to Dr. King as "my dear friend, Martin," once news of the "Beyond Vietnam" sermon broke, the president was reportedly so enraged that he referred to King using a racist slur (Sitkoff, 2010, p. 207; Dyson, 2008, p. 163).

Confronted with such deep wells of negativism, King did not run from his stance. Rather, he doubled down by arguing that not only was ending the war in Vietnam morally imperative but also that ending the overall dependence of American culture on violence and militarism was the solution to all its social ills. "The war in Vietnam," he said, "is but a symptom of a far deeper malady within the American spirit," and to ignore this would only result in next generation leading anti-war marches to protest future wars. What was required was nothing short of a radical revolution, which many conservative critics ostensibly saw as catering to all things un-American. However, in King's view, America had problematically taken on:

> [T]he role of those who make peaceful revolution impossible by refusing to give up the privileges and the pleasures that come from the immense profits of overseas investments. I am convinced that if we are to get on the right side of the world revolution, we as a nation must undergo a radical revolution of values. We must rapidly begin . . . we must rapidly begin the shift from a thing-oriented society to a person-oriented society. When machines and computers, profit motives and property rights, are considered more important than people, the giant triplets of racism, extreme materialism, and militarism are incapable of being conquered.
>
> *(King, 2013, pp. 92–93)*

This was, then, as it is now, a controversial stance. While King's fame as a leader and activist is based on how he courted controversy politically, it is equally important to note how his prominence was also based on his opposition to key figures who espoused the virtues of violence, including Malcolm X and Frantz Fanon.

As both dual and often dueling leaders of the modern African American freedom struggle, Martin Luther King Jr. and Malcolm X had a complex relationship. While they rose to national and international notoriety during the late 1950s and early 1960s, they were viewed by virtually everyone—by supporters and detractors, by many whites, by most Blacks, and even themselves—as ideological antagonists and diametrically opposed nemeses on all important questions involving racial equality, political possibilities, and the role of violence in America. This knee-jerk polarized juxtaposition of King and Malcolm as opposites persists in the narrative surrounding 1960s' radicalism, particularly in "mainstream" (read white-dominated) public opinion, while becoming considerably less so among African Americans nowadays (Baldwin, 1989). The dualistic view portrays these two world-historical figures as "adversaries in a great Manichaean [cosmic] contest," as Peter Goldman states, pitting "the forces of light against the forces of darkness, with the future course of black protest at stake" (Goldman, 2013, p. 79).

There are reasons, some good, to support this narrative. Strongest is that, throughout most of their careers, they had very substantial disagreements, including on interracial integration versus racial separatism qua Black cultural nationalism as the preferred paradigm for racial equality. More to the theme of this volume, they disagreed on whether to use nonviolent civil disobedience as a means of becoming "free at last" or armed self-defense as means to achieve Black liberation from oppression "by any means necessary." While these divisions were quite stark, it has now become increasingly more conventional wisdom to recognize, the role of violence aside, that there was significant convergence of their political ideologies in the latter phases of their lives. In his dual biography of both leaders, James Cone contends that Martin Luther King and Malcolm X "complemented and corrected each other" (Cone, 1991, p. 246).

A more appropriate contemporaneous foil for King's pacifism might be the Afro-Caribbean psychiatrist and revolutionary theorist Frantz Fanon, also discussed in this volume. Hailed as "the Handbook for the Black Revolution," Fanon's *The Wretched of the Earth* is widely celebrated as a probing examination of colonization, a compelling description of the process of decolonization, and a prophetic analysis of global anti-imperialist movements. Decades after its initial publication, literary scholar Henry Louis Gates Jr. refers to Fanon as "both totem and text" for academic theorists of postcolonial, postmodern, and subaltern studies (Gates, 1991, pp. 457–458). The book's first chapter, "Concerning Violence" establishes the premise that "decolonization is always a violent phenomenon" (Fanon, 2005 [1961], p. 35). Fanon is very clear in his message: the struggle for power in contested colonial states will be resolved *only* through violent struggle. Because the colonized states were created and are maintained using violence, it is a logical inevitability that it will take violence by the colonized to reverse these existing power relationships. Although Fanon's primary interest remains squarely focused on the decolonization process in Africa, the Caribbean, and other parts of the Global South, he also discussed the distinct parallels with the struggles of African Americans, in which King was so involved.

In sharp contrast, King was adamantly opposed to the perspective that violence was an inevitable and irrevocable fact of human existence, regardless of the causes or circumstances. This did not mean he was apologetic for the ills of imperial violence. His "Beyond Vietnam" speech is peppered with critiques of the West which, in the ambience of the Cold War, inevitably stirred controversy. The Vietnamese, he opined,

> [M]ust see Americans as strange liberators. . . . Even though they quoted the American Declaration of Independence in their own document of freedom, we refused to recognize them. Instead, we decided to support France in its reconquest of her former colony. Our government felt then that the Vietnamese people were not ready for independence, and we again fell victim to the deadly Western arrogance that has poisoned the international atmosphere for so long. For nine years following 1945 we denied the people of Vietnam the right of independence. For nine years we vigorously supported the French in their abortive effort to re-colonize Vietnam.
>
> *(King, 2013, p. 85)*

Sewing doubts about the effectiveness of "our" violence, King nevertheless understood, without justifying, theirs:

> How can they trust us when now we charge them with violence after the murderous reign of Diem and charge them with violence while we pour every new weapon of death into their land? Surely we must understand their feelings, even if we do not condone their actions. Surely we must see that the men we supported pressed them to their violence.
>
> *(King, 2013, p. 87)*

But any parallels with Fanon are ultimately irreconcilable, for King rejects violence as a liberatory means. That said, he levies a strident moral condemnation of Western imperial war that tracks with Fanon's general positionality. King specifically targets "Western arrogance of feeling that it has everything to teach others and nothing to learn from them," which is his view, "is not just" (King, 2013, p. 93).

For both those who advocate just war to combat evil and spread democracy as well as those who legitimize anti-colonial violence, King's moral vision is a challenging alternative to embrace. And it is a moral stance for which he ultimately paid the highest price—his life.

Legacy

The impact of Martin Luther King Jr.'s work extended far beyond his untimely death. King's stature as a representative hero of social justice remains firmly intact. In addition to referencing the importance of the just war tradition in his 2009 Nobel Prize acceptance speech, President Obama also lauded King:

> I make this statement mindful of what Martin Luther King Jr. said in this same ceremony years ago: "Violence never brings permanent peace. It solves no social problem: it merely creates new and more complicated ones." As someone who stands here as a direct consequence of Dr. King's life work, I am living testimony to the moral force of non-violence.
>
> *(Obama, 2009)*

The merger of King's teachings and just war concepts within Obama's foreign policy outlook is a troubling legacy but suggests the former President was at the very least operating with the knowledge that even purportedly justifiable uses of force tarnished the truth of American ideals, trust in American global leadership, and transparency (Floyd-Thomas, 2018). It also speaks to other facets of King's legacy. Kingian nonviolence is not simply a counter paradigm to that of just war. Rather, it is a call to recognize the limits of the just-war-for-the-sake-of peace dictum so often repeated by just war thinkers who apply their theories to the latest conflicts but so rarely achieved in the long run. And it is a reminder to count the costs of war not simply in terms of civilians and soldiers who die abroad but also in terms of the social programs that die at home because of lack of funding and political support, given the debt incurred by waging war.

Just war scholars might balk at King's views as ultimately naive or embrace the crude caricature of King as a meek, mild-mannered mouthpiece for moderation, but that would elide his more radical side (Smith, 1986). It is when we look beyond the comfortable, convenient depiction of King to see his more complex and confrontational face that the potency of his more radical viewpoints shines. Historian Vincent Harding has advanced the thesis that it is the radical King whose prophetic voice speaks most tellingly to the challenges and woes of the contemporary world, especially the woes caused by imperial violence. The greatest obstacle to understanding the meaning of King's leadership, he argues, "is our apparent determination to forget—or ignore—the last years of his life" (Harding, 2008, p. ix). America suffers from severe selective amnesia (partly innocent, partly willful) about King. Harding claims that the image of King that arose in the post-1965 era is that of the "inconvenient hero" whom most Americans would prefer (and pretend) never existed. Furthermore, King's bravura "I Have a Dream" speech delivered during the "March on Washington" ironically has trapped King within the public memory and popular culture of the nation, as if he were frozen in amber since 1963. Likewise, ethicist and cultural critic Michael Eric Dyson (2000) argues that the mature later leader who surfaced as a radical critic of both U.S. domestic injustice and global militarism is the "true" King. This is the King of the "Beyond Vietnam" speech, who spoke truth to power regardless of the consequence.

If Martin Luther King's life and legacy will continue to hold relevance in a world of intensifying religious conflicts and racial-ethnic hatred, then we owe it to ourselves and posterity to look at King's ideas and radical activism as a serious interlocutor with, and maybe alternative to, the ethics of a just war tradition that has failed to move the global arc of justice via the American wars it has justified in the twenty-first century. King asks us to speak in a different moral language:

> Even when pressed by the demands of inner truth, [people] do not easily assume the task of opposing their government's policy, especially in time of war. Nor does the human spirit move without great difficulty against all the apathy of conformist thought within one's own bosom and in the surrounding world. Moreover, when the issues at hand seem as perplexing as they often do in the case of this dreadful conflict, we are always on the verge of being mesmerized by uncertainty; but we must move on. . . . We must speak with all the humility that is appropriate to our limited vision, but we must speak.
> *(King, 2013, p. 80)*

Works Cited

Baldwin, Lewis V. 1989. A Reassessment of the Relationship between Malcolm X and Martin Luther King Jr. *Western Journal of Black Studies* 13(2), pp. 103–113.

Barnes, Harry Elmer (ed.). 1953. *Perpetual War for Perpetual Peace: A Critical Examination of the Foreign Policy of Franklin Delano Roosevelt and Its Aftermath.* Caldwell, ID: Caxton Press.

Branch, Taylor. 2007. *At Canaan's Edge: America in the King Years, 1965–68.* New York: Simon & Schuster.

Brunstetter, Daniel R. 2014. Trends in Just War Thinking during the U.S. Presidential Debates 2000–12: Genocide Prevention and the Renewed Salience of Last Resort. *Review of International Studies* 4(1), pp. 77–99.

Brunstetter, Daniel R. 2020. The Quest for Peace Revisited. In Eric D. Patterson and Marc LiVecche. eds. *Responsibility and Restraint: James Turner Johnson and the Just War Tradition.* Middletown, CT: Stone Tower Press, pp. 67–95.

Carson, Clayborne (ed.). 1998. *The Autobiography of Martin Luther King Jr.* New York: Warner Books.

Cone, James H. 1991. *Martin & Malcolm & America: A Dream or Nightmare?* Maryknoll, NY: Orbis Books.

Darby, Henry E. and Margaret N. Rowley. 1986. King on Vietnam and beyond. *Phylon* 47(1), (1st Qtr.), pp. 43–50.

Dyson, Michael Eric. 2000. *I May Not Get There with You: The True Martin Luther King Jr.* New York: Free Press.

Dyson, Michael Eric. 2008. *April 4, 1968: Martin Luther King, Jr.'s Death and How It Changed America.* New York: Basic Civitas Books.

Eig, Jonathan. 2023. *King: A Life.* New York: Farrar, Straus and Giroux.

Fairclough, Adam. 1984. Martin Luther King, Jr. and the War in Vietnam. *Phylon* 45(1), pp. 19–39.

Fanon, Frantz. 2005 [1961]. *The Wretched of the Earth.* Translated by Richard Philcox. New York: Grove Press.

Floyd-Thomas, Juan M. 2011. More Than Conquerors: Just War Theory and the Need for a Black Christian Antiwar Movement. *Black Theology: An International Journal* 9(2), pp. 136–160.

Floyd-Thomas, Juan M. 2018. Black Prophetic Discourse and Just War Theory in the Age of Obama. In Juan M. Floyd-Thomas and Anthony B. Pinn. eds. *Religion in the Age of Obama.* London: Bloomsbury Press, pp. 108–127.

Gates, Henry Louis, Jr. 1991. Critical Fanonism. *Critical Inquiry* 17(3), pp. 457–470.

Goldman, Peter. 2013. *The Death and Life of Malcolm X.* Urbana: University of Illinois Press.

Harding, Vincent G. 2008. *Martin Luther King: The Inconvenient Hero.* Maryknoll, NY: Orbis Books.

Johnson, James Turner. 1981. *Just War Tradition and the Restraint of War: A Moral and Historical Inquiry.* Princeton, NJ: Princeton University Press.

King, Martin Luther, Jr. 1958. *Stride toward Freedom: The Montgomery Story.* New York: Harper & Row.

King, Martin Luther, Jr. 1967. *Beyond Vietnam—A Time to Break Silence.* [Accessed 4 May 2024]. Available at: www.americanrhetoric.com/speeches/mlkatimeto-breaksilence.htm.

King, Martin Luther, Jr. 1968. *Where Do We Go from Here: Chaos or Community?* Boston: Beacon Press.

King, Martin Luther, Jr. 2013. *A Time to Break Silence: The Essential Works of Martin Luther King, Jr. for Students.* Edited by Walter Dean Myers. Boston, MA: Beacon Press.

Lewis, David L. 1970. *King: A Critical Biography*. New York: Praeger.
Obama, Barack. 2009. *Remarks by the President at the Acceptance of the Nobel Peace Prize*. [Accessed 5 May 2024]. Available at: https://obamawhitehouse. archives.gov/the-press-office/remarks-president-acceptance-nobel-peace-prize.
Sitkoff, Harvard. 2010. *Toward Freedom Land: The Long Struggle for Racial Equality in America*. Lexington, KY: University Press of Kentucky.
Smith, Kenneth L. 1986. The Radicalization of Martin Luther King, Jr.: The Last Three Years. *Journal of Ecumenical Studies* 26(2), pp. 270–288.
Walzer, Michael. 2006. *Just and Unjust Wars: A Moral Argument with Historical Illustrations*. 4th Edition. New York: Basic Books.
Watley, William D. 1985. *Roots of Resistance: The Nonviolent Ethic of Martin Luther King Jr.* Valley Forge, PA: Judson Press.

17

JOHN RAWLS (1921–2002)

Yvonne Chiu[1]

Contexts

John Bordley Rawls was born in Baltimore, Maryland, as the second of five sons to a successful lawyer father and a politically engaged mother. After attending a strict religious boarding high school, he entered Princeton University in 1939, shortly after Germany invaded Poland. At Princeton, his thesis critiqued naturalism in addressing Christian problems of sin and faith. He considered going to seminary to study for priesthood after university but then accelerated his studies like many in his cohort in order to join the war effort. After graduating summa cum laude in philosophy in January 1943, he enlisted in the U.S. Army, served in the Pacific Theater for two years, then returned to the Ph.D. program at Princeton in 1946 on the GI Bill, where he wrote his dissertation under Walter Stace's supervision. After a variety of positions (Cornell University; Fulbright fellowship at Christ Church, Oxford; MIT), he spent the rest of his career in Harvard University's philosophy department (1962–1995).

As to his philosophic career, John Rawls needs no introduction. Rawls is often credited with reviving Western normative analytic political philosophy (Douglass, 2012). His writings on just war theory are limited, however, and he is mostly associated with a highly abstract, contemporary form of domestic social contract theory (*A Theory of Justice*, 1971, revised 1999; *Political Liberalism*, 1993, expanded 2005), which he then extended to the international realm (*The Law of Peoples*, 1999). Furthermore, as an exemplar of abstract and idealized analytic political philosophy, Rawls's approach in his limited writings on war is dramatically different from that of his near-contemporary and colleague—and preeminent just war thinker of the time—Michael Walzer,

DOI: 10.4324/9781003428688-18

who takes a more historically infused approach to the ethics of warfare. Indeed, it is significant that Rawls's most concrete engagement with the ethics of war comes in his essay "50 Years After Hiroshima" (1995)—one of four essays (along with those by Ronald Takaki, Walzer himself, and Jean Bethke Elshtain, another of the contemporary era's most significant just war theorists)—written for *Dissent* magazine at Walzer's behest, although Walzer has said that he had wanted something a little less abstractly philosophical from Rawls. After 9/11, analytic philosophers would challenge the "traditional" school of just war thinking represented by Walzer (2015 [1977]), but Rawls's own analytical forays into just war theory come to very different conclusions compared to analytical philosophers such as Jeff McMahan, David Rodin, and Helen Frowe, who champion a now-dominant "revisionist" approach. Using the same analytic approach, Rawls provides an alternative set of assumptions and conclusions about warfare ethics that merit our reflection.

It was not Rawls's intention to develop a full doctrine of just war theory; rather, his purpose was to extend particular principles of domestic justice to the international realm, which required some engagement with the nonideal reality of international coercion, violence, and warfare. Doing so with his analytic philosophy method demonstrates that one need not follow revisionist methods with revisionist assumptions to arrive at revisionist conclusions; instead of reducing the international framework of morality to the ethics of daily individual relations in liberal society writ large, one can—as Rawls does—grapple with norms and laws where there is no enforcement (i.e., in the international realm), what Walzer calls the legalist paradigm, and work through the challenge of pursuing ideal theory in a nonideal world.

A recent trend in just war and broader political thought has been to focus on the impact of lived experience on ethical theorizing (O'Driscoll, 2021), and as one delves into Rawls's thoughts on just war, it may enrich our understanding of his philosophy to know how his experience with war—before he became *John Rawls the philosopher*—influenced his theorizing. One would never know from any of Rawls's writings on war that after completing his A.B. in philosophy at Princeton in 1943, he enlisted in the Army and served two years in the World War II Pacific Theater in New Guinea, Philippines, and occupied Japan with the 128th Infantry Regiment (32nd Infantry Division). He saw a fair amount of combat during the war and received a Bronze Star for his radio communications from behind enemy lines in Luzon, Philippines (Pogge, 2007, p. 12). Nor would one know that Rawls's experiences in the war not only altered the path of his life—away from orthodox Christianity and possible priesthood—but also likely significantly shaped his political philosophy. Were it not for a private essay written in the 1990s, "On My Religion," found among his papers after his death, Rawls's life-changing experience with war would have been abstracted away and readers consigned to interpreting only his more analytical texts. But doing so would miss something essential about his foray into just war thinking.

Texts and Tenets

Rawls writes on just war theory briefly in *A Theory of Justice* (§58) [1971, rev. 1999], then in essays "50 Years After Hiroshima" (1995) and "The Law of Peoples" (1999a [1993]), the latter of which was later expanded into the book *The Law of Peoples, with "The Idea of Public Reason Revisited"* (1999). In each case, he extends particular arguments about domestic justice to the international realm, to "the law of nations," and, in the course of doing so, addresses selected aspects of *jus ad bellum*, *jus in bello*, and *jus post bellum*. Taken as a set and interpreted as just war theory, however, Rawls's various writings on war and warfare can seem unsatisfying and overly conventional in their conclusions, despite his radical theoretical framework, which makes peoples the relevant unit of normative analysis and places their representatives in an international original position in order to derive principles of international justice. In the international extension of Rawls's domestic original position, which anchors *A Theory of Justice*, parties (peoples) must come to an agreement on the moral structures of international society while knowing nothing about their relevant capacities and vulnerabilities. Notably and controversially, Rawls does not propose a global original position.[2]

Given the premises of *A Theory of Justice*, Rawls could easily have started in the same place—with individuals—and made the original position with representative individuals a global one, but he found this inadequate and deliberately starts with representative peoples at the international level. The reason is important to his understanding of just war. Rawls notes that liberal societies tolerate a fair amount of difference between themselves and that is because decent hierarchical societies have "certain institutional features that deserve respect"—even if they are not liberal, they should be "similarly tolerant" (*LP* §11.2, p. 84). Thus, one should not assume *a priori* that all non-liberal societies are unacceptable: only by developing a "reasonable liberal Law of Peoples" could one determine whether liberal democracy truly is the only acceptable form of society (*LP*, §11.1, pp. 82–83). This may seem a shocking declaration of ignorance for the man who proposed the priority of liberty, fair equality of opportunity, and the difference principle, but one must remember that Rawls titled his book *A Theory of Justice*, not *The Theory of Justice*. Rawls's vision of the international original position is philosophically consistent with his domestic one, but not in the way his critics would like him to be consistent. The aim of Justice as Fairness in *A Theory of Justice* is the justice of societies rather than the well-being of individuals (*LP*, §16.3, p. 119), and he carries this concern with societies over to the international realm, contra the reductive individualist approaches championed by cosmopolitanism and revisionism.

With this in mind, it is important to note that over time, Rawls modifies the hypothetical participants in the international original position from representatives choosing principles for states (*TJ* §58) to representatives of peoples (*LP* §§2–3) and posits that in this two-stage original position, first liberal peoples and then, separately, decent peoples would affirm the following principles, what he calls The Law of Peoples (LP §4.1, p. 37):

1 Peoples are free and independent, and their freedom and independence are to be respected by other peoples.
2 Peoples are to observe treaties and undertakings.
3 Peoples are equal parties to the agreements that bind them.
4 Peoples are to observe a duty of nonintervention.
5 Peoples have the right of self-defense but no right to instigate war for reasons other than self-defense.
6 Peoples are to honor human rights.
7 Peoples are to observe certain specified restrictions in the conduct of war.
8 Peoples have a duty to assist other peoples living under unfavorable conditions that prevent their having a just or decent political regime.

Rawls does a lot of philosophical heavy lifting in order to reach conclusions largely consistent with *jus gentium* as understood by Grotius, Pufendorf, and Vattel, although Rawls is not polemical in the way that they sometimes might have been. Furthermore, his arguments on just war seem to, over time, largely track Walzer's. One could certainly do much worse than to follow in the footsteps of those stalwarts of the just war tradition, and for any other philosopher, this outcome would be unremarkable. For Rawls, however, this is surprising because of the sharp contrast with his domestic philosophy, which can be quite radical—for example, the difference principle. More so than the contrasts in subject matters (individuals in societies versus peoples in the world), it is the different characters of the desired outcomes—the radical domestic program versus the conservative international agenda—that is most striking. His particular approach and conclusions are less conservative than they initially appear, however. As such, Rawls occupies an important place in the just war theory ecosystem, as some elements of his thought break new ground beyond merely reflecting existing tensions in liberalism.

Contra cosmopolitans and revisionists, Rawls refuses to collapse the international into the domestic, believing that the moral significance of the international cannot be ignored or wished away. But simultaneously, despite his own idealized social contract apparatus at the domestic level, he would also reject a more thoroughgoing contractarian approach to international relations and warfare that might argue that states can come to "war by agreement" and thus give themselves rights to engage in war under the constraint of widely accepted, reciprocal, and presumably mutually beneficial principles and rules of warfare (e.g., Benbaji and Statman, 2019, whose social contract approach leads to a set of norms and laws largely reflected by the Geneva Conventions and UN Charter). Such an approach would give even decent societies a right to war (Finlay, 2022), but Rawls insists on placing more constraints on permissible content for the international contract. Decent societies would not agree to a set of principles that permitted war, except in self-defense, says Rawls, because no society "has a right to war in pursuit of its *rational*, as opposed to *reasonable*, interests" (*LP* §13.2, p. 91).

So, Rawls sits closer to Walzer, and somewhere between McMahan and Benbaji and Statman. Does he occupy a mushy middle, then? By making peoples instead of states the normative unit of analysis at the international level, Rawls de-emphasizes administrations, governments, or territorial boundaries; this has the potential to inject chaos into the international system by muddling the proverbial boundaries for intervention. On the other hand, the focus on the background conditions of society, rather than ad hoc or immediate developments—even though questions of military intervention would still be triggered by an immediate development—emphasizes the importance of evaluating the underlying political structure and conditions to determine whether intervention is warranted.

In his typology of societies, Rawls distinguishes between liberal, decent, and outlaw societies, as well as benevolent absolutisms. A decent society protects its members' human rights, provides adequate political representation through consultation, has rule of law, respects the Law of Peoples, and is non-aggressive internationally (*LP* introduction, 3*n*2; §12.2, p. 88).

Liberal societies should tolerate decent societies despite their shortcomings, argues Rawls. This pushes him toward a more conservative position, that perhaps a humanitarian crisis alone might not warrant intervention if the societal structure and conditions are such that the crisis might eventually be resolved internally. This is both a prudential concern about stability and unintended consequences, which are important for just war theory's probability of success criterion, as well as a consideration of the normative aims of just war theory. Military action must aim at peace, and, like other military actions, humanitarian intervention must not make peace impossible later. This means that one must set the bar for intervention quite high, as to do otherwise would open the door to international chaos and make future peace impossible. This is morally unsatisfying, however, because more often than not, that leaves current injustices uncorrected and unpunished.

That said, Rawls allows for two major exceptions: supreme emergency (discussed later) and outlaw states. Because the "political (moral) force [of human rights] extends to all societies," neither liberal nor decent societies should tolerate outlaw states, which systematically violate the human rights of their people. At the very least, outlaw states should be "condemned and in grave cases, may be subject to forceful sanctions and even to intervention" (*LP* §10.3, pp. 80–81). The bar for intervention is still rather high, however, because while Rawls says "liberal and decent peoples . . . simply do not tolerate . . . [and] have the right, under the Law of Peoples, not to tolerate outlaw states," he later says that well-ordered peoples undertake war "only when they sincerely and reasonably believe that their safety and security are seriously endangered by the expansionist policies of outlaw states." Thus, even outlaw states must still be tolerated, unless, in addition to their domestic failings, they have expansionist ambitions such that they fail to adhere to a "reasonable Law of Peoples" and they believe that war is an acceptable instrument for pursuing their regimes' "rational (not reasonable)" geopolitical interests (*LP* introduction, 5; §13.1, pp. 90–91). In reality, few outlaw states will refrain from threatening other societies, not even "hermit kingdoms," because their rulers usually use the specter of an external threat to help justify the oppression of their own people, which in turn leads them to actually pose threats to other societies. Both aggressive outlaw states and the manifestation of true evil (such as the Nazi threat to the liberal world order) that constitutes a supreme emergency would justify intervention, which means that military options and ensuing exemptions to justice would sometimes be permitted in the nonideal world.

Warfare has historically had the capacity to produce major political shifts by virtue of demonstrating the efficacy of the victor's underlying political system (Chiu, 2019, pp. 224–225). Are those prevailing political systems necessarily better morally, for example well-ordered or at least decent in Rawls's typology? The historical record in its entirety is less than clear, but between 1816 and 1990, democracies won 76% of the wars they fought, while dictatorships and oligarchies won just less than half of their wars. Although all regime types won more often when they initiated the war, democracies won a staggering 93% of wars they initiated and only democracies also won the majority of wars when they were attacked (63%) (Reiter and Stam, 2002, pp. 28–29). There could be a number of explanations for this, but Reiter and Stam (2002) point to causal factors inherent in democracy (political accountability built into institutional structures and emphasis on individual rights and responsibilities) that echo Kant's arguments for more pacific republics. Insofar as democracy is a loose proxy for a well-ordered or at least decent society, one might hope that contemporary war sometimes incentivizes positive moral developments internationally; even if that possibility is only sometimes realized, that would be a significant improvement over the greater historical record.

To that end, Rawls's proposed international system that allows for separate societies is not meant to permit injustice, even if it tolerates injustice. Rawls argues that use of war should be limited and directed only toward achieving an ideal state which is a peaceful Society of Peoples, which requires particular (participatory) political structures for societies whose interactions would be governed by a Confederation of Peoples. As he states in the "50 Years After Hiroshima" essay: "The aim of a just war waged by a decent democratic society is a just and lasting peace between peoples, especially with its present enemy" (see also: *LP*, §1.4, pp. 22–23; *MM*, 6: 354–5; Roberts, 2018, pp. 666–667). His view echoes contemporary international relations scholarship about democratic peace theory, namely the empirical claim that democracies do not war against each other, which has its philosophical roots in Immanuel Kant.

In much of his political philosophy, Rawls is undoubtedly heavily influenced by Kant, even though Kant's place in the just war tradition is disputed (Roberts, 2018; Williams, 2012; Orend, 2018), and this carries over to Rawls's writings on just war theory. This influence is evident in Rawls's broad view of international relations, which self-consciously echoes Kant's *foedus pacificum* from his "Toward Perpetual Peace" essay (*LP*, §12.2, p. 86). In both "The Law of Peoples" (*CP*, p. 543) and "50 Years After Hiroshima" (1995) essays, Rawls makes the empirical claim that "democracies" or "democratic peoples do not wage war against each other." Four years later, in *The Law of Peoples* book, he says instead that "well-ordered peoples do not wage war against each other" (*LP*, §14.1, p. 94) and where he continues to mention "democracies" or "democratic peoples" in the same vein, he adds important qualifiers such as "liberal" or "major established" or "whose basic institutions are well-ordered by liberal conceptions of right and justice" (*LP*, §5.4, pp. 52–53).

This is obviously a version of the democratic peace theory, and his revision to "well-ordered peoples" better captures the complex state of the field of democratic peace studies today (Reiter, 2017). Kant's original formulation of "perpetual peace" pointed to republican forms of governance as the mechanism by which peace would gradually pervade international relations; empirical political science studies of this phenomenon initially focused simply on democracies as being more peaceful and have since yielded a complex and often contradictory set of findings, as well as possible exceptions for illiberal democracies, young democracies, democratizing countries (still in transition), or democracies surrounded by non-democracies.

One key question in democratic peace theory is what is actually driving the phenomenon of greater peace between (certain types of) democracies. Does democracy explain the democratic peace phenomenon? Is it the domestic democratic structure itself exerting internal influence, or the interactive effect between democratic systems, or perhaps not democracy itself but something that is correlated with, captured by, or contributes to (some) democracy? Rawls's thoughts on just war may offer some clues.

Rawls's criteria for a well-ordered society are many and complex (*LP* §§8–9, pp. 63–78), and both liberal and decent societies qualify. Orderliness could be summarized as *at minimum* protecting human rights, having rule of law, and requiring the government to consult the people in some institutionalized way. This crosscuts political structures, excluding some types of democracies while including some nondemocratic societies whose governments are institutionally bound in some way (not merely out of *noblesse oblige*) to consider the needs and opinions of their populations. In many ways, this formulation is better than mere democracy at capturing Kant's identification of republicanism as the causal mechanism for peace, and it incorporates many of the qualifiers that empirical democratic peace studies have found, including established institutions and stability.

Rawls did not endorse waging wars to spread democracy—only to halt the encroachment of outlaw states on liberal and decent societies. A (foreseen?) consequence of this kind of limited, defensive war may require rolling back the outlaw state's capabilities and perhaps thus making the world a little more democratic, but on Rawls's understanding, democratic evangelism would not justify the military ventures of a more muscular foreign policy.

War waged justly against "outlaw" states could, however, teach values that are important to democracies—namely respect of human rights—to nondemocratic peoples. Following Kant, Rawls argued that the democratic state's conduct in war plays a role in creating the conditions of a stable peace. Once a liberal people are reluctantly drawn into war against an outlaw state, the way it wages war should be guided by the purpose of educating the outlaw state in specific ways—demonstrative and instructive—which imposes a moral constraint separate from *jus in bello* principles of warfare.

Because "the aim of war is a just peace," war must not be waged in ways that preclude that end; therefore, "just peoples by their actions and proclamations are to foreshadow during war the kind of peace they aim for and the kind of relations they seek between nations" (*TJ* §58, p. 332; "50 Years"). Here, Rawls follows Kant, who argues that "some trust in the enemy's way of thinking must remain even in the midst of war" in order to retain possibilities for future perpetual peace.

Rawls adds that demonstrating the desired state of international relations will require the additional step of diplomacy (by statesmen) to translate those ethical wartime actions—for example, not using weapons of mass or indiscriminate destruction that salt the earth, so to speak, such that the society's future livelihood is destroyed—into the narrow geopolitical message that just peoples only wish to halt the outlaw state's expansion and not to overthrow it.

There is also a second, instructive component to how just peoples should wage war constrained by the values that democracies hold dear: "to teach enemy soldiers and civilians the content of those [human] rights by the example of how they hold in their own case" ("Fifty Years"). For example, if just peoples distinguish between how they hold the outlaw state's leadership, warfighters, and general population accountable for wartime acts, they would offer a model for how a just society might distinguish between individuals' responsibility and accountability while acknowledging the communal constraints of organized society (Chiu, 2011).

While the demonstrative purpose tells the outlaw state that just peoples have limited geopolitical ambitions, the instructive aspect is actually quite ambitious. It implies the expectation that the individuals in outlaw states not merely passively receive just treatment from other states but that they also participate in, that is, join acceptable global society by meting out just treatment in turn, once they have learned what this entails. Ignorance would no longer excuse injustice. This instructive education carries a potentially transformative aspect, such that the targeted society will become more just—and pacific.[3]

There is a danger that taking advantage of the necessity for war to also demonstrate and instruct in liberal values will open the door to sliding down the slope toward the perceived necessity of instruction (and correction) itself, or that outlaw peoples will be perceived as inferior and barbaric (see Chapter 1 of this volume on Aristotle, for example). These demonstrative and instructive requirements also sit in tension with supreme emergency exceptions, as will be seen later in the chapter.

Controversies

Many questions have been raised about Rawls's international theory as philosophy. For example, one might ask how nonideal principles that guide *jus ad bellum* and *jus in bello* could exist in the world of ideal theory—even ideal international theory. One might probe the implications of grounding *jus in bello* obligations of selective conscientious objection in an orientation toward the common good rather than in individual conscience or natural duties. One might be skeptical of why liberalism must be tolerant of "other reasonable ways of ordering societies" and whether that hinges on how we understand "liberalism" or the ambiguous meaning of the word "reasonable." One might question why "free and democratic peoples" (as representatives of liberal peoples) in the international original position would not adopt a principle declaring their respect for not just *human* rights (*CP*, pp. 546–547, 552) but also liberal rights (*TJ* §11, p. 53; *CP*, p. 540). One might consider Rawls's conception of human rights at the international level as too thin to be of actionable value for military ethics.[4]

Engagement with the ethical challenges of real-world wars highlights several controversial aspects in Rawls's just war thinking, particularly selective conscientious objection and the supreme emergency justification, which are worth exploring at greater length here.

Conscription and Conscientious Objection

The Vietnam War's transformative effects on American society hit the academy hard; across the spectrum, just war thinkers (like Walzer) and pacifists (such as Martin Luther King Jr., discussed in Chapter 16 of this volume) opposed the war, but philosophers and activists alike wrestled differently with its many vexing moral and political challenges, including the draft.

Rawls's opposition to the Vietnam War seems hypocritical to many, given that he (and other notable Harvard faculty) publicly opposed the "2-S" draft deferments for college students at the time, but he did so on narrow grounds, consistent with his concern about unequal distributions of the burdens of society and the intersection of those unequal burdens with race. Rawls was not a pacifist and recognized that societies needed militaries. True to his lexical priority of liberty, he opposed conscription except in emergencies, only when absolutely necessary for national security. At the same time, in the republican tradition, he was uneasy about a volunteer military's threat to individual liberties and about a professional military's stratifying effects on society. This was one reason why he argued that insofar as there would be conscription, its burdens must be distributed equally. In unpublished work, he toyed with the idea of citizen armies, and he signed a public proposal for a "two-stage" defense policy that maintained a volunteer military in peacetime that would be supplemented or replaced by a "conscript force raised by a universal draft (or possibly by a lottery)" in wartime. The proposal argued that wartime conscription—assuming genuine necessity for such— was needed because "unless our people, acting through our representative in Congress, are themselves prepared to fight the war and to share justly the burdens of its hardships, it should never under any circumstances be fought" (Forrester, 2019, pp. 50–51, 77, 95–96; Forrester, 2014, pp. 790–791). Here, Rawls is trying to create the mechanism that Kant identifies will bring about perpetual peace: not only government's accountability to its people but also the constraints that people will place on their governments when they realize that they themselves run risks from war.

Given that Rawls held a high standard for conscription and the Vietnam War did not meet it, why did he not allow greater room for conscientious objection to the draft? Rawls conceived of civil disobedience as a last resort and thought that it would be effective only against societal practices that fell short of just standards—not against unjust arrangements reflecting unjust standards (Forrester, 2019, pp. 52–53, 70). In *TJ* §58 (The Justification of Conscientious Refusal), Rawls extends his discussion in the preceding section of the rights, duties, and bounds of civil disobedience to the realm of foreign policy to address conscientious refusal to serve in the military or engage in particular acts of war. Unlike just war revisionist thinkers, such as McMahan (2009), who argue that soldiers should refuse on moral grounds to participate in unjust wars, for Rawls, conscientious objection should not be rooted in individual conscience, religious belief, or a moral philosophy such as pacifism. Rather, conscientious objection should be not simply grounded in objections to the injustices of *ad bellum* and *in bello* decisions but also oriented toward "political principles conceiving the *common good.*" The purpose of permitting selective conscientious objection is to "improve the workings of the constitutional system to conform more closely with the requirements of [justice as fairness]" (Forrester, 2019, p. 78; Forrester, 2014, p. 790). Despite the lexical priority of liberty in his theory of justice, Rawls's focus on justice at the structural and institutional levels led him to trade off individual liberties of conscience against community interests in the case of conscientious objection.

Supreme Emergency and Statesmanship

Rawls follows Walzer in adopting exceptions for supreme emergencies that in *The Law of Peoples* are narrowly tailored toward "set[ting] aside—in certain special circumstances—the strict status of civilians that normally prevents their being directly attacked in war" (*LP* §14.3, p. 98). He acknowledges that is a difficult position to hold, because what justifies suddenly throwing out traditional just war theory and contemporary customary international law's fundamental rule of civilian distinction? Only survival of a way of life, namely "civilized life everywhere"—and, in the case of the threat from Nazi Germany, "the nature and history of constitutional democracy and its place in European history"—could permit this (*LP* §14.3, p. 99).

In warfare, one must distinguish between the outlaw state's leaders, warfighters, and civilians. Supreme emergency permits violating those distinctions by targeting civilians, in order to force the outlaw state into surrender. Yet, the nature of an outlaw state is such that civilians are precisely the group that outlaw state leaders are willing to sacrifice. Beyond the moral problems with violating a foundational *jus in bello* principle, this is a practical one: use of this exception is unlikely to be effective.

Indeed, this is reflected in Rawls's opposition to a historical application of the supreme emergency exception: he claims that both fire-bombing and using atomic weapons against Japan were unjustified. Historical reexamination of the case may support Rawls here, as it is likely that the Soviet entry into war against Japan and its rapid progress in invading Manchuria—not the mass airpower killings of Japanese civilians—played a larger role in the Japanese leadership's decision to surrender.[5]

Furthermore, the demonstrative and instructive purposes of just warfare that Rawls argues for generate additional difficulties for his supreme emergency exception. How could just peoples demonstrate and instruct of liberal values through warfare, when they can choose when to violate them? Indeed, the American use of atomic weapons in World War II continues to generate charges of hypocrisy and skepticism that liberal democracies have the moral authority to teach others about human rights, as charged by the likes of Charles Mills (2009), who is discussed in Chapter 20 of this volume.

In addition to the difficulty of moral justification, supreme emergency exceptions are, of course, ripe for abuse. What is to be the bulwark against such exploitation? Rawls points to statesmen, who not only translate the demonstrative purpose of warfare into diplomatic communication (as discussed earlier) but also make difficult decisions in selflessly discerning their society's fundamental long-term interests (*LP* §14.2, pp. 97–98). This means that statesmen must be able to set aside their own thoughts, feelings, and interests to guide both the commencement and waging of warfare in ways that promote the possibility for long-term peace. But they must also make excruciating decisions to violate the principle of discrimination and target civilians if supreme emergency arises. This requires both extraordinary judgment, leadership, "strength, wisdom, and courage" (*LP*, §14.2, p. 97) as well as a willingness to "dirty" one's hands, per Walzer's dilemma (1973). Unfortunately, violating the rules of war in the name of supreme emergency is a far easier task—and thus much more common—than guiding the conduct of war in ways that will promote long-term peace, meaning the demands on statesmen are asymmetric.

Rawls acknowledges that the statesman is "an ideal, like that of the truthful or virtuous individual" (*LP*, §14.2, p. 97). Given his philosophical rigor, this seems a somewhat strange and uncharacteristic appeal to a *deus ex machina* to save people in times of crisis—perhaps reminiscent of Rousseau's Lawgiver, along with all its attendant concerns. There may be historical precedent, however, perhaps in Montaigne's focus on Epaminondas, discussed in Chapter 6.

Legacy

Rawls's contribution to just war theory lies in how he philosophizes about urgent and intractable political problems in the nonideal world with an analytical methodology and, given his starting point, how one might analogize from the domestic to the international level when it comes to questions of justice and warfare.

One of the great challenges for liberal theories of any kind is how to make domestic and international ethics consistent with each other. One way to do that is simply to collapse them, to make the international domestic by extending the domestic to the entire world. Contra both the cosmopolitan (Fabre, 2012) and the reductive individualist (McMahan, Rodin, Frowe) approaches to just war, Rawls demonstrates another way to try to reconcile domestic with international relations and thereby to situate just war theory within an ideal all-encompassing moral framework. Doing so means, ultimately, subsuming the domestic within the international, but Rawls thinks that this does not require philosophically eliminating morally relevant global subunits such as societies or peoples.

Despite building his domestic and international philosophies on abstract hypotheticals, Rawls understands this essential aspect of human nature: there is only so far one's sympathies can realistically extend. This imposes limits on the scope of society, for however much one abstracts philosophically, ethical beings can never escape the fact that they are part of a community. That community provides a human good; Rawls understands that attachment to a particular culture and participation in that culture's civic life are goods, and he argues that therefore "proper patriotism" should be promoted (*LP*, §15.4, pp. 111–112). As such, he not only accepts a pluralism of community types but also recognizes the human reality of separate communities.

This is an important lesson for just war theory as well, which is necessarily nonideal: it should not ignore the significant practical and moral constraints on individual ethics generated by politics and polities. The reality and complexity of human societies mean that epistemological limitations plague every military and ethical decision in war. This is something which a revisionist like McMahan acknowledges but does not have a mechanism for handling, beyond *de facto* pacifism. Revisionist approaches to just war theory ignore the reality of navigating war from within political society and its complex web of relationships (civil–military, political–military, intra-military) and accompanying moral questions. Absent a grappling with those political realities, the ethical questions, answers, and guidance for warfare that revisionist just war thinking supplies are ultimately severely limited. However one might criticize Rawls' own answers, he attempts to make sense of the realities of the separate but connected domestic and international realms by navigating the messy moral minefield of international relations and war with analytic rigor.

Why would an exemplar of abstract hypothetical philosophizing accept such messiness in his international theorizing? Surely, Rawls's experiences with the war—an inescapably political activity—and its arbitrariness shoved his thinking onto a different path, shaped the content of his philosophy (e.g., justice as fairness), and changed his understanding of community. In his private essay "On My Religion," Rawls identifies three incidents that ultimately prompted his loss of Christian faith: a Lutheran pastor's sermon during World War II that invoked God's favor and divine providence in wartime outcomes, a friend killed in the war on a reconnaissance mission that he was chosen for instead of Rawls because Rawls happened to have the blood type needed for an urgent donation at the time, and learning about the Holocaust (2009, pp. 262–263). Rawls recounts these in the context of his loss of religious faith, but it is not hard to see how his experiences with and knowledge of the misuse of religion, sheer happenstance, and human evil would influence his political thinking about the structure of overlapping consensus, the arbitrariness of one's starting point in life and childhood, and various features of justice as fairness such as the basic liberties, priority of liberty, and fair equality of opportunity.

One would never know from his published works the extent to which Rawls was shaped by his life experiences. For example, it is clear from the tone of

Rawls's essay on the atomic bombings of Hiroshima and Nagasaki that it is a very personal subject, but he makes no mention that when he served in occupied Japan, his troop train passed through the recently devastated city of Hiroshima. In that essay, he argues vehemently against the use of nuclear weapons against Japan, giving no hint that he likely would have had to participate in an invasion of the Japanese home islands had nuclear weapons not been used.

It mattered to Rawls that he be impartial in his theorizing. The personal is not always political, and readers should not reflexively look to the personal to explain a thinker's political arguments—even when the personal might inform the political to good effect. While understanding Rawls's philosophy is enriched by knowing the ways in which it was deeply informed by his experiences in childhood and war, his work should be approached on the terms he intended. Those terms are not merely an exemplar of abstract, idealized analytic political philosophy, however.

In his own way, Rawls repeatedly draws on historical and scientific study throughout his philosophical oeuvre, and by accounting for social and political realities while striving for impartiality, Rawls provides a strong model for how to engage in theorizing about challenging political and ethical problems. It is an important lesson for just war theorists to tangle with the political, no matter how difficult and messy that will be.

Ultimately, what is just war theory about? Although Rawls's primary method is moral philosophy, when it comes to international ethics, Rawls appears to think that it is about international relations and not about moral philosophy—perhaps this is why his international political thought is so dissatisfying to moral philosophers. In this area, at least, Rawls belies his reputation as an abstract, decision-theoretical moral philosopher.

Rawls's philosophical positions and principles are influenced by an intense awareness of the unavoidable nature of human society, namely its arbitrariness and dangerousness, accompanied by limited human ability to control those features. Thus, he writes in "On My Religion" that "the content of the judgments of practical reason depends on social facts about how human beings are related in society and to one another." These facts should not be abstract concepts but taken "as they undeniably are in our social world" (2009, p. 268). As such, Rawls sees the reality of separateness of persons while retaining the social communities they live in. This requires navigating the muddy contours of international politics by mixing and constraining the ideals of abstract philosophy with the nonideals of the real world in order to map the philosophical contours of just war.

Notes

1 Views are author's own and do not represent those of the U.S. Naval War College, Department of Navy, Department of Defense, or the U.S. government.
2 Rawls considered this global original position in 1969 but deemed that in the nonideal world, the psychological support needed for global institutions and their

global redistribution were too implausible (Forrester, 2014, p. 789). In doing so, he rejected later liberal cosmopolitan theories of justice.

3 Noticeably, Rawls ignores the possibility that warfighting itself might develop virtues in the warfighter necessary for long-term peace. Unlike Kant, who believes that the activity of warfighting offers the possibility of developing essential human capabilities for reason, rational thought, control of the passions, and self-regulation—all virtues necessary for lawful autonomy and future peace (*CJ* §5.263, §5:432; Chiu, 2019, pp. 239–40)—Rawls keeps his purview at the societal level in *The Law of Peoples.*

4 See (Childress, 1978; Martin, 2007; Martin and Reidy, 2006; Moellendorf, 2013; Roberts, 2018).

5 Rawls also notes Truman's motive for positioning the United States vis-à-vis the USSR post-WWII, in already-unfolding Cold War conflict (*LP* §14.4, p. 100).

Works Cited

Benbaji, Yitzhak and Daniel Statman. 2019. *War by Agreement: A Contractarian Ethics of War.* Oxford: Oxford University Press.

Childress, James F. 1978. Just-War Theories: The Bases, Interrelations, Priorities, and Functions of Their Criteria. *Theological Studies* 39(3), pp. 427–445.

Chiu, Yvonne. 2011. Liberal Lustration. *The Journal of Political Philosophy* 19(4), pp. 440–464.

Chiu, Yvonne. 2019. *Conspiring with the Enemy: The Ethic of Cooperation in Warfare.* New York: Columbia University Press.

Douglass, R. Bruce. 2012. John Rawls and the Revival of Political Philosophy: Where Does He Leave Us? *Theoria: A Journal of Social and Political Theory* 59(133), pp. 81–97.

Fabre, Cécile. 2012. *Cosmopolitan War.* Oxford: Oxford University Press.

Finlay, Chris J. 2022. Ethics, Force, and Power: On the Political Preconditions of Just War. *Law and Philosophy* 41, pp. 717–740.

Forrester, Karina. 2014. Citizenship, War, and the Origins of International Ethics in American Political Philosophy, 1960–1975. *The Historical Journal* 57(3), pp. 773–801.

Forrester, Karina. 2019. *In the Shadow of Justice: Postwar Liberalism and the Remaking of Political Philosophy.* Princeton: Princeton University Press.

Martin, Rex. 2007. Walzer and Rawls on Just Wars and Humanitarian Interventions. In S. P. Lee. ed. *Intervention, Terrorism, and Torture: Contemporary Challenges to Just War Theory.* Dordrecht: Springer, pp. 137–153.

Martin, Rex and David A. Reidy. (eds.). 2006. *Rawls's Law of Peoples: A Realistic Utopia?* Malden, MA: Blackwell Publishing.

McMahan, Jeff. 2009. *Killing in War.* Oxford: Oxford University Press.

Mills, Charles W. 2009. Rawls on Race/Race in Rawls. *The Southern Journal of Philosophy* 47, pp. 161–184.

Moellendorf, Darrel. 2013. Just War. In Jon Mandle and David A. Reidy. eds. *A Companion to Rawls.* Hoboken, NJ: Wiley-Blackwell, pp. 378–393.

O'Driscoll, Cian. 2021. *How* and *Why* to Do Just War Theory: A Critical Exchange. *Contemporary Political Theory* 20, pp. 858–889.

Orend, Brian. 2018. Immanuel Kant. In Daniel R. Brunstetter and Cian O'Driscoll. eds. *Just War Thinkers: From Cicero to the 21st Century.* New York: Routledge, pp. 168–179.

Pogge, Thomas W. 2007. *John Rawls: His Life and Theory of Justice.* New York: Oxford University Press.

Rawls, John. 1995. 50 Years After Hiroshima. *Dissent*, Summer. Available at: www.dissentmagazine.org/article/50-years-after-hiroshima-2/.

Rawls, John. 1999a [1993]. *Collected Papers [CP].* Edited by Samuel Freeman. Cambridge, MA: Harvard University Press.

Rawls, John. 1999b. *The Law of Peoples, with 'The Idea of Public Reason Revisited' [LP]*. Cambridge, MA: Harvard University Press.

Rawls, John. 1999 [1971]. *A Theory of Justice [TJ]*. Revised Edition. Cambridge, MA: Harvard University Press.

Rawls, John. 2005 [1993]. *Political Liberalism [PL]*. New York: Columbia University Press.

Rawls, John. 2009. *A Brief Inquiry into the Meaning of Sin and Faith with 'On My Religion'*. Edited by Thomas Nagel. Commentaries by Joshua Cohen, Thomas Nagel, and Robert Merrihew Adams. Cambridge, MA: Harvard University Press.

Reiter, Dan. 2017. Is Democracy a Cause of Peace? *Oxford Research Encyclopedia of Politics*. Available at: https://doi.org/10.1093/acrefore/9780190228637.013.287.

Reiter, Dan and Allan C. Stam. 2002. *Democracies at War*. Princeton, NJ: Princeton University Press.

Roberts, Peri. 2018. War and Peace in the Law of Peoples: Rawls, Kant and the Use of Force. *Kantian Review* 23(4), pp. 661–680.

Walzer, Michael. 1973. Political Action: The Problem of Dirty Hands. *Philosophy and Public Affairs* 2(2), pp. 160–180.

Walzer, Michael. 2015 [1977]. *Just and Unjust Wars: A Moral Argument with Historical Illustrations*. 5th Edition. New York: Basic Books.

Williams, Howard. 2012. *Kant and the End of War: A Critique of Just War Theory*. Basingstoke: Palgrave.

18

JUDITH BUTLER (1956–)

Rosemary Kellison

Introduction

Judith Butler (born in 1956) is an American scholar of gender studies and philosophy, currently serving as Distinguished Professor in the Graduate School at University of California, Berkeley. Their work in gender studies and feminism, critical theory, and ethics has been widely influential and recognized through a number of prestigious fellowships and lectureships.

Butler is best known for their contributions in authoring some of the most influential theoretical works related to the development of third-wave feminism and queer theory beginning in the 1990s. This essay instead focuses on their work relevant to the ethics of war. Like the contemporary thinkers explored in *Just War Thinkers* (Michael Walzer, James Turner Johnson, Jean Elshtain, and Jeff McMahan), as well as those covered in this volume (Charles Mills, Pope Francis), their thoughts on war have been impacted by extant wars, especially those of the post-9/11 context, as well as developments in the Israel–Palestine conflict over the past few decades. However, I will argue that we understand Butler's ethical commentary on war and violence best when we observe its strong theoretical connections to their earlier writings on gender.

Texts and Tenets

Butler is not a just war thinker. As will become clear, their contributions to the ethics of war pose a serious challenge to many aspects of just war reasoning, and they rarely directly engage with just war thinkers in their work. That said, Butler's work on violence and war offers a powerful critique of ideas

DOI: 10.4324/9781003428688-19

and practices associated with the just war tradition and ought to be taken seriously by those engaged in the practice or scholarly study of just war reasoning. Importantly, this strand of Butler's scholarship is not unconnected to their earlier work on gender.

Butler's 1990 book *Gender Trouble: Feminism and the Subversion of Identity* constituted a major contribution to feminist thought. In that text, Butler develops a theory of gender performativity. Reversing the order implied in common descriptions of gender as an inherent, stable identity that humans express in their actions, Butler argues that gender is rather constituted through human practices (1990, p. 192). It is more accurate, then, to say that one *does* woman-ness than to say that one *is* a woman. In Butler's words:

> Gender ought not to be construed as a stable identity or locus of agency from which various acts follow; rather, gender is an identity tenuously constituted in time, instituted in an exterior space through a *stylized repetition of acts*. The effect of gender is produced through the stylization of the body and, hence, must be understood as the mundane way in which bodily gestures, movements, and styles of various kinds constitute the illusion of an abiding gendered self.
>
> *(1990, p. 191)*

This illusion is convincing both to others and to the gendered subject herself.

Butler describes the process of repeating the acts that constitute gender as akin to that of an actor being handed a script and acting out their assigned role (1988, p. 526). This metaphor highlights the way in which the content of one's performance comes largely from sources external to oneself. Not only that, but external forces are significant in compelling one to perform as expected, imposing severe punitive consequences on those who fail to follow their assigned script (1988, pp. 522, 528, 1990, p. 190). This is not to say that there is no room for agency on Butler's model, however. After all, actors in the same role may interpret and enact their shared script in somewhat different ways (1988, p. 526); a drag queen may offer a creative parody of gender (1990, p. 187). Indeed, Butler argues, "[p]aradoxically, the reconceptualization of identity as an *effect*, that is, as *produced* or *generated*, opens up possibilities of 'agency' that are insidiously foreclosed by positions that take identity categories as foundational and fixed" (1990, p. 201).

Insofar as human responsibility is understood as central to ethics—and to the degree to which responsibility is understood as connected with agency—this last point is crucial to the preservation of the possibility of ethics in Butler's work.[1] It is an insight that is further developed in their later book *Giving an Account of Oneself*. To say that the subject is produced by the law, rather than existing independently before the law, is to say that persons are irreducibly relational. This relationality is multilayered; it begins with the obvious biological fact that humans depend on others for their physical existence (and its continuation over time) and extends to the way that human identities are forged in and through the social norms, economic systems, juridical structures, and historical contexts in which persons exist. As Butler puts it,

> [T]here is no "I" that can fully stand apart from the social conditions of its emergence, no "I" that is not implicated in a set of conditioning moral norms, which, being norms, have a social character that exceeds a purely personal or idiosyncratic meaning.
>
> *(2005, p. 7)*

But this fact does not mean, as some might assume, that "I" am not morally accountable. Rather, Butler argues, it is precisely in relation to others that accountability comes to exist; it is others who demand the account that I offer of myself—an account that is, necessarily, incomplete insofar as some parts of my relational self remain "opaque" to my own eyes (2005, pp. 11, 20). Thus, to be relational is to be responsible; it is also, Butler insists, to be vulnerable. If I am constituted in part through my relations to others, I depend on others for my own existence, as they do on me. There is a sense in which the fates of human beings are intertwined. While we are autonomous to some degree, our autonomy is "qualified" by "the fundamental sociality of embodied life, the ways in which we are, from the start and by virtue of being a bodily being, already given over, beyond ourselves, implicated in lives that are not our own" (2004, p. 28). When someone with whom I am in relationship is harmed, I too suffer harm; when I lose someone with whom I am in relationship, I lose some part of myself. Thus, to grieve a loss is to experience a kind of "transformation" and "dispossession" of oneself (2004, pp. 21, 28).

Butler's writing on violence and war, especially *Precarious Life: The Powers of Mourning and Violence* (2004), *Frames of War: When Is Life Grievable?* (2010), and *The Force of Nonviolence* (2020), is strongly shaped by their understanding of human relationality and shared vulnerability. After all, shared struggle coming from a place of recognition of mutual vulnerability is not the only way in which relationality may be enacted. Humans may also relate to one another in violent or dominative ways (2020, p. 10). Indeed, to do violence to another is to exploit their embodied precarity (2004, p. 27). Butler emphasizes that to say that all humans are vulnerable and precarious is not to say they are all vulnerable and precarious in the same ways or to the same degree (2010, p. xvii). The global conditions that structure human interrelations render vulnerability unequal, such that some are more vulnerable to violence than others (2004, pp. 16–18).

One of Butler's most significant contributions to ethical conversations around violence is their concept of *grievability*. Within Butler's work, this concept offers a way of thinking through the causes and implications of unequal vulnerability. Recall Butler's argument that the gendered self is produced by the law in relation to other persons. In a similar vein, and building on Hegel's account of mutual recognition, Butler argues that our recognition of other human beings as persons depends on those others being rendered visible by social "frames" that define the human subject (2005, pp. 26–29, 2010, pp. 4–5). In other words, simply to see human beings is not necessarily to recognize them as such; the degree to which others are recognizable is dependent upon how expansive our social frames for humanity are.

Butler contends that one highly effective way of revealing who fits into our social frames is to examine the differential ways in which we mourn (or fail to mourn) the victims of violence. Butler observes that after the 11 September 2001 attacks by Al-Qaeda on the World Trade Center and the Pentagon, American victims were commemorated with long obituaries and other public mourning rituals (2004, p. 12). By contrast, civilians killed by American forces responding to 9/11 in Afghanistan were not publicly mourned and in fact remained nameless and faceless to the American public (2004, p. 32). In Butler's terms, for Americans, fellow Americans are grievable in ways that members of the civilian populations of the countries in which the post-9/11 wars have been fought are not. Throughout their more recent work, Butler analyzes this disparity to identify its causes and implications.

In order to be grievable, Butler argues that one must be perceived by the potential griever as having certain qualities. First, one must be recognized as a *living* being, such that one's death—the catalyst for grief—is visible. Butler speaks of populations of people who are seen as inhabitants of "a lost and destroyed zone; they are, ontologically, and from the start, already lost and destroyed, which means that when they are destroyed in war, nothing is destroyed" (2010, p. xix). In the eyes of Western powers who send their militaries into those zones, these people may be perceived as "socially dead"; through this frame, their violent deaths are not recognized (2010, p. 42).[2] Butler describes this frame as one that enables perpetual violence. Through this lens, populations are rendered "unreal" and "spectral"; they can be "negated" again and again without ever inflicting grievable death (2004, p. 33). Relatedly, moral prohibitions on killing do not apply in cases in which there is no living person to kill (2020, pp. 57–58). Moreover, one cannot be grieved if one's death is not noted. When civilian deaths that result from war are not investigated and accounted for, those civilians' lives cannot be fully grieved (2020, p. 74).

In addition to being recognized as living and as having been killed, to be grievable, one must be recognized as familiar in some sense. We grieve for those who are "like me"—for those who are part of the "we" for which we fight (2010, p. 36). As a result, frames of humanity that exclude those of different religious, cultural, linguistic, or racial backgrounds greatly circumscribe those who are grievable to a particular person. Like Charles Mills, explored in the final chapter of this volume, Butler identifies racism as a major contributor to the shaping of these frames and asks whether, for contemporary Americans, Afghans, Arabs, and Muslims have been excluded altogether (2004, p. 32, 2020, pp. 4, 11). That is not to say that members of these populations have been absent from public consciousness of the post-9/11 wars. But Butler argues that when they show up, it is often not as persons but rather as instruments or symbols of particular American interests. For example, when American troops toppled the Taliban government in Afghanistan after 9/11, images of Afghan women's faces, "freed" from the burqas they had been forced to wear under Taliban rule, were visible in American newspapers and television programming. Yet, Butler argues that these images did not serve the purpose of humanizing the women who were depicted; rather, they served as a representation of what *Americans* had won in the war (2004, p. 143).

Thus, much like vulnerability, grievability is unequally distributed among human beings. A significant implication of Butler's description of grievability is that those who are rendered ungrievable as a result of racism and other oppressive systems that shape our social frames are simultaneously made more vulnerable to violence. As Butler puts it, "violence operates as an intensification of social inequality" (2020, p. 142). The flipside of this point is that to be recognized as grievable is to be protected from harm to some degree. In other words, grievability operates not only after death, when a grievable person is mourned, but even more significantly while a person is living, when grievability communicates a normative mandate to protect the grievable person:

> To say that a life is grievable is to claim that a life, even before it is lost, is, or will be, worthy of being grieved on the occasion of its loss; the life has value in relation to morality. One treats a person differently if one brings the sense of the grievability of the other to one's ethical bearing toward the other. If an other's loss would register as a loss, would be marked and mourned, and if the prospect of loss is feared, and precautions are thus taken to safeguard that life from harm or destruction, then our very ability to value and safeguard a life depends upon an ongoing sense of its grievability—the conjectured future of a life as an indefinite potential that would be mourned were it cut short or lost.
>
> *(2020, pp. 75–76)*

To recognize someone as grievable is to recognize their humanity and thus their precarity—and to act in ways that safeguard them. It is a status that brings with it a normative demand to be treated as someone worth protecting.

In other words, grievability is a *moral* quality: "If and when a popula-tion is grievable, they can be acknowledged as a living population whose death would be grieved if that life were lost, meaning that such loss would be unacceptable, and even wrong—an occasion of shock and outrage" (2020, p. 105). In their recent book on nonviolence, Butler argues that a "radical equality" of grievability is therefore central to an inclusive nonviolent ethic (2020, pp. 28, 56). In the absence of equal grievability, prohibitions on kill-ing or causing harm—often communicated through the *jus in bello* principle of distinction, or conjectures about noncombatant immunity—protect only some (2010, pp. 180–181).

Violence can not only exploit and deepen the precarity of those against whom it is wielded, but it also often serves as a means by which, accord-ing to Butler, those who employ it attempt to deny their own vulnerabil-ity. The turn to violence is usually a response to some perceived injury to oneself or to others about whom one cares, what just war thinkers talk of in terms of *jus ad bellum* just cause, meaning that it arises in reaction to an experience that reveals (to oneself and to others) one's vulnerability to injury, to loss, and to grief. A violent response to injury is, then, at least in part a means of attempting to deny that revelation by demonstrating one's strength instead—to "banish" grief through powerful action (2004, p. 29, 2005, pp. 99–100). Paradoxically, however, given the reality of human inter-relationality, doing violence to others involves doing a certain kind of violence to oneself (2020, p. 9). Butler endorses a cyclical view of violence: violence begets violence, meaning that attempts to deny one's vulnerability through violence will end up re-exposing oneself sooner or later.[3] Build-ing on Walter Benjamin's "Critique of Violence," Butler argues that "[w]hen any of us commit acts of violence"—even when we do so with the aim of achieving justice—"we are, in and through those acts, building a more violent world" (2020, p. 19).

Butler challenges their readers to envision some other possibility for responding to the experience of vulnerability or grief: "If we are interested in arresting cycles of violence to produce less violent outcomes, it is no doubt important to ask what, politically, might be made of grief besides a cry for war" (2004, p. xii). Butler suggests that a profound ethical resource is lost when we refuse the experience of vulnerability. It is this experience, after all, that might bring us the sense of familiarity that Butler describes as necessary to consider another person grievable. To recognize one's own vulnerability or to experience grief is to see and feel the ways in which one is a relational being, the ways in which harms to other persons implicate oneself, and the fundamental similarities between all humans as relational persons. To allow oneself to fully experience loss and precarity, then, is an important step in the development of a moral worldview that recognizes all persons as equally grievable.

Butler advocates for an ethic of nonviolence that emerges from this experience of grief and vulnerability as fundamentally shared. Importantly, this nonviolence is not a passive form of nonaction. Following Levinas, Butler describes an encounter with the face of the Other in which the face serves as a symbol of our shared ability to harm one another as well as our shared vulnerability to that harm (2012, p. 56). For Levinas, this encounter generates tension as we grapple with our moral commitment not to harm the Other, our fear that they will harm us, and our own temptation to harm them (2012, pp. 58–59). An ethic of nonviolence is one possible response to this experience. What this means is that nonviolence is best understood as involving a struggle: "a struggle not to kill the other, a struggle to encounter and honor the face of the other" (2012, p. 61).

Butler contrasts their understanding of nonviolence as a struggle with models that describe nonviolence as a principle (2010, p. 171, 2020, p. 23). Far from passive, the nonviolent actor engages in practices of struggle against violence (their own and others') and against the frames that enable it. Butler emphasizes that violence is always possible in a relational world. It will not finally be eliminated by a principle or rule against it (2020, pp. 59–61). Nonviolence thus requires what might be characterized as an "aggressive" commitment (2020, pp. 180–181). It is enacted not in the endorsement of a principle but in the practice of solidarity and resistance (2020, p. 202).

Thus, even as Butler acknowledges a certain inevitability of violence as a feature of human relationality, they also insist on the possibility of doing things otherwise—of enacting a nonviolent form of relationality that builds solidarity rather than exploiting vulnerability through violence. In this sense, Butler's endorsement of the normative possibilities open to human beings with respect to violence mirrors their earlier work regarding the qualified autonomy of humans in relation to gender. Additionally, their argument that social frames generate and limit our ability to see members of various populations as living, grievable human beings echoes their earlier claim that the subject is produced by the law. Finally, Butler's argument that the recognition of human relationality and (unequally) shared vulnerability can serve as the starting point for an ethic of nonviolence builds on their earlier work concerning the normative implications of relationality. Thus, there is a strong throughline in Butler's philosophy from their work on gender to their more recent work on ethics and violence.

Controversies

As evidenced by the fact that they include very little engagement with just war thinkers, Butler's work on war and violence is not best understood as a direct critique of just war reasoning. That said, their work is clearly a response to dominant understandings of the ethics of war as well as many aspects of the contemporary practice of war by powerful states, both of which are strongly shaped by the just war tradition. Moreover, they do sometimes comment specifically on language and policies associated with that tradition. In this section, I outline some of the ways in which Butler's work poses a challenge to just war reasoning. I also consider some of the ways Butler has distinguished themself from a prominent critic of just war, Talal Asad.

Before considering some of the specific content of Butler's critical analysis of contemporary warfare, I want to note their emphasis on the importance of maintaining space for dissent on these issues. The strong social and other sanctions that often meet critique during times of war can have a chilling effect on the expression of alternative views. Butler objects to the way that criticisms of U.S. foreign policy or military actions after 9/11 were construed as offering a kind of support or justification for terrorism (2004, pp. 9–10, 15). Similarly, they reject the way that criticisms of Israeli policy toward Palestine are regularly described as antisemitic or as a denial of Israel's right to exist (2004, p. 111); they even devoted an entire book, *Parting Ways* (2012), to making the case that Jewish resources can be marshaled in an effective critique of the Israeli state. Moreover, they endorse a capacious account of dissent that is not exhausted by speaking out against particular policies. Building on their account of the unequal grievability of various populations, Butler argues that simply mourning for someone whom others consider ungrievable can itself be a powerful form of protest (2020, p. 106). In such a case,

> [T]he community that mourns also protests the fact that the life is considered ungrievable, not only by those responsible for taking the life, but also by those who live in a world where the presumption is that such lives are always vanishing, that this is simply the way things go.
>
> *(2020, p. 74)*

Grief here poses a challenge to the very frames that enabled the exploitation of the grieved person's vulnerability.

Given Butler's interest in how juridical structures form subjects and present us with scripts we are socially compelled to act out, it is unsurprising that they are concerned with how the legal framework of just war shapes and authorizes certain violent acts. First, they express concerns about the breadth of circumstances in which going to war is deemed "justified." Sometimes, they note, to be justified in committing violence appears to mean little more than that one is responding to violence committed by someone else (2004, p. 16). Other times, the term is used in an instrumentalist manner to convey the idea that a particular use of violence is strategically useful (2010, p. xv). Butler suggests that the concept of "justified" in respect to war would be better deployed in a more normatively evaluative manner to indicate when going to war is not simply legally justifiable but also "responsible" (2004, p. 16,

2010, pp. xv–xvi). This is a point on which many just war thinkers themselves would agree with Butler, though they might use a term like "prudent"
rather than "responsible."

It is significant that compared to their use of the term "violence," Butler
rarely uses the term "war" and uses the phrase "just war" only when referring to the claims of others. Butler draws our attention to the ways that
states strategically deploy these terms. In particular, states use their power
to label—and thus delegitimate—the actions of their opponents and critics
as "violent" and therefore unjust (2020, pp. 2–3). Even nonviolent forms of
protest are sometimes reinterpreted and renamed as violent by the state; in
some cases, voicing a critique of war is itself framed as an aggressive or violent act (2020, p. 145). All of this means that *"violence is always interpreted"*
(2020, p. 14). To name a protest march as violent, or a destructive bombing
as just war, is to engage in a powerful act of framing.

In contrast, states refer to their own uses of armed force as "war." For
Butler, the concept of "just war" has certain implications that obfuscate the
reality of war. Butler is concerned about how war can be depicted as a means
of spreading democracy or as a way of imposing order when, in their view,
it in fact is inherently undemocratic and represents a failure of order (1995,
p. 44, 2010, pp. 36–37). Most significantly, according to Butler, the concept
of "just war" attempts to transform acts of violence and killing into legal and
permissible acts:

> The idea of a legal war or, indeed, a just war, relies on the controllability
> of instruments of destruction. But because uncontrollability is part of that
> very destructiveness, there is no war that fails to commit a crime against
> humanity, a destruction of civilian life.
>
> *(2010, p. xviii)*

The result, according to Butler, is wars that are considered "legal" but that
unavoidably involve the crime of harming civilians.

In other words, as Butler interprets them, the laws of just war have built-
in exceptions. These exceptions are revealing: "The exception to the rule is
important, perhaps more so than the rule itself" (2020, p. 52). The notion
of just war itself is an exception, insofar as it names the circumstances in
which a state (and only a state) may licitly violate the general prohibition on
the use of violence. As much as the laws of war restrain violence, then, they
also authorize it by naming exceptions in which war is justified in particular
circumstances and to be carried out by particular parties.[4] These exceptions
are of course what just war thinkers refer to as just war criteria.[5] Specifically,
here we might say that Butler is referring to the *jus ad bellum* criteria of
right authority and just cause, while in their discussion of exceptions related
to the killing of civilians, they are referring to the *jus in bello* criterion of

discrimination. However, Butler's argument differs from a typical presentation of just war reasoning in their emphasis on how the application of these criteria, or the appeal to these exceptions, is inescapably shaped by our social frames for humanity.

In the contemporary context, the most widely accepted just cause for war is self-defense. In a world in which not all persons are equally recognized as "selves," Butler argues, this reasoning for war is not as neutral as it may first appear. Quite often, according to Butler, those I recognize as part of my "self" and therefore as worthy of my defense, as well as those I regard as having the right to defend themselves, are those who share some aspect of my identity or my status in the world (2020, pp. 11–12). In other words, the just cause of self-defense in practice often operates to extend the right to use violence—or the right to have violence deployed on one's behalf—only to certain populations who are recognized as human. Butler argues that the notion of self-defense leads to a "war logic" in which I am willing to defend those who are like me, who share some sort of group identity with me, by harming those whom I do not recognize as like me. They conclude that as a result, "there is a moral justification for violence that emerges precisely on a demographic basis" (2020, p. 55).

It is this same basic distinction (between self and Other, between grievable lives and ungrievable lives) that Butler argues frames considerations of what just war thinkers would call the *jus in bello* criteria. Echoing Charles Mills, discussed in the final chapter in this volume, Butler suggests that the concept of "collateral damage"—the prevailing term for innocent persons who are unintentionally killed as a side effect of uses of war aimed at legitimate military targets—rests on racist distinctions between populations (2020, p. 62). They also question the way the requirement of proportionality implies that the value of certain lives is capable of being calculated while that of those more like oneself is deemed incalculable (2020, p. 107).

All of this leads to Butler's central moral argument that nonviolence rests on equal grievability: "[w]hat we might call the 'radical equality of the grievable' could be understood as the demographic precondition for an ethics of nonviolence that does not make the exception" (2020, p. 56). In what might be considered a solution to the dilemma faced by Pope Francis, as discussed in the next chapter, namely, how to hold onto a presumption against war without endorsing pacifism or falling back on just war principles, Butler emphasizes that they are not here endorsing a principle of absolute pacifism but a commitment to the struggle to equally safeguard and preserve all lives (2020, p. 56). Regardless of its limits, however, this is clearly a normative commitment. In making this moral claim explicit, Butler separates themself from other scholars who have offered somewhat similar critiques of just war reasoning.

In particular, Talal Asad has also critically interrogated the frames that shape our understandings of political violence. Asad's focus is on the ways that we have employed the concepts of "just war" and "terrorism" to name an essential moral difference between forms of violence. Focusing primarily on Michael Walzer's account of just war, Asad argues that the term "just war" is employed to characterize violence committed by liberal states as both necessary and well intentioned—that is, morally acceptable or even good—while the label of "terrorism" denotes violence by non-state actors that is deemed morally bad due to evil intentions (2007, pp. 2, 38, 2010, p. 4). These frames, according to Asad, blind us to the reality that the real difference between violence committed by states and violence committed by non-state actors is that state violence is far more destructive (2007, p. 4, 2010, p. 19). Asad emphasizes that a good deal of this destruction is endured by civilians. He is skeptical of dominant paradigms around unintentional collateral damage (2010, pp. 12, 15). Like Butler, Asad suggests that contemporary discourse around just war is inescapably framed by modern Western liberals' implicit sense that certain lives are valued more than others and thus that certain people's deaths matter more than others' (2007, pp. 94–95).

Butler's critique of Asad is aimed not at the substance of his argument but at his claim that this analysis is not normative or evaluative, but simply descriptive (Butler, 2010, p. 150). Butler responds that Asad's questions about just war and terrorism "make sense only on condition that reference is made to a horizon of comparative judgment" (2010, p. 157). We care about Asad's description *because* of the evaluative and normative reactions it engenders:

> Presumably, we want to know about these differential ways of defining and experiencing death dealing because they are consequential for why and how wars are waged, and we are trying to shed light on these differential modes in order, in whatever way, to counter and undo them with the hope of ending or ameliorating such wars. And we want to end them, if we do, because we think they are wrong, unjust, contemporary forms of conquest, racist and destructive. [. .]
>
> *[I]t is only on the condition that we do, in fact, oppose violence and the differential ways it is justified that we can come to understand the normative importance of the comparative judgment that Asad's work makes available to us.*
>
> *(2009, pp. 106–107)*

Butler here clarifies the way that our critical assessment of our social frames and the unequal ways in which they are applied makes sense only within some larger evaluative framework. Much as they insisted that a critical analysis of the norms and juridical structures that generate the subject does not preclude moral agency, here they show that revealing the ways in which our prevailing moral discourses are beholden to deeply embedded frames of humanity does not require us to abandon a moral perspective on violence. To the contrary, it is precisely in illuminating these frames that an evaluative judgment is made possible, and a normative orientation is suggested; it is by seeing the destructive impact of the grievable/ungrievable binary that the moral urgency of dismantling that binary is made clear.

Legacy

Judith Butler is a giant in the fields of gender studies, feminism, and critical theory. Their work is a major influence on many thinkers in those areas. Within the tradition of just war reasoning, however, their legacy is still unclear. Perhaps unsurprisingly, among scholars working on the just war tradition in some way, they have been cited most frequently by feminist scholars who are themselves engaged in various critiques of just war reasoning (Gentry, 2013; Kellison, 2019; Mann, 2014; Sjoberg, 2016). One explanation for their relative absence in just war conversations is the simple fact that Butler's work is not itself framed as a critique of the just war tradition and does not engage in close reading of just war sources, nor do they focus on the kinds of questions that often preoccupy just war thinkers. That said, as the preceding discussion has made clear, Butler is a worthy conversation partner for those studying just war; and their work suggests some serious implications for just war reasoning.

There are several issues on which just war thinkers might productively turn to Butler's work. As noted before, Butler's arguments pertain to both the *jus ad bellum* and *jus in bello* criteria of just war reasoning. Butler suggests that in the *ad bellum* context, just war thinkers have much work to do in confronting the influence of implicit bias in the notion of self-defense. Specifically, Butler argues that we tend to include only those we consider like ourselves among those for whom we are willing to use defensive force and for whom we recognize the right to use force to defend themselves. Just war thinkers who take this argument seriously would be more aware of the patterns according to which certain kinds of states' claims of self-defense are often presumptively taken to be just, while similar claims of other states and non-state actors are regularly dismissed. Similarly, in the *in bello* context, Butler pushes just war thinkers to reject modes of calculating according to which members of certain populations are deemed acceptable "collateral damage" in ways that citizens of Western liberal states would never be. These are, in my view, quite significant challenges to just war reasoning that those who are committed to justice must take seriously.

Perhaps most importantly, Butler's body of work on violence pushes us to recognize that war itself—even war perceived as just according to just war reasoning or international humanitarian law—is violent and destructive. Moreover, the harms inflicted by that violence and destruction are disproportionately borne by the human beings Butler describes as ungrievable. Butler reveals that the language of "just war" can work to hide that violence, destruction, and harm—and its unequal impact—from view. Just war thinkers who understand their work as aiming at a just and lasting peace must confront this unavoidable fact.

Notes

1 See Allen (1999), Chapter 3.
2 As an example of people who inhabit this ambiguous state, Butler describes references to Palestinian children as "human shields" (2010, p. xviii).
3 As Butler says with reference to the war in Iraq, "Tragically, it seems that the US seeks to preempt violence against itself by waging violence first, but the violence it fears is the violence it engenders" (2004, p. 149).
4 On this point, Butler is strongly influenced by Walter Benjamin's understanding of the violence of law (2020, chapter 3).
5 Importantly, some just war thinkers would reject the characterization of the just war criteria as "exceptions" to a general rule against the use of force. In just war circles, this debate is carried out in terms of whether there is a "presumption against war." For a characteristic view in favor of a presumption against war, see Childress (1982), Chapter 3, as well as the chapter on Pope Francis in this volume; for a rejection of that presumption, see Johnson (2011), Chapter 2.

Works Cited

Allen, Amy. 1999. *The Power of Feminist Theory: Domination, Resistance, Solidarity*. New York: Avalon Publishing.
Asad, Talal. 2007. *On Suicide Bombing*. New York: Columbia University Press.
Asad, Talal. 2010. Thinking about Terrorism and Just War. *Cambridge Review of International Affairs* 23(1), pp. 3–24.
Butler, Judith. 1988. Performative Acts and Gender Constitution: An Essay in Phenomenology and Feminist Theory. *Theatre Journal* 40(4), pp. 519–531.
Butler, Judith. 1990. *Gender Trouble: Feminism and the Subversion of Identity*. New York: Routledge.
Butler, Judith. 1995. Contingent Foundations: Feminism and the Question of 'Postmodernism'. In Seyla Benhabib, Judith Butler, Drucilla Cornell, and Nancy Fraser. eds. *Feminist Contentions: A Philosophical Exchange*. New York: Routledge, pp. 35–58.
Butler, Judith. 2004. *Precarious Life: The Powers of Mourning and Violence*. London: Verso.
Butler, Judith. 2005. *Giving an Account of Oneself*. New York: Fordham University Press.
Butler, Judith. 2009. The Sensibility of Critique: Response to Asad and Mahmood. In Talal Asad, Wendy Brown, Judith Butler, and Saba Mahmood. eds. *Is Critique Secular? Blasphemy, Injury, and Free Speech*. Berkeley and Los Angeles: University of California Press, pp. 101–136.
Butler, Judith. 2010. *Frames of War: When is Life Grievable?* London: Verso.
Butler, Judith. 2012. *Parting Ways: Jewishness and the Critique of Zionism*. New York: Columbia University Press.
Butler, Judith. 2020. *The Force of Nonviolence: An Ethico-Political Bind*. London: Verso.

Childress, James. 1982. *Moral Responsibility in Conflicts: Essays on Nonviolence, War, and Conscience*. Baton Rouge: Louisiana State University Press.

Gentry, Caron E. 2013. *Offering Hospitality: Questioning Christian Approaches to War*. Notre Dame, IN: University of Notre Dame Press.

Johnson, James Turner. 2011. *Ethics and the Use of Force: Just War in Historical Perspective*. Farnham, Surrey: Ashgate.

Kellison, Rosemary. 2019. *Expanding Responsibility for the Just War: A Feminist Critique*. New York: Cambridge University Press.

Mann, Bonnie. 2014. *Sovereign Masculinity: Gender Lessons from the War on Terror*. New York: Oxford University Press.

Sjoberg, Laura. 2016. *Women as Wartime Rapists: Beyond Sensation and Stereotyping*. New York: New York University Press.

19

POPE FRANCIS (1936–)

Christian Nikolaus Braun

Introduction

At first look, a chapter on the current pope in a book on thinkers at the margins of just war might seem like an odd fit.[1] After all, the Catholic Church has been at the very center of thinking about the ethics of war and peace for many centuries. Key figures of the just war tradition, like Augustine of Hippo, Thomas Aquinas, or Francisco de Vitoria, were ordained members of the Church. In our own time, the Church continues to play a role in the debates about the rights and wrongs of war, reaching from the grappling with ongoing armed conflicts to moral arguments about particular weapons systems. It is also worth noting that although the popes no longer command their own armed forces, the Holy See remains a full subject of public international law and has been active in peace diplomacy, including in the ongoing war in Ukraine.

While the above seemingly stands against the purpose of this book, however, a closer look at the trajectory of Catholic thinking on war and peace reveals a noticeable change in its general outlook on the justifiability of armed force. Commonly characterized as a change from a relatively permissive "presumption against injustice" toward a restrictive "presumption against war" as starting point of analysis, contemporary Catholic thought considers the use of armed force justifiable in exceptional circumstances only. While this trend toward restraint has a long history whose origins can be traced to the advent of industrialized warfare in the mid-nineteenth century, the popes of the twentieth and twenty-first centuries have developed this skepticism further. One illustrative aspect of this skepticism has been the treatment of the very term just war itself. Many Catholic voices today reject using the language of just war. Pope Francis is arguably the Church leader who has put forward the most far-reaching skepticism toward the justifiability of war of any pope to date.

DOI: 10.4324/9781003428688-20

While Francis, like his predecessors, does not adopt a pacifist position, he has used very powerful language in his condemnation of the condition of war and, clearly, feels uneasy about the just war framework which, in his opinion, has time and again been an enabler of war. Therefore, this chapter seeks to portray Francis as a "reluctant just war thinker" who affirms states' right to self-defense but seeks to transcend the use of armed force in world politics. As a result, considering the remarkable journey of Catholic thinking on war and peace toward this very restrictive outlook, and Francis' powerful oratory skills as arguably its hitherto culmination, the 266th successor of Saint Peter deserves a place in this volume on the margins of just war.

Contexts

So who is this man, who, in his first greeting as newly elected pope told the faithful that his "brother Cardinals have gone to the ends of the earth" to find a new Bishop of Rome (Pope Francis, 2013)? With this reference to the margins, Francis pointed to the fact that he is the first pope from the Western Hemisphere. Considering the history of the European-dominated Catholic Church, Francis' election was remarkable indeed. In further firsts, he is the first Jesuit pope, and there had never been a pope before who chose the name Francis. Francis was born as Jorge Mario Bergoglio in Buenos Aires, Argentina, in 1936. His family history is one of immigration. His father's family emigrated from Italy to Argentina in 1929, and his mother, although born in Argentina herself, was the daughter of Italian immigrants, too.

Throughout his life, Bergoglio has repeatedly encountered war, indirectly at first but directly in a leadership position later in life. His father and uncles were veterans of the First World War and would talk about their experiences during family gatherings (Vallely, 2015, p. 16). Bergoglio's early childhood coincided with the Second World War, and one biographical account notes that the Italian community in Buenos Aires celebrated the end of the war and shared news of relatives who had remained in war-torn Europe (Ivereigh, 2014, p. 15). Doubtlessly, the two World Wars have influenced Bergoglio's thinking about war. Although he did not experience the Second World War as directly as his two immediate predecessors John Paul II (1978–2005) and Benedict XVI (2005–2013), who respectively grew up in Poland and Germany, Bergoglio, as pope, has repeatedly spoken as a member of an older generation who seeks to remind the younger generations about the horrors that resonate from this past. In fact, in the context of the war in Ukraine, Francis returned to his early childhood memories:

> What we are witnessing is yet another barbarity and unfortunately we have a short memory. Yes, because if we had a memory, we would remember what our grandparents and our parents told us, and we would feel the need for peace just as our lungs need oxygen.
>
> *(2022a, p. 2)*

Bergoglio joined the Jesuit order in 1958 and first studied humanities in Chile before he returned to Argentina, where he completed a degree in philosophy in 1963. In 1970, he added a theology degree to his CV. He was ordained a priest in 1969 and made his final profession with the Jesuit Order in 1973. In addition to numerous positions within the order, Bergoglio would also hold academic positions, including the role of Rector of the Colegio Máximo de San José. In 1986, Bergoglio finished his doctorate after studies in Germany. His rise through the ranks of the Church included time as superior of the Jesuit province of Argentina (1973–1979), his appointment as an auxiliary bishop of Buenos Aires (1992), promotion to be the city's archbishop (1998), becoming a cardinal (2001), and eventual election as pope (2013).

Of all of these roles, it was during his time as head of the Jesuit Order in Argentina that Bergoglio experienced war most directly, when his role as superior coincided with Argentina's so-called Dirty War (1976–1983) during which the country's military dictatorship waged a brutal campaign against those it claimed were left-wing extremists. Estimates of the number of casualties vary from 10,000 to 30,000. Many fates remain unknown until today, as large numbers were "disappeared," meaning that they were kidnapped, tortured, and their bodies oftentimes thrown into the sea. Without having the space here to go into detail, Bergoglio's time as superior has been discussed controversially with regard to his dealings with the regime. It seems that he put a premium on protecting the Jesuit Order and the Church. That is arguably the reason why he, like most Argentinian bishops, did not publicly denounce the regime.[2] Interestingly, for students of history, the role of the Argentinian Church in the Dirty War provides curious parallels with the German Church in the Second World War, which, likewise, did not publicly condemn Hitler's military aggression (Braun, 2023). While, in that sense, the Argentinian experience is not radically new, it nonetheless speaks to the fact that for Francis, the experience of war and the moral problems it provokes are by no means a philosophical question only. This becomes readily apparent in his thinking on war, in both his oratory and written work.

Texts and Tenets

Having occupied the *Cathedra Petri* for more than a decade now, Francis has made his voice heard vis-à-vis questions of war and peace numerous times. A noncomprehensive list of issues he has engaged with includes what he sees as a moral obligation to build a just peace, the condition of war generally, as well as the many particular wars that have occurred during his pontificate. Francis has also commented on the morality of specific weapons systems, including nuclear weapons and emerging weapons technology such as lethal autonomous weapons systems.

Without seeking to pull the reader into the intricacies of papal teaching authority, it is important to note that some papal documents and remarks are more authoritative than others. For the purpose of this chapter, a basic distinction should be made between various prepared statements including his encyclical *Fratelli Tutti* (Pope Francis, 2020), in which he provides his most elaborate discussion of war, and off-the-cuff remarks. While the latter are illustrative of Francis' revulsion for the condition of war, they carry less authority than the former. Crucially, Francis' extemporaneous comments have at times been contradictory and opened the door for his critics to accuse him of abandoning the established Church teaching, which, clearly, he has not done. Such misinterpretation has arguably been of the pope's own making, as he has sought to avoid remarks that might be seen as him affirming directly the use of armed force, which has created a certain tension with the Church's established teaching on licit self-defense.

All this to say, when analyzing his thinking on war and peace, it is important to bear in mind that what Francis said during, say, an in-flight press conference, does not carry the same significance as one of his encyclicals or an official letter. That said, there can be no doubt that the ethics of war and peace has been an important issue on Francis' agenda. One illustration of this can be found in a book Francis (Pope Francis, 2022a) published in the context of the war in Ukraine entitled *Against War: Building a Culture of Peace*. This book, to which the pope contributed the introduction, is a collection of his various statements and excerpts from official documents, which provides an all-encompassing picture of Francis' thinking on war and how to build a just peace. A comprehensive analysis of what Francis has said and written about war, as the next section will demonstrate, reveals that some of the controversies his remarks have caused, especially those on the war in Ukraine, are grounded in a reading that focuses mostly on his off-the-cuff remarks, without paying due attention to the pope's official statements and the established teaching of the Church.

Controversies

Francis' thinking on war and peace has repeatedly been headline news, certainly within the Catholic world but also beyond. This section concentrates on two main aspects with regard to the justifiability of armed force. First, Francis' powerful condemnation of the condition of war, expressed via his obvious uneasiness toward the just war tradition and, second, his remarks on the war in Ukraine for which, at least initially, he received considerable criticism.

Before turning to war, however, there needs to be an engagement with Francis' thinking on peace. Indeed, separating the pope's calls to build a just peace and his ideas about how to achieve it would not only do the pope an injustice but also stand against the overarching horizon of just war thinking, namely, the goal just war ostensibly ought to play to facilitate a just peace. Furthermore, Francis' peace ethics has given an impetus to an emerging school of thought working within the Catholic tradition that has adopted the name of "just peace." This heterogeneous group of thinkers, which includes both paci- fists and non-pacifists, hopes to increasingly marginalize the inheritance of the just war tradition and give more emphasis to nonviolent peacebuilding (Cahill, 2019). It is worth comparing how his thinking differs not only from Catho- lic thinkers covered in this volume (Sepúlveda, Vera Cruz) but also from the canonical thinkers (Aquinas Vitoria, Suárez) and contemporary proponents (Elshtain, Johnson) covered in the original (2017) *Just War Thinkers* volume.

Without having the space here to provide a detailed discussion of the the- ological reasoning that undergirds Francis' thinking, the pope takes Jesus' example of love as the maxim that should determine human interaction in the temporal realm.[3] Francis firmly believes that a just peace on earth can be achieved, which is why he asks everyone, those with and without political authority alike, to overcome the reliance on and faith in violent means. As this chapter will demonstrate in its final section on Francis' legacy, this way of arguing has not been an original contribution of the current pope but a long-established feature of papal thought that has foregrounded the papacy's role as peacemaker. That said, however, Francis has added distinctive ideas, such as the frame of fraternity, which he considers to be "the foundation and pathway to peace" (Pope Francis, 2014). Derived from the highest theologi- cal virtue, *caritas*, fraternity, according to Francis, reminds humankind of its relatedness and thus "helps us look upon and to treat each person as a true sister or brother; without fraternity it is impossible to build a just society and a solid and lasting peace" (Pope Francis, 2014). It is in the context of *caritas* and fraternity that Francis has also engaged with the role of nonviolence:

> May charity and nonviolence govern how we treat each other as individu- als, within society and in international life. When victims of violence are able to resist the temptation to retaliate, they become the most credible promoters of nonviolent peacemaking. In the most local and ordinary situ- ations and in the international order, may nonviolence become the hall- mark of our decisions, our relationships and our actions, and indeed of political life in all its forms.
>
> *(Pope Francis, 2017)*

In other words, Francis' long-term hope is that humankind, through *caritas* and the awareness of its fraternity, will be able to transcend the use of vio- lence and, thus, accomplish a just peace. For the time being, alas, the use of armed force within strict limits, as will be discussed next, continues to be morally acceptable. In Francis' own words, until humankind succeeds in the abolition of war, "[p]eacebuilding through active nonviolence is the natural and necessary complement to the Church's continuing efforts to limit the use of force by the application of moral norms" (Pope Francis, 2017).

Another aspect of Francis' thinking on how to build a just peace has to do with what has been described as the "growing edges of just war theory" (Allman and Winright, 2012). This concept has in mind the rationale to allocate more importance to considerations of *jus ante bellum* (right before war) and *jus post bellum* (right after war), considerations that, respectively, might prevent the outbreak of, or the return to, war. Essentially, *jus ante bellum* and *jus post bellum* ideas can be imagined as contributions to the debate about how to overcome war and build a just peace. Not surprisingly, aspects of this emerging stream of thought can be found in Francis' writing:

> However, as Christians we remain deeply convinced that the ultimate aim, that most worthy of the person and of the human community, is the abolition of war. We must therefore always commit ourselves to building bridges that unite rather than walls that separate; we must always help to find a small opening for mediation and reconciliation; we must never give in to the temptation of considering the other as merely an enemy to destroy, but rather as a person endowed with intrinsic dignity, created by God in his image.
>
> *(Pope Francis, 2015)*

Having considered the importance Francis lends to peacebuilding as a way to overcome war, the chapter now turns to Francis' thinking on the circumstances where the use of armed force may be, regrettably, justifiable. Francis (Pope Francis, 2022a, p. 2) leaves no doubt about his rejection of the condition of war:

> War is madness, war is a monster, war is a cancer that feeds on itself, engulfing everything! What is more, war is a sacrilege that wreaks havoc on what is most precious on our earth: human life, the innocence of the little ones, the beauty of creation. Yes, war is a sacrilege!

Francis (Pope Francis, 2022a, p. 4) has also repeatedly described contemporary international affairs as an amalgam of various wars that, put together, "risks becoming the scene of a unique Third World War." Importantly, this powerful renunciation of war and the warning against another global conflagration are related to his attitude toward the just war tradition. Clearly, Francis has an uneasy relationship with just war, with both the semantics of the term and the framework of moral analysis. However, as already indicated earlier, Francis does not adopt a pacifist outlook. As will emerge shortly in the context of his remarks on the war in Ukraine, the seemingly tension between rejecting war as a condition and accepting the right to self-defense that informs Francis' thinking has at times been difficult to entangle and made him the subject of critique.

The best starting point to assess Francis' take on just war can be found in his 2020 encyclical *Fratelli Tutti*, in which he elaborated on his skepticism toward the just war framework:

> War can easily be chosen by invoking all sorts of allegedly humanitarian, defensive or precautionary excuses, and even resorting to the manipulation of information. In recent decades, every single war has been ostensibly "justified". The Catechism of the Catholic Church speaks of the possibility of legitimate defence by means of military force, which involves demonstrating that certain "rigorous conditions of moral legitimacy" have been met. Yet it is easy to fall into an overly broad interpretation of this potential right. . . . We can no longer think of war as a solution, because its risks will probably always be greater than its supposed benefits. In view of this, it is very difficult nowadays to invoke the rational criteria elaborated in earlier centuries to speak of the possibility of a "just war". Never again war!
>
> *(Pope Francis, 2020, § 258)*

This excerpt provides an important insight into Francis' thinking.[4] Essentially, Francis takes issue with the just war framework because in his eyes it has time and again not just failed to prevent war but, in fact, taken on an enabling function. At the same time, however, he refers to the Catechism's section on legitimate defense, which proves that he does not rule out circumstances that may justify the resort to armed force. In fact, he even cites the "rigorous conditions of moral legitimacy" that the Catechism requires in order for defensive force to be justifiable. Interestingly, Francis does not cite the phrase the Catechism (Catholic Church, 1993, § 2309), in the same section, employs to summarize these conditions: "These are the traditional elements enumerated in what is called the "just war" doctrine" (Catholic Church, 1993). It thus seems fair to assume that Francis seeks to distance himself from the just war framework without seeking to adopt a pacifist stance. However, inevitably, the affirmation of legitimate defense takes him back to just war reasoning, which provides the criteria to assess the moral defensibility of defensive uses of armed force.

The need for criteria of assessment, which any non-pacifist ethical outlook will require, arguably leads Francis to his final sentence in which he states that it would be "very difficult" to employ the just war criteria today. Note that he is not using a stronger wording such as "impossible." It seems that the "very difficult" is as far as he can go without adopting a pacifist point of view.

Not surprisingly, perhaps, the complexity of Francis' general outlook on the moral justifiability of armed force, and his particular uneasiness toward the term just war and its historical baggage, resulted in some considerable debate after the start of the war in Ukraine. What exactly, many observers asked, was the pope's moral judgment on the war in Ukraine? Francis himself, frequently in unprepared remarks, sparked confusion about his position, and he also contradicted himself at least once on the question of the possibility of a just war in response to the Russian invasion. Consider, for example, the following remark the pope (as cited in Vatican News, 2022) is reported to have made in a video call with Russian Orthodox Patriarch Kirill in March 2022, a few weeks after the start of the war: "There was a time, even in our Churches, when people spoke of a holy war or a just war. Today we cannot speak in this manner. A Christian awareness of the importance of peace has developed." Only a few days later, Francis (Pope Francis, 2022b), in a prepared address, used an even stronger wording when he declared:

> A war is always—always!—the defeat of humanity, always. We, the educated, who work in education, are defeated by this war, because on another side we are responsible. There is no such thing as a just war: they do not exist!

These two remarks generally followed the position Francis had laid out in his official teaching document *Fratelli Tutti* two years earlier. However, three months later, in July 2022, the pope (as cited in Mares, 2022) arguably backtracked slightly from his ostensibly crystal-clear conviction that war can never be just when he stated, in an interview, the following: "I believe it is time to rethink the concept of a "just war." A war may be just, there is the right to defend oneself. But we need to rethink the way that the concept is used nowadays." In this particular remark, perhaps without intending it, Francis confirmed the possibility of a just war imagined as a war of legitimate defense as described in the Catechism. Essentially, what had happened was that he did not apply the very careful semantic distinction between just war and legitimate defense he made in *Fratelli Tutti*.

It is of interest to note that Francis' remarks revived a debate about the state of just war thinking within Catholicism. This debate, as I will explain in more detail in the following section on Francis' legacy, has been a long-standing one, but clearly, both critics and advocates of just war felt encouraged to weigh in and, respectively, welcomed or rejected the pope's argument. On the one hand, writing for *Pace e Bene*, a U.S.-based non-governmental organization committed to nonviolence, Ken Butigan (Butigan, 2022) had the hope that the view Francis had expressed in his conversation with Patriarch Kirill might open the door toward transcending just war and embracing nonviolence: "Just war theology remains enshrined in the Catholic Catechism. Perhaps the pope's firm declaration this week will open the possibility of an historic shift there." On the other hand, long-time critics of the Catholic "presumption against war," such as George Weigel (Weigel, 2022), did not hide their irritation about what in their eyes amounted to problematic reasoning on the side of the pope:

> It is simply not the case that serious Christians can no longer use the categories of "just" and "unjust" in thinking about warfare. There are, in truth, just and unjust wars. Russia's war in Ukraine is unjust and ignoble. Ukraine's war is just and noble. Informal papal comments do not change that reality. They can, unfortunately, obscure it.

Besides this general conversation about the changed character of war and the possibility of a just war, Francis' specific remarks on the war in Ukraine subjected him to even stronger critique. While the pope, from the start, had forcefully condemned the war and called for an end to the fighting, other aspects of his approach caused considerable controversy. For example, Francis showed an initial hesitancy to single out Russia as the aggressor and to think through what this meant for Ukraine's right to self-defense and the international community's responsibilities in assisting Ukraine.[5] Reporting about what some labeled as "the pope's ambiguities about the war in Ukraine" (Chambraud, 2022) featured prominently in media outlets around the world.[6] At times there even seemed to be some confusion within the Holy See's senior hierarchy itself. For example, the pope's "foreign minister," Archbishop Gallagher (as cited in O'Connell, 2022), when asked about whether his remarks on Russia as the aggressor and if the Holy See's support for the territorial integrity of Ukraine were being made in the name of the pope, answered: "I was speaking in the name of the Holy See, and the Holy Father hasn't corrected me so far on what I've said on his behalf."

In fairness to Francis, some of the critics who expected a clearer moral language from him tend to forget that the popes have historically sought to play an active role in conflict mediation. Indeed, Francis would start a peace initiative during which a special envoy, Cardinal Zuppi, traveled both to Kyiv and to Moscow in 2023. The (modern) Holy See generally does not enter into alliances and has historically been neutral in armed conflicts, partly because it seeks to play the role of honest broker capable of facilitating peace deals. It seems that this role of the papacy has had an impact on at least some of Francis' remarks on the war in Ukraine. At the same time, it should be observed that since the 1929 Lateran Treaty, neutrality in international affairs has been a legally binding obligation for the Holy See, meaning it can only engage in mediation when the interested parties welcome such efforts.[7]

Despite these early "ambiguities," however, in September 2022, the pope did make a clear statement on Ukraine's right to self-defense. In response to the question asked during an in-flight press conference on whether Ukraine should be given weapons to defend herself against Russian aggression, Francis (Pope Francis, 2022c) replied:

> This is a political decision, which can be moral—morally acceptable—if it is done according to the conditions of morality, which are manifold, and then we can talk about it. But it can be immoral if it is done with the intention of provoking more war or selling weapons or discarding those weapons that are no longer needed. The motivation is what largely qualifies the morality of this act. To defend oneself is not only lawful but also an expression of love of country. Those who do not defend themselves, those who do not defend something, do not love it; instead, those who defend, love.

Taking a closer look at this response, Francis' answer is striking in several ways and essentially leads back to the conundrum discussed earlier, namely that any non-pacifist moral framework requires criteria to assess the justifiability of particular uses of force or, in this case, acts that enable the use of force.

Francis points not only to "the conditions of morality" that are needed to assess the morality of weapons deliveries but also to conditions, which it is safe to assume, are what the Catechism refers to as the "traditional elements enumerated in what is called the 'just war' doctrine." What is more, he directly engages with the right intention criterion of just war as taking on a fundamental role in the moral evaluation of the decision. Finally, Francis invokes the old Christian just war idea of giving one's life for others as an expression of love when he states that legitimate defense is an act of love of country. All in all, without mentioning the term just war, Francis provided a basic just war argument caged in the Catechism's language of legitimate defense. This suggests Francis is very much a "reluctant just war thinker," who seeks to avoid speaking in the language of just war but who is inevitably drawn back to the framework and its categories to assess the morality of specific cases of legitimate defense. In a sense, the "triumph of just war theory," to employ Michael Walzer's (Walzer, 2004, ch. 1) claim that just war reasoning has become an inescapable part of how we make sense of the rights and wrongs of war, has been so overwhelming that even a clear skeptic of any use of armed force like Francis has been affected by it. Moreover, as the next section on his legacy will argue, while Francis has arguably used the most powerful rhetoric of any pope to condemn the condition of war, his actual position on the ethics of war and peace is firmly in line with his immediate predecessors and, in fact, the trajectory the Catholic Church has been on for roughly a century and a half.

Legacy

As an institution that allocates a prominent role to what it refers to as Sacred Tradition, it will come as no surprise that the conversation about the status of just war in Catholic thought, and especially changes to it, has been heated at times. Having presided over a home to ethical reflection on war and peace that has endured several millennia, Francis stands in a long line of popes who argued about these matters. Arguably he has, due to his pronounced uneasiness toward the just war framework and his hesitancy to affirm any use of armed force, revived ethical reflection on war and peace. The war in Ukraine has further elevated the interest in just war within Catholicism.

Francis' teaching has provided a rich ground for both critics and advocates of just war to explore. For just war's critics, the pope has not gone far enough in his distancing from the just war framework, while the advocates of just war continue to point to what they see as an undiminished role of just war. However, this polarized state of affairs within contemporary Catholic ethical thought on war and peace is not solely attributable to the Francis papacy. Rather, the conversation about the diminished permissibility and augmented restraint of armed force has been dominant in the past two centuries. Therefore, to grasp Francis' legacy, the debate his musings raise about just war needs to be situated within Catholicism's long history, and especially recent history, of ethical reflection on war and peace.

While Francis' thinking can be perceived as speaking to the general historical fault line between pacifism and just war that has inspired Catholic thought since its early days, the key to understanding the modern popes' thinking on war and peace, according to Gregory Reichberg (Reichberg, 2017, p. 264), can be found in the pontificate of Pius IX (1846–1878).[8] Pius' reign coincided with several key developments, in terms of both the changing character of war and the role of the papacy. Regarding the former, the American Civil War (1861–1865) and the Franco-Prussian War (1870–1871) were powerful illustrations of the destructiveness of modern warfare that convinced Pius and his successors that war could no longer be waged with restraint. Later developments, such as the twentieth century's two world wars (1914–1918; 1939–1945) and the advent of nuclear weapons (1945), gave further clout to this conviction.[9] Regarding the latter, the loss of the papal territories, and with it the papacy's ability to wage war, let the popes emphasize their role of peacemakers. Put differently, and perhaps slightly provocatively, losing the ability to wage wars may be seen as an explanatory factor for why the popes started to move away from the framework that for centuries had been a helpful tool to legitimize the violent elements of their statecraft.

A century later, the U.S. Catholic Bishops (National Conference of Catholic Bishops, 1983) would proffer that a "presumption against war" was the starting point of Catholic moral analysis about war and peace, a phrasing which gives a succinct name to a process that can be traced back to Pius, and arguably with antecedents in the sixteenth-century priest Bartolomé de las Casas (Brunstetter, 2020). The Catholic thinking on war and peace that has evolved over the last century and a half, as critics have noted repeatedly, differs from earlier versions of just war that were more permissive and did not shy away from speaking in the language of just war.[10] James Turner Johnson (Johnson, 1996) introduced the term of a "presumption against injustice" to capture the gist of this earlier understanding and suggested that supporters of a "presumption against war" were presiding over a "broken tradition." In this critique, Johnson has been echoed by prominent Catholic thinkers including George Weigel (Weigel, 2002, p. 691) who has spoken of the "great forgetting of the classic Catholic just war tradition." Supporters (see, e.g., Joblin, 1988; Coste, 1962; Hehir, 2000) of the changed general Catholic outlook on war and peace, on the other hand, have welcomed the "presumption against war" as a laudable development in Catholic teaching.

Bearing in mind this brief nod to the rich history of Catholic reflection on war and peace, it emerges that most of the controversies associated with Francis' remarks and writings on just war have their place within a long-standing Catholic conversation. The war in Ukraine has required specific answers to practical ethical questions arising on the battlefields. This has put Francis, perhaps more so than his immediate predecessors, in the challenging position of having to apply the Church's teaching on legitimate defense in the context when holding to a strong "presumption against war" approaches its limits. In his remarks, Francis, due to his uneasiness to legitimize any use of armed force, has arguably gone the farthest of any pope to date in condemning the condition of war.

It should be expected that the precedent upon which he himself has built will also be the horizon that shapes the thinking of his successors. There might be some alterations in terms of style, for example, regarding use of the term "just war," but future popes are likely to continue the trajectory that the Church has been on for a century and a half, namely, to elaborate on a deeply rooted "presumption against war." Unless history makes that position untenable.

Conclusion

When the current pope chose the name Francis, many observers interpreted this choice as indicative of his intent to not just care for those at the margins of society but also, like Francis of Assisi, to bring radical change to the Church as an institution. And, indeed, for his conservative critics, Francis has been exactly that—a radical pope who has not given due attention to the Church's role as the "pillar of truth." On the other hand, many of his liberal supporters, while welcoming the changes the pope has implemented, had hoped for a much more radical approach toward Church reforms. Francis' thinking on just war mirrors this general picture. On one side of the spectrum, one finds critics of the pope who, building on a rejection of the Catholic "presumption against war" that predates Francis, have continued to rebut this position as a betrayal of Christian thought on just war, as well as Francis' remarks on particular current manifestations. On the other side, critics of just war, including the emerging school of "just peace," would like the pope to go further in his embrace of nonviolent peacebuilding and his distancing from the just war framework.

The war in Ukraine, by highlighting some of the tensions in contemporary Catholic thinking on war and peace, has further revived a conversation that has been present in Catholicism for many centuries. Francis is only the latest successor of Peter who has weighed in on these matters. As this chapter has sought to demonstrate, while the pope has certainly added a characteristic "Francis flavor" to the debate, his thinking follows a long-standing, though relatively recent in the scope of the just war tradition, trend in Catholic Social Teaching and, against the charges of his critics, follows the established Church teaching. Francis, therefore, is a "reluctant just war thinker."

Notes

1 The author is grateful for very helpful feedback from Gregory Reichberg and Paul Vallely.
2 For a nuanced assessment of Bergoglio's time as superior of the Jesuit province of Argentina, see Vallely (2015, Chapter 4). For a more favorable take, see Ivereigh (2014, Chapter 4).
3 For detailed analyses of Francis' peace ethics, see Braun (2018); De Volder (2023); and Smytsnyuk (2023).
4 For an insightful discussion of *Fratelli Tutti* before the horizon of the just war thinking of Thomas Aquinas, see Reichberg (2022).
5 For an in-depth analysis of the Holy See's position in the early phase of the war in Ukraine, see Smytsnyuk (2023).
6 A similar pattern could be observed in October 2023 when Francis was criticized by the Israeli Embassy to the Holy See for "linguistic ambiguities" and "parallelisms" that seemingly equated the aggressor Hamas with Israel—the party that invoked its right of self-defense. Similar to his remarks on the war in Ukraine, Francis would follow up on his initial comments with a clear affirmation of Israel's right of self-defense, for which he received a publicly stated appreciation from the Israeli ambassador. See Allen (2023).
7 See Ernesti (2022) for a book-length study of the Holy See's foreign policy since 1870.
8 Reichberg draws this argument from the work of Joseph Joblin. See Joblin (1988).
9 This is the argument that James Turner Johnson (see, e.g., 1996) has made in the context of the "presumption against war" versus "presumption against injustice" debate. For a recent reassessment of this debate, see Braun (2020).
10 There is some debate about how radically Catholic teaching on war has actually changed since the times of the scholastics. Reichberg (2017, Chapter 11), for example, argues that there is much more continuity than critics tend to admit.

Works Cited

Allen, Elise Ann. 2023. Israel's Vatican Envoy Says Pope's Remarks on Self-Defense, Hostages 'Fill a Vacuum'. *Crux*, 12 October. [Accessed 11 November 2023]. Available at: https://cruxnow.com/vatican/2023/10/israels-vatican-envoy-says-popes-remarks-on-self-defense-hostages-fill-a-vacuum.

Allman, Mark J. and Tobias L. Winright. 2012. Growing Edges of Just War Theory: *Jus ante bellum, jus post bellum*, and Imperfect Justice. *Journal of the Society of Christian Ethics* 32(2), pp. 173–191.

Braun, Christian Nikolaus. 2018. Pope Francis on War and Peace. *Journal of Catholic Social Thought* 15(1), pp. 63–87.

Braun, Christian Nikolaus. 2020. The Catholic Presumption against War Revisited. *International Relations* 32(4), pp. 583–602.

Braun, Christian Nikolaus. 2023. *Quo Vadis?* On the Role of Just Peace *within* Just War. *International Theory* 15(1), pp. 106–128.

Brunstetter, Daniel. 2020. The Quest for Peace Revisited. In Eric D. Patterson and Marc LiVecche. eds. *Responsibility and Restraint: James Turner Johnson and the Just War Tradition*. Middletown: Stone Tower Press, pp. 67–95.

Butigan, Ken. 2022. Is a Great Shift in the Human Journey at Hand? *Pace e Bene*, 21 March. [Accessed 13 November 2023]. Available at: https://paceebene.org/blog/2022/3/21/is-a-great-shift-in-the-human-journey-at-hand.

Cahill, Lisa Sowle. 2019. *Blessed Are the Peacemakers: Pacifism, Just War, and Peacebuilding*. Minneapolis, MN: Fortress Press.

Catholic Church. 1993. *Catechism of the Catholic Church*. Citta del Vaticano: Libreria Editrice Vaticana.

Chambraud, Cécile. 2022. Pope Francis's Ambiguities about the War in Ukraine. *Le Monde*, 13 May. [Accessed 14 November 2023]. Available at: www.lemonde.fr/en/opinion/article/2022/05/13/pope-francis-s-ambiguities-about-the-war-in-ukraine_5983338_23.html.

Coste, René. 1962. *Le problème du droit de guerre dans la pensée de Pie XII*. Paris: Aubier.

De Volder, Jan. 2023. Pope Francis's Contribution to Catholic Thinking and Acting on War and Peace. *Theological Studies* 84(1), pp. 30–43.

Ernesti, Jörg. 2022. *Friedensmacht: Die vatikanische Außenpolitik seit 1870*. Freiburg im Breisgau: Herder.

Hehir, J. Bryan. 2000. In Defense of Justice. *Commonweal* 127(5), pp. 32–33.

Ivereigh, Austen. 2014. *The Great Reformer: Francis and the Making of a Radical Pope*. New York: Henry Holt & Company.

Joblin, Joseph. 1988. *L'Église et la guerre*. Paris: Desclée de Brouwer.

Johnson, James Turner. 1996. The Broken Tradition. *The National Interest* 45, pp. 27–36.

Mares, Courtney. 2022. Pope Francis: 'I Believe It Is Time to Rethink the Concept of a Just War'. *Catholic News Agency*, 1 July. [Accessed 14 July 2023]. Available at: www.catholicnewsagency.com/news/251691/pope-francis-i-believe-it-is-time-to-rethink-the-concept-of-a-just-war.

National Conference of Catholic Bishops. 1983. The Challenge of Peace: God's Promise and Our Response. Washington, DC, 3 May. [Accessed 19 July 2023]. Available at: www.usccb.org/upload/challenge-peace-gods-promise-our-response-1983.pdf.

O'Connell, Gerard. 2022. Interview: Archbishop Gallagher on Vatican Diplomacy, Ukraine and the Threat of World War III. *America: The Jesuit Review*, 18 July. [Accessed 16 November 2023]. Available at: www.americamagazine.org/politics-society/2022/07/18/archbishop-gallagher-ukraine-war-243376.

Pope Francis. 2013. First Greeting of the Holy Father Pope Francis. *Rome*, 13 March. [Accessed 12 July 2023]. Available at: www.vatican.va/content/francesco/en/speeches/2013/march/documents/papa-francesco_20130313_benedizione-urbi-et-orbi.html.

Pope Francis. 2014. Message of His Holiness Pope Francis for the Celebration of the World Day of Peace. *Rome*, 1 January. [Accessed 15 July 2023]. Available at: www.vatican.va/content/francesco/en/messages/peace/documents/papa-francesco_20131208_messaggio-xlvii-giornata-mondiale-pace-2014.html.

Pope Francis. 2015. Address of His Holiness Pope Francis in the Fourth Course for the Formation of Military Chaplains on International Humanitarian Law. *Rome*, 26 October. [Accessed 15 July 2023]. Available at: www.vatican.va/content/francesco/en/speeches/2015/october/documents/papa-francesco_20151026_cappellani-militari.html.

Pope Francis. 2017. Message of His Holiness Pope Francis for the Celebration of the Fiftieth World Day of Peace. *Rome*, 1 January. [Accessed 15 July 2023]. Available at: www.vatican.va/content/francesco/en/messages/peace/documents/papa-francesco_20161208_messaggio-l-giornata-mondiale-pace-2017.html.

Pope Francis. 2020. Encyclical Letter *Fratelli Tutti* of the Holy Father Francis on Fraternity and Social Friendship. *Assisi*, 3 October. [Accessed 12 July 2023]. Available at: www.vatican.va/content/francesco/en/encyclicals/documents/papa-francesco_20201003_enciclica-fratelli-tutti.html.

Pope Francis. 2022a. *Against War: Building a Culture of Peace*. Maryknoll, NY: Orbis Books.

Pope Francis. 2022b. Address of His Holiness Pope Francis to Participants in the International Congress Promoted by the Pontifical Foundation Gravissimum Educationis. *Rome*, 18 March. [Accessed 14 July 2023]. Available at: www.vatican.va/content/francesco/en/speeches/2022/march/documents/20220318-fondazione-gravissimum-educationis.html.

Pope Francis. 2022c. *Press Conference on the Return Flight to Rome*, 15 September. [Accessed 14 July 2023]. Available at: www.vatican.va/content/francesco/en/speeches/2022/september/documents/20220915-kazakhstan-voloritorno.html.

Reichberg, Gregory M. 2017. *Thomas Aquinas on War and Peace*. Cambridge: Cambridge University Press.

Reichberg, Gregory M. 2022. Note doctrinale sur la guerre juste. À propos de *Fratelli tutti*. *La Revue Thomiste* 122(3), pp. 439–466.

Smytsnyuk, Pavlo. 2023. The Holy See Confronts the War in Ukraine between Just War Theory and Nonviolence. *ET-Studies* 14(1), pp. 3–24.

Vallely, Paul. 2015. *Pope Francis: Untying the Knots: The Struggle for the Soul of Catholicism*. London and Oxford: Bloomsbury.

Vatican News. 2022. Pope to Russian Patriarch: 'Church Uses Language of Jesus, Not of Politics'. *Vatican News*, 16 March. [Accessed 14 July 2023]. Available at: www.vaticannews.va/en/pope/news/2022-03/pope-francis-calls-patriarch-kirill-orthodox-patriarch-ukraine.html.

Walzer, Michael. 2004. *Arguing about War*. New Haven: Yale University Press.

Weigel, George. 2002. The Just War Tradition and the World after September 11. *Catholic University Law Review* 51(3), pp. 689–714.

Weigel, George. 2022. No 'Just Wars'? *First Things*. 30 March. [Accessed 16 June 2023]. Available at: www.firstthings.com/web-exclusives/2022/03/no-just-wars.

20

CHARLES W. MILLS (1951–2021)

Jessica Wolfendale

Introduction

Charles Mills' work was (and continues to be) extremely influential in political philosophy, political science, and philosophy of race, but few authors working in philosophical just war theory and military ethics have discussed his work. Yet, as this chapter demonstrates, Mills' work on the Racial Contract and white supremacy as a political system has profound implications for just war theory. Recognizing the colonialist origins of many Western states challenges important *jus ad bellum* criteria, including legitimate authority, and the use of ideal theorizing in just war thinking has masked yet replicated colonialist tropes about the nature of war and shaped the language of just war theory in ways that reflect histories of colonialism and white supremacy.

Context

Charles Mills (1951–2021), Distinguished Professor of Philosophy at the City University of New York Graduate Center, was born in the UK to Jamaican parents and grew up in Jamaica. He studied physics at the University of the West Indies in the early 1970s during a time of radical political upheaval. This experience fueled his growing interest in radical politics, colonialism, and racism and led him to pursue graduate studies at the University of Toronto, where he completed his Ph.D. in Philosophy in 1985. His doctoral work focused on Marxism because no one at the University of Toronto at that time worked on issues of race-the philosophy department had no non-white faculty, and the University of Toronto had no Africana Studies department or program. After graduating, Dr. Mills held faculty positions at the University

DOI: 10.4324/9781003428688-21

of Oklahoma, the University of Illinois at Chicago, and Northwestern University, before joining the CUNY Graduate Center in 2016. By the time of his death, he had published six books and approximately 100 articles and book chapters. Mills was an authoritative political philosopher all but ignored by just war thinkers, and applying his insights to the idea of civilizational war offers a powerful counter-narrative to the opening chapter on Aristotle and those that follow in the Western philosophical tradition.

Texts and Tenets

Mills' primary research explored the intersection between white supremacy, colonialism, and Western political philosophy. He challenged political philosophers and political theorists in the Western liberal tradition to move beyond the highly abstract theorizing that dominated discussions of justice and liberal political philosophy since the publication of John Rawls' *A Theory of Justice* (1971).

Instead, he emphasized the importance of engaging with and theorizing about the practices of structural oppression that characterized the formation of many, if not most, of the world's nations. He argued that much philosophical liberal thought was molded by unacknowledged conceptions of whiteness and colonialism and failed to grapple with the realities of a world dramatically shaped by profound racial injustice and other forms of oppression. For example, in an article published on the eve of the U.S. withdrawal from Afghanistan entitled "Race and Global Justice," Mills argued:

> [R]ace has provided the theoretical and normative rationale for reconciling egalitarianism and inegalitarianism, differentiating the human population into those deserving and those underserving of equal treatment. . . . the key concepts of liberalism *as a theory* are shaped by . . . imperial logic.
> *(2019, pp. 106–107,)*

In his most famous work, *The Racial Contract* (1997), Mills critiqued Rawls' theoretical device of a hypothetical social contract between rational, self-interested individuals choosing principles of justice from behind a "veil of ignorance." According to Rawls, individuals behind the veil of ignorance would not know their race, class, gender, age, natural talents, or place in society. Nor would they know "[t]he political system of the society, its class structure, economic system, or level of economic development" (Wenar, 2021). Rawls argued that, in those circumstances, such individuals would choose two principles to be used to structure social and political arrangements: a primary principle of equality and a secondary principle stipulating that deviations from the principle of equality would only be justified if all have a fair chance to compete for available opportunities and only if inequalities (e.g., economic inequality) benefit the worst-off in society (Rawls, 2001a, pp. 42–43).

Mills criticized Rawls' approach on the grounds that Rawls falsely assumed that it would be possible to derive principles of justice that could be used to structure a just society without any knowledge of "the political system of the society." Excluding knowledge of actual political systems from behind the veil of ignorance meant excluding knowledge of the history of racialized oppression that was central to the construction of much of the Western world and Western political thought. As Mills put it, "we live in a world which has been *foundationally shaped for the past five hundred years by the realities of European domination and the gradual consolidation of global white supremacy*" (Mills, 1997, p. 20, italics in original). He used the idea of the Racial Contract to describe these practices of oppression and the political system they created.

The Racial Contract demonstrates how a "racial polity" was founded in settler and colonial states, which portioned and transformed "human populations into 'white' and 'nonwhite' men." (Mills, 1997, p. 13). The Racial Contract thus:

> [E]stablishes a racial polity, a racial state, and a racial juridical system, where the status of whites and nonwhites is clearly demarcated, whether by law or custom. And the purpose of this state . . . is, inter alia, specifically to maintain and reproduce this racial order, securing the privileges and advantages of the full white citizens and maintaining the subordination of nonwhites.
>
> *(1997, pp. 13–14)*

White supremacy operates as a *political* system, that is, "a particular power structure of formal or informal rule, socioeconomic privilege, and norms for the differential distribution of material wealth and opportunities, benefits and burdens, rights and duties" (1997, p. 3). Yet, despite decades of anti-racist and anti-colonial theorizing from non-white scholars, the (white-dominated) Western political philosophy tradition ignored and marginalized the study of the Racial Contract and the patterns of injustice it creates and so failed to produce theories of justice that were relevant to creating a more just world. For example, Rawls' rational agents choosing principles of justice from behind the veil of ignorance have no knowledge of the "political system of society" (and so no knowledge of the Racial Contract), and so the principles of justice that Rawls claims they would choose would provide little guidance about how to remedy *actual* injustices in the world. While Rawlsian ideal theory might be "raceless" in principle,

> [A] society where race has been created and then dismantled through the appropriate public policy measures is not the moral equivalent of a society

where race never came into existence in the first place. So, the principles of justice necessary for bringing about the former will in key respects be different from Rawlsian principles, since they will be predicated on the need to correct past oppression, whereas Rawls's principles are not.

(2019, pp. 113–114)

In addition to critiquing Rawlsian political theory, this quote also reflects Mills' criticisms of ideal theorizing in political philosophy (and moral theory more broadly). As he argued in "Ideal Theory as 'Ideology,'" ideal theory approaches to questions of justice and morality abstract "away from realities that are crucial to our comprehension of the actual workings of injustice in human interactions and social institutions . . . thereby guaranteeing that the ideal-as-idealized-model will never be achieved" (2017, p. 77).

Additionally, Mills argued that the use of ideal theorizing in moral and political philosophy is *ideological* because it masks and distorts the backgrounds of group privilege that shape the norms and assumptions that inform such theorizing (2017, p. 73). Ideal theorizing:

[I]nvolves the modeling of what people should be like (character), how they should treat each other (right and good actions), and how society should be structured in its basic institutions (justice). . . . What distinguishes ideal theory is the reliance on idealization to the exclusion, or at least marginalization, of the actual.

(Mills, 2017, p. 75)

Thus, ideal theorizing often relies on unstated and unexamined assumptions about the ideal human agent (the "abstract and undifferentiated equal atomic individuals of classical liberalism") and idealized human capacities, such as rational self-interest (Mills, 2017, p. 75). Because of this conception of the ideal human agent as an abstract rational agent, ideal theorizing involves a corresponding "silence on oppression . . . little or nothing will be said about actual historic oppression and its legacy in the present or current ongoing oppression" (Mills, 2017, p. 76). So, far from challenging or enabling a greater understanding of the structures of contemporary society, "opting for 'ideal' theory has served to rationalize the status quo" (Mills, 2017, p. 89). Only by understanding *actual* structures of oppression and injustice can progress toward a truly just and equal society be achieved: "the best way of realizing the ideal is through the recognition of the importance of theorizing the *non*-ideal" (Mills, 2017, p. 73).

But Mills' goal in *The Racial Contract* and later works such as *Contract and Domination* (authored with Carole Pateman, 2013) and *Black Rights/ White Wrongs: The Critique of Racial Liberalism* (2017) was not just to critique the Rawlsian and ideal theory tradition in political philosophy. Deeply committed to the liberal values of equality and justice, Mills believed that it was possible to reconstitute the social contract tradition to better serve the ends of achieving a just and equal society by foregrounding the importance of addressing racial injustice. As he writes in *Black Rights/White Wrongs*, while Western liberal theorizing often failed to "accommodate an ontology of groups in relations of domination and subordination," this failure was not "an intrinsic feature of liberalism's conceptual apparatus" (Mills, 2017, p. 16) but flowed instead from the privileged stance of most theorists working within this tradition. Thus, he advocated not for the complete *rejection* of political liberal philosophy but rather the radical *revision* of the Rawlsian tradition to explicitly address racial justice. What is needed, he argued in "Race and Global Justice," is a "conceptual shift from distributive to corrective justice, from the distributive norms of a well-ordered society to the rectificatory norms appropriate for an ill-ordered society" (2019, p. 115).

To this end, he argued that political philosophy as a discipline must do three things. First, political philosophers must engage directly with "the intimate historical connection between liberalism and race" and give "racial discourse . . . the theoretical centrality to moral and political philosophy it deserves" (Mills, 2019, p. 109). Second, political philosophy must cease to marginalize non-Western and non-white thinkers and "open the conceptual door for the admission to the canon of thinkers in the black, anti-colonial, and Third World traditions of political philosophy" (Mills, 2019, p. 109). Finally, "the ghettoization of contemporary political philosophers who work on race" must stop, to create "a conceptual space . . . for a rethinking of descriptive and normative political philosophy in the light of this unacknowledged history" (Mills, 2019, p. 110). Only once these steps are taken will political philosophy be able to offer the theoretical and practical guidance needed to create a more just world that lives up to the ideals of the Western liberal tradition.

Controversies

While Mills wrote extensively on global white supremacy and its impact on Western political philosophy, he did not write directly on the ethics of war, despite Rawls (the target of many of his critiques and a subject of a chapter in this volume) having written on this topic in the *Law of Peoples* (2001b). Similarly, few philosophers working in the just war tradition have engaged directly with Mills' work. This is unfortunate because Mills' ideas have profound implications for the controversies that populated, and continue to populate, just war theorizing. Mills' critiques of Western political philosophy apply to past and contemporary Western just war theorizing, which is overwhelming white.[1] Canonical and contemporary just war theorists rarely discuss, let alone theorize, the historical and contemporary role that just war concepts and arguments play in white supremacist colonialism and Western imperialism. This becomes clear in the controversies that have, as some scholars claim, advanced just war thinking.

Colonial Controversies and the Repetition of the Myth of Civilizational War

During the sixteenth, seventeenth, and eighteenth centuries, the Spanish discovery of the New World and the conquest and decimation of indigenous peoples that followed led the Spanish Crown to sponsor a series of debates in which renowned thinkers discussed the "the relationship between war and justice in an effort to provide guidelines for Spain's political relationship with the 'barbarians of the New World, commonly called Indians'" (Brunstetter and Zartner, 2011, p. 734).

Known as the Affair of the Indies, these debates shaped the thought of canonical just war thinkers including Francisco de Vitoria (Bellamy, 2018), Hugo Grotius (Lang, 2018), and Emer de Vattel (Christov, 2018). While some participants, such as Juan Gínes de Sepúlveda (discussed in this volume), argued for the expansion of *jus ad bellum* and the right of imperial powers to engage in conquest, others, including Vitoria and Bartolomé de las Casas, defended the rights and sovereignty of the "barbarians" (Brunstetter, 2018a). Indeed, Vitoria's arguments, and the debates themselves, have been praised by some contemporary just war theorists as examples of rare scholarly courage—a "heroic moment in the history of the academic world," in Michael Walzer's words (quoted in Brunstetter and Zartner, 2011, p. 734). However, as Brunstetter and Zartner point out, praising Vitoria for his stance on indigenous rights ignores the fact that he did not repudiate war against indigenous peoples and instead offered "ambiguous" conclusions "regarding the possibility of regime change, the scope of saving the innocent and the durability of the sovereignty of the barbarians" (Brunstetter and Zartner, 2011, p. 734). It is therefore misleading to frame Vitoria as a champion of indigenous rights. And it is equally misleading to see the commissioning of the Affair of the Indies as evidence of the Spanish Crown's desire to constrain imperial conquest by moral and legal principles. Instead, these debates legitimized racialized practices of conquest, even when some participants supported indigenous sovereignty. The process of legitimization occurred in three interrelated ways.

First, the fact that the rights of indigenous peoples against invasion and extermination were literally "up for debate" during the Affair of the Indies reveals tacit acceptance of the view that indigenous peoples had lesser moral status compared to Europeans (whose rights were *not* up for debate). All participants, including Vitoria and Las Casas, accepted the description of indigenous peoples as "barbarians" with debatable rights, even when they differed on the meaning of that term (Brunstetter and Zartner, 2011, p. 734). The exclusion of the subjects of the debate from the debates themselves reflected and reinforced their exclusion from moral equality by treating them as objects whose basic moral standing was in question.

Second, the staging of these debates signaled to all involved that arguments defending imperialism and colonialism and denying indigenous peoples' moral and legal rights were worthy of a public platform and defense by prominent scholars. As a result, such arguments had *prima facie* legitimacy even when some debate participants argued against them. These debates are thus an early example of what is known today as "both sides-ism" (Merriam-Webster, 2024). It is hardly surprising, therefore, that the outcome of the debates had a limited impact on the practices of colonialism and imperialism in the long run. Consider the chapter in this volume on Vera Cruz, which showcases some of the on-the-ground challenges of applying Vitoria's just war theory to colonial problems. Once a group of people are described as "barbarians" whose rights are "up for debate," and once arguments in favor of conquest are treated as legitimate positions for debate, there is an important sense in which the debate is already lost.

Finally, the debates created a theoretical and imaginative space for the idea that conquest and colonialism could be conducted within moral constraints. Thus, while some participants in the debates—Las Casas (discussed in *Just War Thinkers*) in particular—condemned the "evil lives . . . monstrosities . . . savagery . . . and pride" of the Spanish conquistadors (Brunstetter and Zartner, 2011, p. 742), the debates made possible the idea of the "just" colonizer. This imaginary figure allowed the audience and participants who defended imperial conquest to align themselves with "good" forms of colonialism, against the excesses of "bad" conquistadors.

This imaginative distinction between "good" and "bad" forms of conquest and colonialism made no real difference to the actual brutality of colonialism. Colonialism is, of its nature, characterized by absolute domination over others and the use of arbitrary violence effectively unconstrained by legal or moral norms (Buffachi, 2017, p. 207). As with the institution of slavery (which often accompanied colonialism), attempts to draw distinctions between "good" and "bad" colonizers (or slaveowners) are based on the false belief that institutions founded on the imposition of absolute domination over others can, nonetheless, be reconciled with humane values (Wolfendale, 2022, pp. 242–243). Indeed, those engaged in colonialism or slavery—both institutions founded on the presumption of a right of absolute domination over others—cannot in practice recognize any moral constraints on their actions because any such recognition would threaten the legitimacy of that total domination.[2]

The above discussion illustrates the intimate relationship between the reality of colonial brutality, debates about just war thinking, and the legitimization of colonialism. While the Affair of the Indies shaped the thought of canonical just war thinkers such as Vitoria, it also provided justificatory moral narratives that created theoretical and imaginative space for the illusory idea of morally decent colonialism. The failure of most contemporary philosophical just war theorists to understand or theorize this history contributes to their failure to recognize how such narratives (and illusory ideas) are replicated in justifications of current conflicts—an issue discussed in the Legacy section of this chapter. Arguably, the same civilized–barbarian dichotomies, first drawn from Aristotle as discussed earlier in this volume, explicitly represented by Sepúlveda, and then explicated in Mills' Racial Contract, implicitly still inform the practice of just war theorizing in contemporary philosophy. Contemporary philosophical discussions of just war theory, far from serving to limit the wrongful uses of military force in the contemporary world, may effectively *promote* and reinforce white supremacist and colonialist tropes and justify the seemingly endless and unconstrained use of military force.

The debates on just war theory that took place during the era of Spanish colonialism of the New World legitimized permissive conceptions of *jus ad bellum* and *jus in bello* against indigenous peoples—"barbarians"—whose customs were supposedly evidence of their uncivilized and savage nature. While canonical just war thinkers debated the legitimacy of waging war to eliminate these customs (Vitoria, Las Casas, Grotius, Pufendorf, and Vattel, e.g.), doing so created a tradition of inherited ideas that accepted the legitimacy of the underlying idea that "civilized" states might have a moral duty to use violence to impose the "right" ideas for the good of the uncivilized. This acceptance entrenched the illusory image that colonialism could be motivated by noble aims and not by the desire for domination over others, à la Sepúlveda. As international relations theorist Kimberly Hutchings—one of the few theorists to cite Mills' work—explains, "[c]olonial war was expressly justified in terms of its purpose of bringing civilization into benighted areas of the world" (2019, p. 225). Hutchings goes on to argue, by being described as "barbarians," inhabitants of invaded and colonized states are presented as "either morally or practically incapable, in need of protection, education, or punishment" (2019, p. 212). The use of military force is then framed as the moral duty of the "civilized" state (who alone has "the authority to use violence for good ends") to rescue, punish, or liberate such peoples (Hutchings, 2019, p. 212). Because this task is, of its nature, never-ending, this narrative justifies an expansive, even unlimited, conception of war—an "interventionist, proactive *jus ad bellum* that recognizes moral righteousness and power as a political license to enforce justice against evil" that (as discussed later in the chapter) can be mapped onto the American War on Terror (Brunstetter and Zartner, 2011, p. 746). Hutchings identifies the issue at hand:

[I]n colonial warfare . . . a whole range of policing and pre-emptive military action across borders can become subsumed under the concept of war. This position reflects the . . . assumption in which the world is already characterised as belonging to the coloniser, whose just violence has to be imagined as a permanent possibility.

(2019, p. 212)

This narrative of civilizing warfare has been replicated many times since the Affair of the Indies, to justify colonialism and even the extermination of indigenous peoples. For example, Daniel Brunstetter explains how, after the American Revolutionary War, wars of extermination against Native Americans were justified by the claim that Native Americans "did not abide by (European) rules of war but rather waged merciless warfare that ignored all civilized constraints. . . . And within the norms of European warfare, different standards were justified when dealing with such peoples" (2018b, p. 297). These *different standards* evaded the supposedly restraining influence of just war thinking lauded by just war theorists.

A similar narrative was used to justify the U.S. invasion of the Philippines and the use of torture against Filipino soldiers and civilians. When a 1904 Senate Report found evidence of the widespread use of torture by U.S. troops, President Roosevelt defended the invasion by claiming that it represented "the triumph of civilization over forces which stand for the black chaos of savagery and barbarism" (Kramer, 2006, p. 169). Members of the Roosevelt Administration also suggested that, if torture had occurred, it "might at times be justified by the frequent violations of the rules of 'civilized warfare' committed by a 'barbaric and treacherous' enemy" (Kramer, 2006, p. 169).

This same narrative appeared yet again after the terrorist attacks of 11 September 2001, when the Bush Administration deployed the language of "civilization vs. barbarism" to justify the invasion of Afghanistan and Iraq (Wagner-Pacifici, 2009) and the use of torture (Wolfendale, 2022, pp. 249–254). For example, on 24 September 2001, Attorney General John Ashcroft claimed that:

[T]he attacks of September 11th drew a bright line of demarcation between the civil and the savage. . . . On one side of this line are freedom's enemies, murderers of innocents in the name of a barbarous cause. On the other side are friends of freedom; citizens of every race and ethnicity, bound together in quiet resolve to defend our way of life.

(quoted in Esch, 2010, p. 382)

These examples demonstrate that any serious ethical analysis of contemporary uses of military force by Western states must address how colonialist narratives have been (and still are) used to justify expansive conceptions of *jus ad bellum* and *jus in bello* against perceived "barbaric" enemies. As the above examples illustrate, far from restraining conflict, just war theory has internalized colonialist conceptions of war to justify expansive and ongoing conflicts, such as the post-9/11 wars, that have directly and indirectly caused the deaths, injuries, and displacement of millions of people, many of them civilians (Costs of War Project, 2024).

However, despite the clear historical connection between just war theory and colonialist justificatory narratives, philosophical just war theorists may dispute the claim that they should theorize or investigate this connection. Instead, they might argue, the goal of philosophical just war theory is to clarify the *ideal* theoretical principles governing the use and conduct of war. If so, it is not the role of just war theorists to address how political actors have misused just war concepts in racially and politically motivated ways. Later in the chapter, I explain why this approach to philosophical just war theorizing ignores how ideal theory is shaped by colonialist assumptions and so not only fails to provide concrete answers to questions about the conduct of war but also reflects and reinforces the core elements of the Racial Contract.

New Controversies in Just War Theory and the Dangers of Ideal Theorizing

Mills argued that Western political philosophy's failure to address the Racial Contract left political philosophy without the theoretical resources to redress the history of racial oppression and colonialism and create a more just society. A similar critique can be leveled against the Western philosophical just war tradition today. While some international relations scholars have discussed the impact of colonialism and white supremacy on the conduct and theory of war, the trend in just war thinking has gravitated toward debates between traditionalist just war theory and what is known as revisionist (or reductionist) just war theory—a set of views that reject the assumption that the morality of war has a status separate from ordinary morality (Lazar, 2017). Scholars working in revisionist just war theory tend to use idealized hypotheticals to generate support for different theoretical positions on the liability to lethal force and largely ignore issues of race, racism, or colonialism. But traditionalists (to use the conventional term), and indeed most canonical thinkers and historians of the just war tradition, are also guilty of this omission: few discuss the historical role of "just war" in colonialism and imperialism, and even fewer explore the connection between this history and the development of just war theory, international law, and dominant just war controversies.

For example, Michael Walzer's famous book *Just and Unjust Wars* (2006) aims to make just war theory relevant for the twentieth and twenty-first century by applying it to real-world examples. The examples he discusses include the Allied strategic bombing campaign in World War II, terrorist bombings in the Algerian war of independence, and the Vietnam War massacre at My Lai. But Walzer does not probe the deeper roots and moral trespasses of imperial war, as exposed in the chapters on Fanon and Martin Luther King in this volume. More recent scholarship in philosophical just war theory suffers from a similar lacuna. For example, the 2018 *Oxford Handbook of the Ethics of War* contains only one chapter that discusses colonialism and then only in the context of a discussion of the relationship between territorial rights and rights of defense (Stilz, 2014). There are no chapters or index entries on the topics of race, racism, or white supremacy.

This combined failure illustrates perfectly Mills' point that, after World War II, the West effectively erased "the centrality of race to its rule" (2019, p. 105). As a result, just war theorists have ignored the implications of Mills' claim that the "modern state in general . . . is in fact a racial state" (2019, p. 100). Thus, they have failed to recognize the role of just war theory as a tool of expansion and colonialism—a far cry from the tradition's claim that the just war tradition aims to only justify war for sake of peace.

This failure is problematic for two reasons. First, debates about the relationship between war and colonialism were central to the formation of just war theory and the rationalization of centuries of colonialism. Second, this history continues to shape contemporary debates on (and the practice of) military force in ways that have gone largely unrecognized by philosophical just war theorists. For all the scholarship written and controversies about the revisionist challenge to traditionalist just war thinking, both revisionists and traditionalists have missed how the relationship between the Racial Contract and just war theory shaped, and continues to shape, the world.

Legacy

There is a poetic symmetry that the previous (Brunstetter and O'Driscoll, 2018) *Just War Thinkers* volume ended with Jeff McMahan, while this volume ends with Charles Mills, who eviscerates McMahan's form of ideal theorizing. Just war thinking today is defined by the revisionist turn, which relies on analytical philosophy to discern the morality of violence. But just war thinkers would gain theoretical purchase on the reality of war by recognizing and understanding the long-standing relationship between the Racial Contract and the practices of war.

Revisionist just war theory claims that the moral principles governing the use of force in war should be no different from the moral principles governing the use of force in any other context. Paradigmatic revisionist approaches to just war theory, such as those of Jeff McMahan (2008, 2009), David Rodin (2005), and Helen Frowe (2023), are motivated in part by the perceived need to resolve inconsistencies between the moral principles of traditional just war theory and those of ordinary morality.

For example, Jeff McMahan (2008, 2009) argues that traditional just war theory's separation of *jus in bello* from *jus ad bellum* is morally indefensible. In the traditionalist view (as defended by Walzer (2006)), soldiers are morally and legally accountable for adhering to *jus in bello* constraints in how they fight, but not for the *justice* of the war itself—a view known as the *moral equality of combatants*. This means that soldiers on both sides of a conflict have an equal right to kill each other but are also equally liable to be killed, simply because they pose a threat of harm to each other. Correspondingly, noncombatants are *not* liable to be killed because they do not pose a threat of harm to others, a view known as the principle of noncombatant immunity. The Fanon chapter in this volume controversially rejects this principle, which tracks with some revisionist arguments that argue that noncombatants can be liable to defensive force (McMahan, 2009; Mares, 2021).

More generally, McMahan argues that both the moral equality of combatants and the principle of noncombatant immunity are at odds with basic moral principles regarding liability to defensive force. Even if the laws of war grant combatants on both sides a *legal* right to attack each other because combatants pose a threat of harm to each other,

> [T]he corresponding moral principles are false. It is not true, for example, that one makes oneself liable to defensive force simply by posing a threat to another. . . . The correct criterion of liability to attack in these cases is not posing a threat, nor even posing an unjust threat, but moral responsibility for an unjust threat.
>
> *(2008, pp. 21–22)*

Since revisionists argue that the morality of killing in war is no different from the morality of killing in other contexts, their arguments often (although not always[3]) make use of what Kimberly Hutchings describes as "hypothetical idealized scenarios" to generate intuitions about the conditions under which a person is liable to defensive force. They do so to elucidate the moral principles that should govern just war. Such scenarios typically appeal to what "idealised moral actors" would or would not do in various imaginary situations as a means of generating moral principles to apply to armed conflict (Hutchings, 2019, p. 215). For example, McMahan uses the following scenario to support his argument that unjust combatants have no right to use force against just combatants:

> If a murderer is in the process of killing a number of innocent people and the only way to stop him is to kill him, the police officer who takes aim to shoot him does not thereby make herself morally liable to defensive action, and if the murderer kills her in self-defense, he adds one more murder to the list of his offenses.
>
> *(2009, p. 14)*

Scholars have criticized the use of hypothetical scenarios such as this for failing to consider the "messiness" of actual armed conflicts (Walzer, 2006). As Hutchings puts it, such hypotheticals "have little prescriptive purchase" on the actions of real individuals engaged in combat (2019, p. 217). Mills' critique of ideal theory, informed by the history of the Racial Contract, offers a different perspective on the problems with using ideal theory in the context of the ethics of war.

Mills criticized ideal theorizing in political and moral philosophy not only because it failed to yield practical guidance on real-world issues but because it was shaped by problematic and unexamined assumptions about the nature of ideal theorizing and ideal agents in the first place. For example, he argued that ideal political theory tends to assume that ideal human agents are the "abstract and undifferentiated equal atomic individuals of classical liberalism" (Mills, 2017, p. 75). Hutchings makes a similar point, arguing that the ideal theorizing characteristic of revisionist just war theory often replicates and reinforces the tropes of colonial warfare: that some states have the moral authority to use force "for good ends," that "a whole range of policing and pre-emptive military action across borders can become subsumed under the concept of war," and that "military action [can be] legitimised in civilizational terms" (Hutchings, 2019, p. 212).

In her discussion of McMahan's thought experiments, Hutchings argues that his intended audience is Western liberal states and liberal scholars, who are invited to see themselves reflected in the ideal rational agents who are the actors in the thought experiments, and not the "less than ideal moral agents in the theoretical world . . . as for instance the shadowy non-combatants that the just warrior may or may not foreseeably but unintentionally or intentionally kill" (2019, p. 215). McMahan's thought experiments thereby encourage the idea that rational, liberal actors (the policeman, in the hypothetical example described earlier) may have a moral duty to rescue innocent victims from a savage or irrational enemy (the murderer). Read through Mills' Racial Contract, this scenario effectively replays the colonialist trope of the "civilized" West (the "beneficent powerful" (Hutchings, 2019, p. 217)) taking on the "duty" of protecting the innocent of the world from savage barbarians— a trope, as noted in the previous section, that was used to justify expansionist conceptions of *jus ad bellum* and *jus in bello* and continues to shape the ongoing War on Terror. This expansionist conception of war is given further theoretical legitimacy by revisionists who, by recasting the morality of war in terms of individual liability to defensive force, remove the distinction between killing in war and killing in self or other defense, thereby blurring the lines between policing and war. The result of this unexamined infusion of colonial tropes into revisionist just war ideal theorizing is, as Mills argued in relation to ideal theorizing in political philosophy, that contemporary philosophical just war theory often serves "to rationalize the status quo" (2017,

p. 89)—which is a racialized status quo—whatever the intentions of the theorists themselves.

McMahan and other revisionist just war theorists would no doubt object that it is not their intention to provide justifications for the actual uses of force by Western states. That may be true, but such a response ignores how (as the Affair of the Indies and its continued legacy demonstrate) philosophical arguments can provide *prima facie* legitimacy to political justifications for war that replicate and reinforce colonialist tropes in ways that can and do have devastating impacts on real communities. As Mills puts it, "The abstractions of ideal theory are not innocent" (2017, p. 89). Thus, it is not only naïve, but also arguably morally reckless, for theorists working on the ethics of war to ignore how assumptions that inform their theorizing are shaped by the long-standing and intimate relationship between just war thinking and colonialism.

Conclusion

Incorporating Charles Mills' ideas into just war theorizing offers much-needed insights into the consequences of the failures of contemporary just war thinkers to understand and acknowledge the relationship between the Racial Contract, historical and contemporary uses of military force, and so-called advances in just war thinking. First and foremost, Mills' arguments suggest that just war thinkers must theorize the actual historical role of military force in colonialism. Doing so is necessary not only to understand the empirical facts of this legacy but also to recognize how this history has shaped and distorted the very language, concepts, and narratives through which debates about the ethics of war are conducted. Fundamentally, the Racial Contract is as central to the history of military force as it is to the structure of modern states. Thus, understanding the interrelationship between the Racial Contract and just war theory is crucial if just war thinkers are to develop principles that can guide the future use of military action. This requires that just war theorists, especially analytical philosophers, engage directly with scholars working in colonial and decolonial studies and open the "conceptual door" (Mills, 2019, p. 109) to non-white and non-Western approaches to the ethics of war. Finally, philosophical just war scholars should take seriously Mills' critiques of ideal theorizing and examine closely how the use of idealized scenarios can uncritically reinforce colonialist narratives of war that may serve to expand, rather than curtail, the use of military force. As part of this critical reflection, just war theorists should also reflect on *who* is the presumed audience of ideal theory and whose experiences of war are given priority in discussions of the ethics of war. For far too long just war theorists, from Vitoria to Walzer to McMahan, have prioritized the perspectives of those who *wield* military force rather than those who suffer from war. Shifting the focus of just war theorizing to the victims of military force—who have often been dehumanized and denigrated by the narratives critiqued in this chapter—would further expand the conceptual space of the just war tradition.

Notes

1 As a discipline, philosophy is disproportionately white. For example, one survey found that, in 2021, 81% of recipients of PhDs in philosophy were white, and only 2.8% were Black or African (Schwitzgebel, 2023).
2 This view is made explicit in relation to slavery by the judge in *State v. Mann* (1829), *when the North Carolina Supreme Court overturned the conviction of John Mann for the assault and battery of his slave, Lydia, on the grounds that "'inherent in the relation of master and slave' was the fact that 'the power of the master must be absolute to render the submission of the slave perfect"* (quoted in Wolfendale, 2022, p. 243). Thus, while lip service might be paid (and even laws passed) to protect enslaved people from inhumane treatment, in practice, there were few, if any, constraints on slaveowners' treatment of the enslaved people.
3 As Seth Lazar notes, "there are revisionists who draw deeply on military history . . . and traditionalist who use far-fetched hypotheticals" (Lazar, 2017, p. 35).

Works Cited

Bellamy, Alex. 2018. Francisco de Vitoria (1492–1546). In Daniel Brunstetter and Cian O'Driscoll. eds. *Just War Thinkers: From Cicero to the 21st Century*. Abingdon, UK: Routledge, pp. 77–91.

Brunstetter, Daniel R. 2018a. Bartolomé de las Casas (1484–1566). In Daniel Brunstetter and Cian O'Driscoll. eds. *Just War Thinkers: From Cicero to the 21st Century*. Abingdon, UK: Routledge, pp. 92–104.

Brunstetter, Daniel R. 2018b. Neutrality, Race and Wars of Extermination: Native Americans in the Aftermath of the American Revolution. In Glenn Moots and Phil Hamilton. eds. *Justifying Revolution: Law, Virtue, and Violence in the American War of Independence*. Norman, OK: University of Oklahoma Press, pp. 286–307.

Brunstetter, Daniel R. and Cian O'Driscoll. eds. 2018. *Just War Thinkers: From Cicero to the 21st Century*. Abingdon, UK: Routledge.

Brunstetter, Daniel R. and Dana Zartner. 2011. Just War against Barbarians: Revisiting the Valladolid Debates between Sepúlveda and Las Casas. *Political Studies* 59(3), pp. 733–752.

Buffachi, Vittorio. 2017. Colonialism, Injustice, and Arbitrariness. *Journal of Social Philosophy* 48(2), pp. 197–211.

Christov, Theodore. 2018. Emer de Vattel (1714–1767). In Daniel Brunstetter and Cian O'Driscoll. eds. *Just War Thinkers: From Cicero to the 21st Century*. Abingdon, UK: Routledge, pp. 156–167.

Costs of War. 2024. [Accessed 2 May 2024]. Available at: https://watson.brown.edu/costsofwar/.

Esch, Joanne. 2010. Legitimizing the 'War on Terror': Political Myth in Official-Level Rhetoric. *Political Psychology* 31(3), pp. 357–391.

Frowe, Helen. 2023. *The Ethics of War and Peace—An Introduction*. 3rd Edition. Abingdon UK: Routledge.

Hutchings, Kimberley. 2019. Cosmopolitan Just War and Coloniality. In Duncan Bell. ed. *Empire, Race, and Global Justice*. Cambridge, UK: Cambridge University Press, pp. 211–227.

Kramer, Paul A. 2006. Race-Making and Colonial Violence in the U.S. Empire: The Philippine-American War as Race War. *Diplomatic History* 30(20), pp. 169–210.

Lang, Anthony F., Jr. 2018. Hugo Grotius (1583–1645). In Daniel Brunstetter and Cian O'Driscoll. eds. *Just War Thinkers: From Cicero to the 21st Century*. Abingdon, UK: Routledge, pp. 128–143.

Lazar, Seth. 2017. Just War Theory: Revisionists vs. Traditionalists. *Annual Review of Political Science* 20, pp. 37–54.

Mares, Gabriel. 2021. Just War Theory after Empire and the War on Terror: Re-Examining Non-Combatant Immunity. *International Theory* 13(3), pp. 483–505.

McMahan, Jeff. 2008. The Morality of War and the Law of War. In David Rodin and Henry Shue. eds. *Just and Unjust Warriors: The Moral and Legal Status of Soldiers*. Oxford, UK: Oxford University Press, pp. 19–43.

McMahan, Jeff. 2009. *Killing in War*. Oxford, UK: Oxford University Press.

Merriam-Webster. 2024. *Looking at 'Bothsidesing'*. Available at: www.merriam-webster.com/wordplay/bothsidesing-bothsidesism-new-words-were-watching.

Mills, Charles W. 1997. *The Racial Contract*. Ithaca, NY: Cornell University Press.

Mills, Charles W. 2017. 'Ideal Theory' as Ideology. In Charles Mills. ed. *Black Rights, White Wrongs: The Critique of Radical Liberalism*. New York: Oxford University Press, pp. 72–90.

Mills, Charles W. 2019. Race and Global Justice. In Duncan Bell. ed. *Empire, Race, and Global Justice*. Cambridge, UK: Cambridge University Press, pp. 94–119.

Mills, Charles W. and Carole Pateman. 2013. *Contract and Domination*. UK: Polity Press.

Rawls, John. 1971. *A Theory of Justice*. Cambridge, MA: Harvard University Press.

Rawls, John. 2001a. *Justice as Fairness: A Restatement*. Edited by E. Kelly. Cambridge, MA: Harvard University Press.

Rawls, John. 2001b. *The Laws of Peoples with the 'Idea of Public Reason' Revisited*. Cambridge, MA: Harvard University Press.

Rodin, David. 2005. *War and Self-Defense*. Oxford, UK: Clarendon Press.

Schwitzgebel, Eric. 2023. Non-Hispanic White (Though a Bit Less So Than 10 Years Ago), *The Splintered Mind*. [Accessed 12 May 2024]. Available at: https://schwitzsplinters.blogspot.com/2023/02/us-philosophy-phds-are-still.html.

Stilz, Anna. 2014. Territorial Rights and National Defense. In Seth Lazar and Helen Frowe. eds. *The Oxford Handbook of the Ethics of War*. Oxford, UK: Oxford University Press, pp. 242–259.

Wagner-Pacifici, Robin. 2009. The Innocuousness of State Lethality in an Age of National Security. In Austin Sarat and Jennifer L. Culbert. eds. *States of Violence: War, Capital Punishment, and Letting Die*. Cambridge, UK: Cambridge University Press, pp. 25–50.

Walzer, Michael. 2006. *Just and Unjust Wars: A Moral Argument with Historical Illustrations*. 4th Edition. New York: Basic Books.

Wenar, Leif. 2021. John Rawls. *The Stanford Encyclopedia of Philosophy*. [Accessed 2 May 2024]. Available at: https://plato.stanford.edu/archives/sum2021/entries/rawls/.

Wolfendale, Jessica. 2022. The Erasure of Torture in America. *Case Western Reserve Journal of International Law* 54(1), pp. 231–257.

CONCLUSION

Heretics and Humanists and Radicals, Oh My!

Daniel R. Brunstetter

> Listen! you hear the grating roar /Of pebbles which the waves draw back, and fling, /At their return, up the high strand, /Begin, and cease, and then again begin, /With tremulous cadence slow, and bring /The eternal note of sadness in.
>
> —Matthew Arnold, "Dover Beach"

Introduction

As I write these words in Spring 2024, the world is experiencing a new wave of violence that threatens to further erode the liberal peace that arguably shaped global order since the Second World War, a world in which violence was supposedly restrained by the tenets of just war and human rights. Of course, terrible wars have long chipped away at the moral veneer of the just war tradition as the standard bearer when it comes to judging the rights and wrongs of war. I have been teaching a class about just war since 2006, and it seems every year I teach it, a new war erupts, ongoing wars take new directions, and old wars reignite. Make your own list; it is humbling. As I do so, I associate the tragic loss of human life that comes with war, that eternal fixture in human affairs, with the verses of Matthew Arnold cited above. When looking at the wars on my post-2006 list, it should come as no surprise that the just war principles were inevitably evoked, applied, stretched, renegotiated, and sometimes overridden.

Now look at a map and pick where the next war, for inevitably there will be at least one, will take place. Imagining the possibility of what might come next—total war raging through Europe, nuclear war, uninhibited decolonial violence in (post)colonial spaces—will leave those who believe in just war

DOI: 10.4324/9781003428688-22

as a moral compass aghast. How can such unhinged violence still be possible in the twenty-first century? The chapters in this volume would suggest that such violence is at par for the course of human history. Indeed, the original *Just War Thinkers* volume starts from the same premise, for each canonical thinker was confronted with the violence of their day (Brunstetter and O'Driscoll, 2018). While that volume was designed to sketch out the contours of the just war tradition by showcasing the perpetual dilemmas and perennial debates that resurfaced across time as key thinkers followed, rejected, or renegotiated inherited authorities, *Just War Thinkers Revisited* takes a different approach. It begins with the same premise, that those who live through war inevitably face the challenge of how to make sense of it, but abandons the supposition that the just war tradition is the source of moral authority to search for the answers. As the subtitle—*Heretics, Humanists, and Radicals*—suggests, while canonical authority can be a guide, it can also be a hindrance when it comes to understanding why humans are prone to violence. In the spirit of the iconic film *The Wizard of Oz*, I would even add an *"Oh my!"* to the subtitle to further accentuate what the book asks of its readers: to step outside the tradition and to not to be afraid of those who would, by appearances alone, give reason to fear them. And maybe even be surprised by what one might learn from fresh perspectives.

As I contemplated that why-do-we-still-wage-unethical-war question and the lessons I learned from the chapters in this volume, I found myself re-reading Simon Weil's penetrating and profound reflection on war: "The *Iliad* or the Poem of Force." I was drawn to it because I was simultaneously seeking the source of the Western understanding of war (perhaps Homer) and a contemporary thinker with whom I could identify to better understand my world clutching at madness (definitely Weil). Penned between 1939 and 1940, begun while Europe and the world cantered toward the devastations that we would come to know as the Second World War and completed after France had inconceivably fallen to the Nazis, Weil's essay attempts to understand why humans run headstrong to their own deaths by waging war. The Second World War was, in important ways, the bedrock foundation of the short-lived triumph of the contemporary just war revival. From the protocols infused with just war insights that came to shape the UN charter and international law, to Michael Walzer's heralding the "triumph" of just war, the just war principles were meant to restrain humanity's impulse to wage war (Walzer, 2002). And yet, as I write this, daily news feeds probe the prospect of nuclear war in Europe or on the Korean peninsula, try to predict the probability of the escalation of already heightened conflicts in the Middle East, and prod the prospects of civil war in the United States, almost as if these were not simply possibilities, but self-fulfilling prophecies.

But it doesn't have to be that way. Instinctively, I turned to Weil because she asks her readers to grapple with violence in a way that does not follow a formulaic philosophical inheritance such as the just war tradition, which perhaps too easily justifies war or, at the very least, sanitizes its gruesome consequences with nice and tidy philosophical formulations. Or nowadays, hands all the moral grappling over to trusted algorithms that make complex moral decisions in a fraction of a split second. Weil, living in war-torn Europe, could not escape grappling with war. The *Iliad* offers no such moral get-out-of-jail cards or ethical helps-me-sleep-at-night pills. Neither does this volume. Rather, its chapters explore a variety of thinkers who offer exploratory avenues to deepen our understanding of why we make war, or why we should refrain from doing so. For those interested in the just war tradition, these avenues reveal blind spots, shortcomings, and crossroads that open alternative Western traditions to probe as we collectively grapple with how to confront an imperfect, but all too human, world of systemic oppression, aggrieved communities, and diverse aspirations for peace and human flourishing.

Bookends

The chronological organization of this volume, like in *Just War Thinkers*, communicates how certain assumptions about war and the human condition have been passed down through the ages, albeit not necessarily as a welcomed tradition. The bookend chapters on Aristotle and Charles Mills illustrate a different lesson, namely, that inherited authority can simultaneously inform and deform the moral frames we—by *we* I simply mean all readers interested in the ethics of war regardless of our differences—use to make sense of warfare.

The volume begins with Aristotle, who admired Homer's epic for its educational value. Plutarch reports that Aristotle's pupil, Alexander the Great, whom Montaigne placed among the three great souls of human history (the other two were Homer himself and Epaminondas), "thought and called the *Iliad* a viaticum of the military art." The famous warrior "took with him Aristotle's recension of the poem . . . and always kept it lying with his dagger under his pillow" (Plutarch, 1914, 8.2). Aristotle's place in Western philosophy and Christian just war thinking as a bridge between ancient Greece, medieval just war via Aquinas, as well as in, contemporary virtue ethics, is foundational. The Homeric epic was part of his moral foundation. As he explains in the *Poetics*, the *Iliad* as a poetic tragedy has educational value. It "reveals moral choice" in its most naked form by highlighting intentions as ethical dispositions that come face to face with complex realities when the poem's protagonists struggle to sustain virtuous behavior in the heat of the moment, sometimes succumbing to the passions, other times falling prey to manipulative behavior, and often missing the mark with too little, or indeed too much, virtue (Aristotle, 1995, p. 53 [ch. 6]). Above all, however, Aristotle remarks that the *Iliad* was "rich in suffering" and that the reader, by entering the mindset of the protagonists, recognizes something universal about this suffering (Aristotle, 1995, p. 19, [ch. 24). Much like walking a mile in another person's shoes, "recognition," the Philosopher writes, "as the very name indicates, is a change from ignorance to knowledge" (Aristotle, 1995, p. 65 [ch. 11]).

There is much to be lauded when it comes to Aristotle's virtue ethics, as McIntyre shows us, but other aspects of his thought have permeated the way we see the world in problematic ways. The Sepúlveda chapter reads like a dark chapter from History that we should have moved on from—the portrait of a humanist who took all the wrong bits from Aristotle. From the beginning of my career, and seemingly at every turn, I've tried to push back against implicit Sepúlvedian undercurrents of just war against "barbarians" that unwittingly infiltrate political and moral discourse (Brunstetter and Zartner, 2011; Brunstetter, 2021, pp. 188–193). But now I wonder if they ever really went away, if we are at a time when once again Sepúlveda might be openly embraced as a just war authority?

The civilized–barbaric dichotomy still colors how violence is viewed. Butler reminds us that it shapes who we grieve. And as the concluding chapter on Charles Mills shows, Aristotelean categories infuse the way we wage war against the Other, then and now. Mills asks his readers to look at the world though a racial lens to see extant hierarchies that deform the values underpinning Enlightenment morals and those of just war, too. Many from the just war tradition explicitly posit these hierarchies in problematic ways. Sepúlveda to be sure. And to a lesser extent, as troublesome caveats, Vitoria, Grotius, and Vattel (Brunstetter and O'Driscoll, 2018, pp. 252, 254). Caveats, but not to those who bore the brunt of Western war, such as the Iroquois at the dawn of the American founding (Brunstetter, 2018). But other just war thinkers—read here the revisionists especially—assume it away. Mills can thus be read as a foil to analytical philosophers such as Jeff McMahan, David Rodin, and Helen Frowe (and Rawls, too), who ignore race in their hypotheticals and thus miss something important about the real world. If we take Aristotle at his word, looking through a Millsian lens is the first step of recognition of the implicit hierarchies that the just war tradition—in all its iterations, from Cicero to the twenty-first-century debates between Walzer and McMahan—imposes on ethical international relations.

Tributaries

The just war tradition has been a source of authority down the ages. Those confronting tumult sometimes turn to it, other times turn away from it, and occasionally seek to compliment it with different paradigms. The volume covers some such tributaries of thought, which are revealing of its lure and lacunae. For those interested in the history of the just war tradition, the Vera Cruz chapter offers a fascinating parallel to the Sepúlveda chapter, showcasing the reach and limits of the on-the-ground influence of the canonical figure, Francisco de Vitoria; it also asks readers skeptical of the tradition to imagine: what if Sepúlveda had really won the day? Vera Cruz was no doubt a cog in the wheels of the Spanish empire, but he used the language of just war to effect change, not radical change, but change that did something to improve the lives of the conquered and halt (albeit only temporarily) some future conquests. Calvo is another example of using the language of the powerful, in this case Western entrepreneurs of international law, to negotiate status in a hierarchical international realm, forged and maintained by Western powers.

What should we make of these thinkers, considering the vestiges of colonial violence and ongoing decolonial violence that continue to shape the world? Are they naive torchbearers of morals that only preserve hierarchy? It is easy to dismiss them as such, but their mixed intentions, flaws, and contradictions are instructive to those of us today who turn to just war or international law: to understand our positionality, to grasp how the ideas we employ contribute to or sustain hierarchy. Or perhaps even change the status quo for the better if visions of justice and the future align.

If just war is flawed, what then? Perhaps the answer is to complement its teaching with insights from other paradigms. I was struck that President Obama and Martin Luther King Jr. both cited Niebuhr as a major influence on their world views, and yet they held such diametrically opposed views on the role war might play in remaking the world. Reading Niebuhr confronts us with the question, what do we do in the face of evil? Can we hold onto the power of moral idealism to guide our ways in the face of enemies seeking our annihilation? Just war thinking has long been a moral answer to confronting evil, from Augustine to Obama and beyond. But the wars of the Obama era failed in that regard, and the ongoing wars do not bode well either: surely there must another way? King and indeed Pope Francis, too, ask us to reconsider the morality of war as a means to confront evil. Perhaps they are asking too much.

Every time I teach my class on just war, I begin with the phrase: "I am a pacifist, except when it is just not to be. This is the tension that frames my just war thinking." Maybe this is why Pope Francis' presumption against war resonates with me. Or why listening to King's "Beyond Vietnam" speech drives me to think about how to better invest time and resources into constructing a better society rather than bombing our enemies, some of whom are real, others perhaps mostly conjured. One does not even need to be a pacifist to put more energy into such endeavors. Equally as important, King gives us further motivation to reconsider where our resources ought to go: just war, sometimes referred to as "military adventures abroad," or perennial domestic concerns?

And yet, there is evil in the world. Evil that needs to be confronted. Rawls, following Walzer, sees Nazism as the ultimate evil. We could probably all agree to that. We might even all agree that using force is justified to destroy Nazism, with Anscombe being a fascinating exception to ponder. But starting from resisting ultimate evil to determine a universal moral language of war only gets us so far. And maybe not as far as we think. More importantly, it misses the point that there are other evils in the world. That who, or what, is considered evil is sometimes a matter of perspective. When Vera Cruz was defending the rights of so-called "barbarians" in the New World against the likes of Sepúlveda, no such rights were afforded to Protestants like Martin Luther who were enemies in an indiscriminate, fratricidal war. Montaigne bore witness to the violent Catholic–Protestant civil wars and turned to the ancient Greeks for insights. In the same essay in which he lauds Epaminondas—the most virtuous of all warriors and the example par excellence for future princes to emulate—he describes Homer as a personal "source of wonder to me, almost above the human condition" (Montaigne, 1943, II.36, 569, my translation). Almost, because Homer portrays the evil that Montaigne saw across his lifetime—the frayed and flawed frenzy of the human passions that engulfed his homeland in civil war.

And there are more evils. For Proudhon, evil was capitalism and the state; for Luxembourg, it was capitalism as well as German industrial militarism. While Rawls justifies Britain's fire-bombing German cities filled with civilians as a supreme emergency to combat the Nazi evil, Fanon justifies total violence against French colonial civilians—liable civilians, to use revisionist just war terminology, if one is being provocative—in Algeria to combat the horrors of colonial evil.

Depending on one's positionality, we give a variety of different names that place such thinkers in the category of the Other, whom we would not dare consult, because they are beyond the pale. Heretics, because they are unorthodox. Humanists, because they are woefully naive about the prospects of humans just getting along. Radicals, because their ideas challenge longstanding traditions. To put things in perspective: Montaigne, the quintessential humanist, the Homer and Epaminondas enthusiast, was put on the Catholic-banned books' list for nearly two centuries because of his unorthodox views. Oh my!

But instead of fearing these thinkers, we could instead turn to them as sources of insight, uncomfortable as they might be.

Crossroads

At the same time Anscombe was reflecting philosophically on how to respond to Germany's bubbling military aggression, Weil was pondering war from a more individual perspective. Force, observes Weil, is a character unto itself in human affairs. The human spirit is "modified by its relations with force, swept away, blinded by the very force it imagined it could handle, and deformed by the weight of the force it submits to" (Weil, 1986, p. 3). The protagonists of the *Iliad*, she notices, think they can possess it, and therein lies the error of their ways.

The observant reader sees a warning in her words. A warning that just war thinkers would do well to heed: that those who think they morally possess just force can be deformed by the feeling of invincibility this gives them.

Reading Weil's essay and knowing the history of what was to come, I was struck by a particular passage that points not only to why humans so willingly drink the poison of war but also to an antidote. "The man who is the possessor of force," she writes, "seems to walk through a non-resistant element; in the human substance that surrounds him nothing has the power to interpose between the impulse and the act." Nothing, except knowledge that war brings terrible suffering to everyone. Between the impulse and the act there is "the tiny interval that is reflection" (Weil, 1986, p. 193).

There is meta-theoretical wisdom in her observation, which captures something about the inspiration of this volume. I have a precise image in my mind when I try to conjure what that moment, the tiny interval that is reflection, might look like: Jacques-Louis David's painting *The Intervention of the Sabine Women*, a version of which adorns the cover of this volume. It depicts Hersilia, the daughter of Titus Tatius (leader of the Sabines), standing between him and her husband, Romulus. The latter, in one of the most notorious stories from the founding mythology of Rome, had orchestrated the abduction of dozens of Sabine women (Hersilia included) and their subsequent forced marriage to Roman soldiers. The former was waging a war to avenge and rescue them. Plutarch describes the scene of the painting in his *Life of Romulus*:

> Here, as they were preparing to renew the battle, they were checked by a sight that was wonderful to behold and a spectacle that passes description. The ravished daughters of the Sabines were seen rushing from every direction, with shouts and lamentations, through the armed men and the dead bodies, as if in a frenzy of possession, up to their husbands and their fathers, some carrying young children in their arms, some veiled in their disheveled hair, and all calling with the most endearing names now upon the Sabines and now upon the Romans.
>
> *(Plutarch, 1914, 19.1)*

A father waging a war against his daughter's kidnapper in the presence of her children: if ever there was an image that captures the quintessential pause, a gesture to calm the passions of force—vengeance, vindication, fury, and hatred—for the sake of peace, this is it. I am captivated by the symbolism of this image, for it asks those who might justify war to consider other pathways despite the justice of their cause, and those already in the fray to ponder arresting the fury of combat to pursue other destinies. Instead of simply hurtling across that "tiny interval" at breakneck speed by running through the *jus ad bellum* and *jus in bello* categories and ticking the moral boxes, to pause. Herselia's arms are spread wide, creating a space between the combatants. She looks at her rapacious husband as if to reprimand his unjust fury, her palm outward facing to us, as if in a gesture of take-my-hand-to-embrace-peace instead. Her other hand, palm raised, is outstretched with a gesture destined to her father who is filled with justified vengeance and rage, as if to say, enough is enough, think of the children. The two gestures are intertwined, interconnected by the possibilities of the future that branches in different directions based on the choices the protagonists make, whether it be all-too-human war or peace based on compromise and calming the bloodlust. Those of us looking at the painting could, of course, take the father or the husband's perspectives, but the painting asks us to take Herselia's, complicated as it may be. She demands from us a pause. Plutarch describes what happened during the "tiny interval" of reflection this way:

> So then both armies were moved to compassion, and drew apart to give the women place between the lines of battle; sorrow ran through all the ranks, and abundant pity was stirred by the sight of the women, and still more by their words, which began with argument and reproach, and ended with supplication and entreaty.
>
> *(Plutarch, 1914, 19.2)*

To pause means thinking about violence and its consequences differently, with less certainty in the power of our moral convictions and less optimism in the power of violence to achieve our goals. Alasdair MacIntyre, who included Weil's essay in an edited collection entitled *Changing Perspectives in Moral Philosophy*, concluded after reflecting on her essay:

> A part of moral philosophy and moral psychology must therefore be concerned with how we come to see things as they are, the variety of ways in which we may fail, the variety of causes of failure, and the kind of discipline that can overcome these obstacles. How do we learn to see things differently?
>
> *(MacIntyre, 1983, p. 13)*

That is the question. This volume is an exercise of the sort: to learn to see things differently by reading those branded as heretics, humanists, and radicals, so as to follow the crossroads of thought thus revealed and explore where these paths might lead. To this end, consider the following.

In hindsight, maybe we don't see the "evil" of Luther or the radicality of John Brown, but back in their day, they were harnessing violence to fundamentally alter the status quo, or at least to try to do so. Advocates and critics of Fanon rarely see him as the humanist he was, or at least, hoped to be for postcolonial peoples across the globe. To read these thinkers is not to endorse their points of view, but to understand the lures of violence from different angles. Weil says it powerfully, with poetic distance that shields us from the human passions that risk blinding us on the battlefield:

> But at the time [of action] their own destruction seems impossible to them. For they do not see that the force in their possession is only a limited quantity; nor do they see their relations with other human beings as a kind of balance between unequal amounts of force. Since other people do not impose on their movements that halt, that interval of hesitation, wherein lies all our consideration for our brothers in humanity, they conclude that destiny has given complete license to them.
>
> *(Weil, 1986, p. 194)*

This veil of invulnerability is lifted when we see, by embracing Aristotle, that we are equally vulnerable, equally fraternal. "Recognition," to repurpose the passage cited earlier, "as the very name indicates, is a change from ignorance to knowledge" (Aristotle, 1995, p. 65 [ch. 11]).

Whenever I fall into a critical mode regarding just war—which is more and more often these days when I see militaries who, in public, aspire to its tenets but, in practice, blatantly betray them—I catch myself being surprised by the levels of violence justified by the likes of Luther or Fanon—for the sake of their cause, in the fight against what they see as evil. Luther, we might say, is passé. But Fanon holds a place of fascination for many who see Western colonialism as the greatest global evil and uninhibited violence as a justifiable response. But Fanon, if one reads beyond the soundbites, was ultimately on a similar page as Weil. Whatever the perceived necessity of violence, it will take its toll and leave a permanent psychological scar on those who wield it for their cause.

The postcolonial humanism that Fanonian violence aimed at never came to fruition, which was predictable if, as Fanon himself knew all too well, one fails to account for the scars left by those enacting violence. And to bring things full circle, the so-called just wars that I have taught about in my classes (Afghanistan, Iraq, Libya, Mali) have not yielded a lasting peace. The scars from these wars, and others, are ubiquitous. This is perhaps the takeaway lesson that applies to not only (de)colonial violence as Fanon observed but also to all violence that someone, somewhere, justifies.

Self-Introspection

The goal of this volume is not to persuade its readers that any of these think-ers have the right answers. Or to trounce and denounce the just war tradi-tion. It is to take that pause, as Weil described, to look at how others justify or reject violence with a more intimate goal in mind. In the spirit of Mont-aigne, to rub minds with those who might hold starkly different assumptions about violence to better sharpen one's own. My own experience in doing so has not always been comfortable. This is the *Oh my!* moment. Navigating the world with the categories of the just war tradition in tow is a comfort-able space, but these chapters have forced me to train different intellectual and moral muscles. Sharpening my soul with Epictetus and Montaigne's Epaminondas demands intellectual openness and curiosity. Reading King and Fanon side by side is troubling and intellectually destabilizing; a long afternoon spent traversing 2,000 years of moral grappling via Aristotle–Sepúlveda–McIntyre–Rawls–Charles Mills is like getting several pairs of new intellectual eyeglasses, none of which are adjusted to my moral eyes. Luxem-bourg and Anscombe confront me with different sides of the anti-militarist coin and push my pacifist sensitivities to the brink. Butler paired with Fanon flips the script to ask whether colonial civilian deaths are grievable. I've mentioned other patterns before, and I challenge the reader to find their own pairings and combinations to think deep and hard about their own assumptions about war.

Wherever that leads the reader on an individual basis, I hope this "pause" can be an intellectual catalyst for new conversations. The world is in a dark place. The time is ripe for new conversations that can lead us out from behind our intellectual barricades in the hope that before we plunge into the next war, we choose to linger a little longer in "the tiny interval that is reflection." If we do, we just might collectively realize the lesson of the *Iliad* that Weil tries to teach her readers. For those who wield force and believe in their own invincibility, a time will inevitably come that reveals this is a false promise, with the only true promise being that war will bring great suffering to every-one: "gone is the armor of power that formerly protected their naked souls; nothing, no shield, stands between them and tears" (Weil, 1986, p. 194).

Works Cited

Aristotle. 1995. *Poetics*. Edited and Translated by Stephen Halliwell and W. H. Fyfe, with Donald Russell and Doreen C. Innes. Cambridge, MA: Harvard University Press.

Brunstetter, Daniel R. 2018. Neutrality, Race and Wars of Annihilation: Native Americans in the Aftermath of the American Revolution. In Glenn Moots and Phil Hamilton. eds. *Justifying Revolution: Law, Virtue, and Violence in the American War of Independence*. Norman, OK: University of Oklahoma Press, pp. 286–307.

Brunstetter, Daniel R. 2021. *Just and Unjust Uses of Limited Force: A Moral Argu-ment with Contemporary Illustrations*. Oxford: Oxford University Press.

Brunstetter, Daniel R. and Cian O'Driscoll (eds.). 2018. *Just War Thinkers: From Cicero to the Twenty-First Century*. New York: Routledge.

Brunstetter, Daniel R. and Dana Zartner. 2011. Just War against Barbarians: Revisiting the Valladolid Debates between Sepúlveda and Las Casas. *Political Studies* 59(3), pp. 733–752.

MacIntyre, Alasdair. 1983. Moral Philosophy: What Next? In Stanley Hauerwas and Alasdair MacIntyre. eds. *Revisions: Changing Perspectives in Moral Philosophy*. Notre Dame: University of Notre Dame Press, pp. 1–15.

Montaigne, Michel de. 1943. *The Complete Essays of Montaigne*. Translated by Donald M. Frame. Stanford: Stanford University Press.

Plutarch. 1914. *Plutarch's Lives*. Translated by Bernadotte Perrin. Cambridge, MA: Harvard University Press. [Accessed 21 June 2024]. Available at: www.perseus.tufts.edu/hopper/collection?collection=Perseus%3Acollection%3AGreco-Roman.

Walzer, Michael. 2002. The Triumph of Just War Theory (and the Dangers of Success). *Social Research: An International Quarterly* 69(4), pp. 925–944.

Weil, Simone. 1986. The *Iliad* or the Poem of Force. In Siân Miles. ed. *Simone Weil: An Anthology*. London: Virago Press Limited, pp. 182–215.

INDEX